Alexander Tikhomirov
Andrej Tyurin (Eds.)

**Algebraic Geometry
and its Applications**

W0055303

Aspects of Mathematics

Edited by Klas Diederich

Vol. E 2:   M. Knebusch/M. Kolster: Wittrings

Vol. E 3:   G. Hector/U. Hirsch: Introduction to the Geometry of Foliations, Part B

Vol. E 5:   P. Stiller: Automorphic Forms and the Picard Number of an Elliptic Surface

Vol. E 6:   G. Faltings/G. Wüstholz et al.: Rational Points*

Vol. E 7:   W. Stoll: Value Distribution Theory for Meromorphic Maps

Vol. E 9:   A. Howard/P.-M. Wong (Eds.): Contribution to Several Complex Variables

Vol. E 10:   A. J. Tromba (Ed.): Seminar of New Results in Nonlinear Partial Differential Equations*

Vol. E 13:   Y. André: G-Functions and Geometry*

Vol. E 14:   U. Cegrell: Capacities in Complex Analysis

Vol. E 15:   J.-P. Serre: Lectures on the Mordell-Weil Theorem

Vol. E 16:   K. Iwasaki/H. Kimura/S. Shimomura/M. Yoshida: From Gauss to Painlevé

Vol. E 17:   K. Diederich (Ed.): Complex Analysis

Vol. E 18:   W. W. J. Hulsbergen: Conjectures in Arithmetic Algebraic Geometry

Vol. E 19:   R. Racke: Lectures on Nonlinear Evolution Equations

Vol. E 20:   F. Hirzebruch, Th. Berger, R. Jung: Manifolds and Modular Forms*

Vol. E 21:   H. Fujimoto: Value Distribution Theory of the Gauss Map of Minimal Surfaces in $\mathbf{R}^m$

Vol. E 22:   D. V. Anosov/A. A. Bolibruch: The Riemann-Hilbert Problem

Vol. E 23:   A. P. Fordy/J. C. Wood (Eds.): Harmonic Maps and Integrable Systems

Vol. E 24:   D. S. Alexander: A History of Complex Dynamics

Vol. E 25:   A. Tikhomirov/A. Tyurin (Eds.): Algebraic Geometry and its Applications

*A Publication of the Max-Planck-Institut für Mathematik, Bonn

Volumes of the German-language subseries "Aspekte der Mathematik" are listed at the end of the book.

Alexander Tikhomirov
Andrej Tyurin
(Eds.)

Algebraic Geometry and its Applications

Proceedings of the 8th Algebraic
Geometry Conference,
Yaroslavl' 1992

A Publication from the
Steklov Institute of Mathematics

Adviser: Armen Sergeev

Alexander Tikhomirov
Department of Mathematics
State Pedagogical Institute
Respublikanskayastr. 108
Yaroslavl' 150000
Russia

Andrej Tyurin
Steklov Mathematical Institute
Vavilova 42
Moscow 117966
Russia

Die Deutsche Bibliothek – CIP-Einheitsaufnahme

Algebraic geometry and its applications: proceeedings of
the 8th Algebraic Geometry Conference, Yaroslavl', 1992;
a publ. of the Steklov Institute of Mathematics /
Alexander Tikhomirov; Andrej Tyurin (ed.). –
Braunschweig; Wiesbaden: Vieweg, 1994
    (Aspects of mathematics: E; Vol. 25)

NE: Tichomirov, Aleksandr S. [Hrsg.]; Algebraic Geometry
Conference ⟨8, 1992, Jaroslavl'⟩; Matematičeskij Institut Imeni
V. A. Steklova ⟨Moskva⟩; Aspects of Mathematics / E

Mathematics Subject Classification:
14C05, 14C30, 14D25, 14E05, 14E07, 14F05, 14F25, 14F45, 14Jxx, 14J10, 81Exx, 10Dxx

ISBN 978-3-322-99344-1   ISBN 978-3-322-99342-7 (eBook)
DOI 10.1007/978-3-322-99342-7

Cover design: Wolfgang Nieger, Wiesbaden

Printed on acid-free paper

Foreword

This volume consists of articles presented as talks at the Algebraic Geometry Conference held in the State Pedagogical Institute of Yaroslavl'from August 10 to 14, 1992. These conferences in Yaroslavl' have become traditional in the former USSR, now in Russia, since January 1979, and are held at least every two years. The present conference, the eighth one, was the first in which several foreign mathematicians participated. From the Russian side, 36 specialists in algebraic geometry and related fields (invariant theory, topology of manifolds, theory of categories, mathematical physics etc.) were present. As well modern directions in algebraic geometry, such as the theory of exceptional bundles and helices on algebraic varieties, moduli of vector bundles on algebraic surfaces with applications to Donaldson's theory, geometry of Hilbert schemes of points, twistor spaces and applications to string theory, as more traditional areas, such as birational geometry of manifolds, adjunction theory, Hodge theory, problems of rationality in the invariant theory, topology of complex algebraic varieties and others were represented in the lectures of the conference.

In the following we will give a brief sketch of the contents of the volume.

In the paper of W. L. Baily three problems of algebro-geometric nature are posed. They are connected with hermitian symmetric tube domains. In particular, the 27-dimensional tube domain T_e is treated, on which a certain real form of E_7 acts, which contains a "nice" arithmetic subgroup Γ_e, as observed earlier by W. Baily. The author discusses an approach to finding the interpretation of T_e/Γ_e as a moduli space of a certain family of polarized algebraic varieties. The approach is based on the relation between Severi varieties of F. Zak and irreducible symmetric tube domains of $\mathbb{R}$-rank 3.

The paper of M. C. Beltrametti, G. M. Besana and A. J. Sommese concerns the dimension of the adjoint linear system $K_X \otimes L^{\otimes(n-2)}$ for quadric fibrations. Namely, let $(X^\wedge, L^\wedge)$ be a smooth n-dimensional projective manifold with a very ample line bundle $L^\wedge$ on it and let (X, L) be its reduction (in the natural sense that there exists a blow up $r : X^\wedge \longrightarrow X$ of a projective manifold X at a finite set B such that $L = (r_* L^\wedge)^{**}$ is ample and $L^\wedge = r^* L \otimes \mathcal{O}_{X^\wedge}(-r^{-1}(B))$). (It is known from the adjunction theory that except for an explicit list of special pairs $(X^\wedge, L^\wedge)$ this reduction exists and is unique up to an isomorphism.) Using the technique, developed by M. C. Beltrametti and A. J. Sommese and the earlier results of G. M. Besana, the authors prove the following main result of this paper (theorems 2.2 and 3.3): if (X, L) is a quadric fibration over a surface, then $h^0(K_{X^\wedge} \otimes L^{\wedge(n-2)}) \geq 2$ with 3 possible exceptions, described explicitly.

D. Butler considers in his paper the left transform M_E of a given vector bundle E with a slope $\mu(E) \geq 2$, generated by global sections, on the curve C, this left transform being defined in a standard way as the kernel of the evaluation map $H^0(E) \otimes \mathcal{O}_C \longrightarrow E$. The main result of the paper (theorem 1) states that the (semi)stability of E implies the (semi)stability of its left transform M_E.

The paper of H. d'Souza concerns the description of the intermediate Jacobian $J(X)$ of a threefold X with a Del Pezzo fibration via Prym-Tyurin variety P associated with a cylinder map for a family of lines on X. The main result of the paper gives the condition under which the principally polarized abelian varieties $(J(X), \Theta)$ and (P, Ξ) are isomorphic, where Θ is the Poincare theta-divisor on $J(X)$ and Ξ the canonical polarization of P: this is the condition that the incidence correspondence for the family of lines on X has no fixed points. The work is essentially based on the results and methods developed by V. Kanev for the description of (P, Ξ) of the Del Pezzo fibrations via Prym-Tyurin variety.

The ramification, decomposition and inertia subgroups of the Cremona groups are studied in the paper of M. H. Gizatullin. The author shows that the simplicity of the Cremona group of the plane is a consequence of a positive solution of a congruence subgroup problem. A representation of Manin's group of minimal cubic surfaces in the Cremona group of the space is given. A nontriviality of some ramification groups is established.

In the first part of the paper of A. L. Gorodentsev a survey of helix theory is given. This theory was developed in order to obtain constructive description of the set of exceptional vector bundles (i.e., bundles E with $\dim \mathrm{Hom}(E, E) = 1$ and $\mathrm{Ext}^i(E, E) = 0$ for $i > 0$). In the second part the author discusses some connections between helix theory and arithmetical properties of nonsymmetrical bilinear forms on lattices and formulates some conjectures about braid group action on the set of semiorthogonal bases of these forms. In the third part there are given examples of calculations with these forms and their groups of isometries.

The moduli space $M_3^{[2]}$ of curves of genus 3 (plane quartics) with a fixed point of order 2 (or, equivalently, with a fixed unramified double covering) is considered in the paper of P. I. Katsylo. This space is closely connected with the moduli space M_3 of curves of genus 3, the classical problem of rationality of which was recently solved affirmatively by Katsylo. In this paper the similar result for $M_3^{[2]}$ is proved: $M_3^{[2]}$ is rational variety. The proof is based on the reduction of the problem to that of rationality of fields of invariants for certian representations of SL_2, stated by the author earlier.

The moduli space $C_n(m)$ of stable weightened ordered n-tuples of points in $\mathbb{P}^1$ under the natural action of $\mathrm{PGL}_2(\mathbb{C})$ is studied in the paper of A. A. Klyachko (here $m = (m_1, m_2, \ldots, m_n)$ is the vector of weights). As it was shown earlier by the author, this space naturally arises in the description of the structure of rank-2 stable sheaves with odd determinant on $\mathbb{P}^2$. In the paper the original method of the study of $C_n(m)$ is introduced, based on the interpretation of this space (more precisely, of its natural compactification by a finite number of points) in terms of the geometry of the spatial polygons in the Euclidean 3-space. This interpretation enables the author to compute the Betti numbers of $C_n(m)$ and give the explicit description of $C_n(m)$ for small n.

In the paper of S. A. Kuleshov vector bundles E with the condition $\mathrm{Ext}^1(E, E) = 0$

on Del Pezzo surfaces are studied. These bundles are called rigid in view of the lack of their infinitesimal deformations. The author proves that any such bundle is a direct sum of exceptional bundles, i.e. rigid and simple ($\mathrm{End}E = \mathbb{C}$), these last being indecomposable.

The fundamental group $G = \pi_1(\mathbb{C}^2 - D)$ of the complement of an algebraic curve D in $\mathbb{C}^2$ is treated in the paper of Vic. S. Kulikov. Certain properties of G and of the Alexander polynomial $\Delta_D(t)$ of the curve D are stated, e.g., it is proved that: if D is irreducible then the commutator $G' = [G, G]$ of G is finitely generated and $\mathrm{rk}(G'/[G', G'])_{free} = \deg \Delta_D(t)$ is an even number; if D is connected, then all the roots of $\Delta_D(t)$ are the roots of unit; if G is an irreducible C-group, then $\Delta_D(1) = \pm 1$. Besides, the description of $\Delta_D(t)$ in terms of the C-corepresentation of G is given.

In the paper of V. V. Nikulin the group $_2Br(X)$ of elements of order 2 in the Brauer group $Br(X)$ of the projective algebraic variety X over $\mathbb{R}$ is considered. Let s be the number of connected components of $X(\mathbb{R})$. The author proves that the canonical map $_2Br(X) \longrightarrow (\mathbb{Z}/2)^s$ is epimorphic if $H^3(X(\mathbb{C})/G; \mathbb{Z}/2) = 0$ and $X(\mathbb{R}) \neq \emptyset$, where $G = \mathrm{Gal}(\mathbb{C}/\mathbb{R})$. This result is then applied to Enriques surfaces.

In the paper of A. D. Popov and A. G. Sergeev there is presented the geometric quantization scheme for the bosonic string theory in twistor terms. Starting from the loop space of a Lie group they define a symplectic twistor bundle over the loop space and reformulate the geometric quantization problem in terms of this bundle. For the standard bosonic string they recover in this way the well known critical dimension condition.

In his list of problems of complex and differentiable varieties F. Hirzebruch (1954) mentioned the classification of all compact complex manifolds V with $b_2(V) = 1$ containing the open analytic subset A such that $V \setminus A$ is biholomorphic to $\mathbb{C}^n$. The complete answer to this problem at the moment is known for $n \leq 3$. The paper of Yu. G. Prokhorov deals with such compactifications (V, A) of $\mathbb{C}^4$ when V is projective. The main result of the paper (theorem 3.1) states that in the case of index 3 (i.e. when $K_V = -3H$ for some $H \in \mathrm{Pic}V$) there exists only four such compactifications. For all of these cases V appears to be the linear section of the grassmanian $G(2, 5)$ in $\mathbb{P}^9$, and they differ only by the divisor A explicitly described in each case. (Note that the case of index ≥ 4, containing projective 4-space and quadric for V, is an easy consequence of the results of Kobayashi and Ochiai.)

It was supposed by A. Gorodentsev and A. Rudakov that for any two exceptional sheaves on a projective plane only one ext-group can be different from zero, which was proved later by A. Bondal and A. Gorodentsev. In the first paper of A. N. Rudakov the similar question is studied for exceptional sheaves on a quadric surface. The result is that this is not longer true, there are exceptional sheaves A, B on a quadric such that $\mathrm{Hom}(A, B)$ and $\mathrm{Ext}^1(A, B)$ are nonzero. But it is proved at the same time that for a subclass of symmetrical exceptional sheaves on a quadric surface the statement holds true: if both A and B are symmetrical, then only one ext-group can be different from zero.

The aim of the second paper of A. N. Rudakov is to summarize known properties of exceptional sheaves and vector bundles on a Del Pezzo surface and to express a set of hypotheses about these sheaves. The main known fact about an individual exceptional sheaf is that if the sheaf has no torsion, then it is a stable vector bundle, and if it has a torsion, then its support is an exceptional curve. But it is worth to study not the individual sheaves but exceptional systems of exceptional sheaves. Here the main fact is that one can define a braid group action on the set of exceptional systems of a given length. It is especially important to study this action when the length is maximal possible for a given surface. In the paper there is a series of conjectures on the properties of this action.

Computation of Segre (respectively, Chern) classes of standard vector bundles on Hilbert scheme of points on a surface, being itself an interesting problem of enumerative geometry, attracted new attention after the recent works of A. N. Tyurin, who interpreted these classes, in particular, top Segre classes δ_d, as new invariants of smooth structure of a fourfold underlying the complex algebraic surface. In the paper of A. S. Tikhomirov it is proved that δ_d is the polynomial of degree d ($=$ degree of 0-cycles on a surface S) of four invariants $x = (D^2)$, $y = (D \cdot K_S)$, $z = s_2(S)$, $w = (K_S^2)$, where D is a given divisor on S. As an example, the formula for $\delta_3(x, y, z, w)$ is given, which coincides with the formula of P. Le Barz obtained earlier in different terms.

In the next paper of A. S. Tikhomirov and T. L. Troshina the original formula for $\delta_4(x, y, z, w)$ is found. The method used here is based on the application of the double point formula for the immersion ("generally") of the chordal variety of a surface S in the space $\mathbb{P}^{10}$ by an appropriate linear subseries of $|D|$ for ample divisor D.

In the last paper of the volume A. N. Tyurin studies the modification of the Gieseker closure of the moduli space $M^H(2; c_1, c_2)$ of H-stable rank-2 vector bundles on the surface S, with given Chern classes c_1, c_2, when the polarization H, lying inside the Kähler cone K^+ of S, deforms in such a way that it passes through the "walls of chambers", these walls being by the definition orthogonal to the vectors $e \in \mathrm{Pic}(S)$ with conditions $e \equiv c_1 \pmod 2$, $c_1^2 - 4c_2 \leq (e^2) \leq 0$. For the description of this modification the method of geometric approximation (GA-procedure) is developed. This procedure enables the author to compute the almost canonical spin-polynomials of S via their geometric approximation. The correction terms of this procedure are treated, and several problems and relations with Donaldson's theory are discussed.

We are very grateful to Tatyana Troshina, and also to Andrej Kazusev and Igor Khomutinnikov, for doing the largest part of the technical work in the preparation of these proceedings for publication.

Alexander Tikhomirov

Andrej Tyurin

Authors' addresses

W.L.Baily Jr., Department of Mathematics, The University of Chicago, 5734 University Avenue, Chicago, Illinois 60637, USA – baily@zaphod.uchicago.edu

M.C.Beltrametti, Dipartimento di Matematica, Universita di Genova, Via L.B.Alberti 4, 16132 Genova, ITALY

G.M.Besana, Department of Mathematics, University of Notre Dame, Notre Dame, Indiana 46556, USA

D.C.Butler, Department of Mathematics, University of Michigan, Ann Arbor, MI 48106, USA

Harry D'Souza, Department of Mathematics, University of Michigan, Flint, MI 48502-2186, USA – dsouza_H@msb.flint.umich.edu

M.H.Gizatullin, Department of Mathematics, Samara State University, 443011, Samara, academician Pavlov str. 1, RUSSIA

A.L.Gorodentsev, Steklov Mathematical Institute, Vavilova 42, 117966 Moscow, GSP-1, RUSSIA

P.I.Katsylo, Moscow Institute of Electronic Machinery, Moscow, RUSSIA

A.A.Klyachko, Department of Mathematics, Samara State University, Pavlova 1, 443011 Samara, RUSSIA

S.A.Kuleshov, Moscow Independent University, Moscow, RUSSIA

V.S.Kulikov, Moscow Institute of Engineers, of Transport, Moscow, RUSSIA

V.V.Nikulin, Steklov Mathematical Institute, Vavilova 42, 117966 Moscow, GSP-1, RUSSIA – slava@nikulin.main.su

A.D.Popov, Steklov Mathematical Institute, ul. Vavilova 42, 117966 Moscow, GSP-1, RUSSIA

Y.G.Prokhorov, Chair of Algebra, Department of Mathematics, Moscow State University, 117234 Moscow, RUSSIA

A.N.Rudakov, Institute for System Analysis, Russian Academy of Science, ul. Avtozavodskaya, 23, 119280 Moscow, RUSSIA – rudal@systud.msk.su

A.G.Sergeev, Steklov Mathematical Institute, Vavilova 42, 117966 Moscow, GSP-1, RUSSIA – armen@sergeev.mian.su

A.J.Sommese, Department of Mathematics, University of Notre Dame, Notre Dame, Indiana 46556, USA

A.S.Tikhomirov, Department of Mathematics, State Pedagogical Institute of Yaroslavl', Respublikanskaya 108, 150000 Yaroslavl', RUSSIA – tikho@delta.yaroslavl.su

T.L.Troshina, Department of Mathematics, State Pedagogical Institute of Yaroslavl', Respublikanskaya 108, 150000 Yaroslavl', RUSSIA

A.N.Tyurin, Steklov Mathematical Institute, Vavilova 42, 117966 Moscow, GSP-1, RUSSIA – tyurin@top.mian.su

Contents

Three Problems on an Exceptional Domain 1
Walter L. Baily, Jr.

On the Dimension of the Adjoint Linear System for Quadric Fibrations 9
M.C.Beltrametti, G.M.Besana and A.J.Sommese

On the Stability of M_E 21
David C. Butler

On a Class of Del Pezzo Fiber Spaces 27
Harry D'Souza

The Decomposition, Inertia and Ramification Groups in Birational Geometry ... 39
M.H.Gizatullin

Helix Theory and Nonsymmetrical Bilinear Forms 47
A.L.Gorodentsev

On the Unramified 2-covers of the Curves of
Genus 3 61
P.I.Katsylo

Spatial Polygons and Stable Configurations of Points in the Projective Line 67
Alexander A. Klyachko

Rigid Sheaves on Surfaces 85
S.A.Kuleshov

The Alexander Polynomials of Algebraic Curves
in C^2 105
Vic. S.Kulikov

On the Brauer Group of Real Algebraic Surfaces 113
Viacheslav V. Nikulin

Symplectic Twistors and Geometric Quantization of Strings 137
A.D.Popov, A.G.Sergeev

Compactifications of C^4 of Index 3 159
Yuri G. Prokhorov

A Note on Cohomologies of Exceptional Bundles on a Quadric Surface 171
Alexei N. Rudakov

Exceptional Vector Bundles on a Del Pezzo Surface 177
Alexei N. Rudakov

Standard Bundles on a Hilbert Scheme of Points on a Surface 183
A.S.Tikhomirov

Top Segre Class of a Standard Vector Bundle E_D^4 on the Hilbert Scheme $Hilb^4 S$
of a Surface S 205
A.S.Tikhomirov, T.L.Troshina

Almost Canonical Polynomials of Algebraic Surfaces 227
Andrej N. Tyurin

Three Problems on an Exceptional Domain

Walter L. Baily, Jr.

We wish to consider three problems of an algebraico-geometric nature connected with hermitian symmetric tube domains, and in particular with the 27-dimensional tube domain $\mathcal{T}_e$ on which a certain real form of E_7 acts.

First some notation. Let $V \cong \mathbb{R}^n$ be an n-dimensional real vector space and $\mathcal{K}$ be a self-adjoint homogeneous convex cone such that the tube domain $\mathcal{T} = V + i\mathcal{K}$ is a complex hermitian symmetric domain whose group G of holomorphic automorphisms is a semi-simple real algebraic Lie group defined over $\mathbb{Q}$. Let $\Gamma \subset G(\mathbb{Q})$ be an arithmetic subgroup of $G(\mathbb{Q})$. Let R be any subring of $\mathbb{C}$ and let $\mathcal{Q}(\Gamma, R)$ be the graded R-algebra of modular forms f on $\mathcal{T}$ with respect to Γ such that all the coefficients of the Fourier expansion of f are in R, and let $\mathcal{Q}(\Gamma) = \mathcal{Q}(\Gamma, \mathbb{C})$.

I. We assume that

i) $\mathcal{Q}(\Gamma) = \mathcal{Q}(\Gamma, \mathbb{Q}) \otimes \mathbb{C}$, and

ii) There exists a *finite* set $\gamma = \{p_1, \ldots, p_s\}$ of primes called "bad" primes such that if $R = \mathbb{Z}[p_1^{-1}, \ldots, p_s^{-1}]$, then $\mathcal{Q}(\Gamma, R)$ is finitely generated as a graded R-algebra.

One has the following results:

1) If $\mathcal{T} = \mathcal{H}_n$, the Siegel upper half-space of degree n, and $\Gamma = \Gamma_n = \mathrm{PSp}(n, \mathbb{Z})$, then $\mathcal{Q}(\Gamma, \mathbb{Z})$ is a finitely generated graded ring. This is classical for $n = 1$, was proved by Igusa for $n = 2$, and proved for all $n > 2$ by Chai and Faltings [CF] (after Chai had proved the finiteness result for $R = \mathbb{Z}[1/2]$).

In fact, if i) holds and $\mathcal{Q}(\Gamma, \mathbb{Q})$ is the integral closure in its quotient field of its graded subalgebra generated by elements whose Fourier coefficients have bounded denominators (e.g., by Eisenstein series), then $\mathcal{Q}(\Gamma)$ is finitely generated as a graded algebra over $\mathbb{C}$ by $\mathcal{Q}(\Gamma, \mathbb{Z})$ (but this does not of course imply of itself that $\mathcal{Q}(\Gamma, \mathbb{Z})$ is itself finitely generated as a graded ring), and one can also show then that there is a finitely generated subring R of $\mathbb{Q}$ such that $\mathcal{Q}(\Gamma, R)$ is finitely generated as a graded R-algebra [Ba2], and this fact, which holds in particular for $\mathcal{T}_e$ and for a certain "nice" arithmetic subgroup Γ_e of $\mathrm{Hol}(\mathcal{T}_e)$, suffices for some number theoretic applications. However, the set of "bad" primes is unclear. Therefore, we pose:

Problem 1. Find a general approach to the problem of finding R as small as possible so that $\mathcal{Q}(\Gamma, R)$ is a finitely generated graded R-algebra. We have mentioned a "nice" arithmetic subgroup Γ_e of $G(\mathbb{R}) = E_{7(-25)}$ acting on $\mathcal{T}_e$ (described in [Ba1]). It is natural to conjecture that $\mathcal{Q}(\Gamma_e, \mathbb{Z})$ is finitely generated. One idea in approaching this problem is to find modular forms φ of low (=critical) weight with respect to Γ_e which might play the role which theta functions play in the case of the Siegel modular group. H.Kim [Ki] has found, using the method of analytic continuation of Eisenstein series (of Maass, Kubota, and Shimura), modular forms φ_4 and φ_8 of weights 4 and 8 with respect to Γ_e, all of whose Fourier coefficients are in $\mathbb{Z}$ and having leading Fourier coefficient $=1$. We hope these might play a role in constructing many $\varphi \in \mathcal{Q}(\Gamma_e, \mathbb{Z})$ of low weights of some pre-assigned initial Fourier expansion to obtain the conjectured result.

Remark. $\varphi_8 = \varphi_4^2$.

II. Suppose now $\xi \in \mathcal{T}$ is a special point, i.e., that there exists a maximal $\mathbb{R}$-anisotropic (compact) torus T of G, defined over $\mathbb{Q}$ such that ξ is the unique fixed point of $T(\mathbb{R})^0$ in $\mathcal{T}$. In principal it is not difficult to prove that if K_ξ is the CM-field of rationality of ξ, and if $K_\xi^\#$ is its reflex field, and if we define L_ξ to be the field

$$L_\xi = K_\xi^\#(\{\varphi(\xi) \mid \varphi : \text{arithmetic modular function finite at } \xi\}),$$

then $L_\xi/K_\xi^\#$ is an abelian extension. One wishes in general to describe the extension $L_\xi/K_\xi^\#$ by describing its reciprocity law, and, more generally, to describe the effect of an arbitrary $\sigma \in \text{Aut}(\mathbb{C})$ on L_ξ. The first part has been done in classical cases (Shimura) [Sh1,2], and a solution to the latter was conjectured by Langlands [LaR] and proved by Milne [Mi] and Borovoi [Bo]. All these solutions depend in principle on the theory of moduli and families of abelian varieties.

Problem 2 is to find an elementary resolution of the above issues. An elementary solution would be one not involving abelian varieties. This M. Karel and I have been able to do for arithmetic Hilbert Modular functions [Ba3] for the case when σ fixes the reflex field [Ba4, Ka1]. This implies the same result for Siegel modular functions [Ba5]. Karel has been able [Ka2] to verify Tate's version [LaS, Ch.7] of Langlands' conjecture in the case of arithmetic modular functions for a modular surface, i.e., in the case of a real quadratic field ($n = 2$). Unfortunately the proof is so simple it is hard to anticipate from it how to handle larger n. Also, it would perhaps be interesting to see what the reciprocity law is at special points of an exceptional domain, e.g., of $\mathcal{T}_e$. I don't know of anyone who has written this down.

III. Again let $\mathcal{T}_e$ be the exceptional 27-dimensional tube domain. Now if $\mathcal{T} = \mathcal{H}_n$ is the Siegel upper half-space and $\Gamma = \Gamma_n$, the Siegel modular group of degree n, then the orbits of Γ_n in $\mathcal{H}_n$ correspond one-to-one to the isomorphism classes of normally polarized abelian varieties of dimension n. Moreover, there is a Γ_n-invariant complex analytic closed subset $\mathcal{J}_n$ of $\mathcal{H}_n$ of which a Zariski-open subset $\mathcal{J}_n^\#$ is in one-to-one correspondence with canonically polarized Jacobian varieties of curves of genus n, hence via Torelli's theorem with the isomorphism classes of non-singular curves of genus n. If $n = 3$, then $\mathcal{J}_3 = \mathcal{H}_3$, and this special case is important for the considerations which follow.

However, there is no known interpretation of the space of orbits of Γ_e in $\mathcal{T}_e$ as the space of moduli of some family of polarized algebraic varieties. **Problem 3** is to seek such a family.

We approach this problem by considering the four Severi varieties of Zak [LaZ] S_n, $n = 1, 2, 3, 4$, where $\dim(S_n) = 2^n$. These are given explicitly as follows ($\mathbb{P}^n$ is the projective space of dimension n):

$$S_1 = \mathbb{P}^2 \hookrightarrow \mathbb{P}^5 \qquad \text{(Veronese imbedding);}$$

$$S_2 = \mathbb{P}^2 \times \mathbb{P}^2 \hookrightarrow \mathbb{P}^8 \qquad \text{(Segre imbedding);}$$

$$S_3 = G(2, 6) \hookrightarrow \mathbb{P}^{14} \qquad \text{(imbedding by Plücker coordinates),}$$

where $G(2, 6)$ is the Grassmannian of planes in 6-space; and

$$S_4 = C\mathbb{P}^2 \hookrightarrow \mathbb{P}^{26},$$

where C stands for the Cayley numbers and $C\mathbb{P}^2$ is the Cayley projective plane, realized as the projective variety of the primitive idempotents in the exceptional 27-dimensional Jordan algebra $\mathcal{J}$ of three-by-three hermitian matrices over C, whose projective space is $\mathbb{P}^{26}$, viewed as a 27-dimensional irreducible module of E_6.

Now observe that the *generic* quadric hypersurface section $Q \cap S_2$ of S_2 as imbedded in $\mathbb{P}^5$, where Q is a quadric in $\mathbb{P}^5$, is a non-singular, non-hyperelliptic plane quartic curve C of genus 3 in $\mathbb{P}^2$. Then, as mentioned above, the moduli of such curves are essentially given, via Torelli's theorem, by the orbits of Γ_3 in $\mathcal{H}_3$. By "essentially" we mean that this correspondence holds on the complement of a divisor on the Satake compactification $(\Gamma_3 \setminus \mathcal{H}_3)^*$ of the moduli space $\Gamma_3 \setminus \mathcal{H}_3$ of normally polarized abelian varieties.

This leads us to consider the four irreducible symmetric hermitian domains of $(\mathbb{R}-)$rank 3, of which $\mathcal{H}_3$ is that of lowest dimension, and of which $\mathcal{T}_e$ is that of the highest dimension, 27. The four domains referred to are:

$D_1 = \mathcal{H}_3 = \mathrm{Sp}(3, \mathbb{R})/K_1$, where $K_1 = \mathrm{U}(3)$, the group of 3 by 3 complex unitary matrices.

$D_2 = H_3$, the 9-dimensional tube domain in $\mathbb{C}^9 \cong M_3(\mathbb{C})$, the 3 by 3 complex matrices $Z = X + iY$, where X and Y are complex hermitian 3 by 3 matrices, and H_3 consists of those for which Y is positive definite, and can be written as $\mathrm{U}(3, 3)/K_2$, where $K_2 = \mathrm{U}(3) \times \mathrm{U}(3)$;

$D_3 = \mathbb{Q}_3$ is the tube domain in $\mathbb{C}^{15}$ analogous to H_3 with quaternion hermitian 3 by 3 matrices in place of complex hermitian, and $\mathbb{Q}_3$ can be written as $\mathrm{SO}^*(6)/K_3$ with $K_3 = \mathrm{U}(3)$. (The notation $\mathrm{SO}^*(2n)$ is from Helgason: "Differential Geometry and Symmetric Spaces", 1st Ed., p.354.)

$D_4 = \mathcal{T}_e$ is the tube domain $\{Z = X + iY \mid X, Y \in$ 27-dimensional real exceptional Jordan algebra $\mathcal{J}$ of 3 by 3 hermitian matrices over C such that Y is positive definite (written $Y > 0$)}. We may write $D_4 = E_{7(-25)}/K_4$, where $K_4 = E_{6(-78)} \times C'$, C' being the unit circle.

We note in each of the above cases that *there is a natural action of K_n on the ambient manifold of S_n*, if we identify that ambient manifold with the appropriate Jordan algebra of hermitian matrices over R, C, H, or C; thus,

$$\mathbb{P}^5 = \mathbb{P}(S_3), \quad S_3 \text{ being the 3 by 3 symmetric matrices;}$$

$$\mathbb{P}^8 = \mathbb{P}(H_3), \quad H_3 \text{ being the hermitian 3 by 3 complexmatrices;}$$

$$\mathbb{P}^{14} = \mathbb{P}(Q_3), \quad Q_3 \text{ being the quaternion hermitian 3 by 3matrices;}$$

$$\mathbb{P}^{26} = \mathbb{P}(\mathcal{J}_3), \quad \mathcal{J}_3 \text{ being the 3 by 3 Cayley hermitian matrices.}$$

Motivated by these considerations, we try the simplest first step. Namely, we examine the configurations of algebraic varieties which arise when for a generic quadric hypersurface $Q \subset \mathbb{P}^8$ we construct the non-singular 3-fold

$$F = F_Q = Q \cap (\mathbb{P}^2 \times \mathbb{P}^2) \subset \mathbb{P}^8.$$

A hyperplane section of F is a $K-3$ surface in two different ways. On the one hand the generic hyperplane section $F^H = F \cap H$ of F (where H is a hyperplane in $\mathbb{P}^8$) is a $K-3$ surface of genus 7. The generic hyperplane section of F^H is a canonical curve of genus 7 in $\mathbb{P}^6$. On the other hand, F is fibered into conics over $\mathbb{P}^2$ with a sextic branch curve. We now explain this more in detail. Explicitly, F is fibered into conics over $\mathbb{P}^2$ as follows:

$$\pi = \pi_Q : Q \cap (\mathbb{P}^2 \times \mathbb{P}^2) \longrightarrow \mathbb{P}^2,$$

where π_Q is the restriction of pr_2 to $F = F_Q$. For $s \in \mathbb{P}^2$, $\pi^{-1}(s) = Q \cap (\mathbb{P}^2 \times \{s\})$, which is a plane conic in the coordinates of the first factor by virtue of the nature of the Segre imbedding $\mathbb{P}^2 \times \mathbb{P}^2 \longrightarrow \mathbb{P}^8$. Let Δ_π be the locus of $s \in \mathbb{P}^2$ such that $\pi^{-1}(s) = F_s$ is the union of two lines. Let $r = [r_0 : r_1 : r_2]$ resp. $s = [s_0 : s_1 : s_2]$ be the coordinates in the first resp. second factor $\mathbb{P}^2$, and t_{ij}, $0 \le i, j \le 2$, be the coordinates in $\mathbb{P}^8$. Suppose the quadric hypersurface Q in $\mathbb{P}^8$ is given by $A(t) = 0$, where A is the quadratic form

$$A(t) = \sum a_{ijkl} t_{ij} t_{kl},$$

so that with $t_{ij} = r_i s_j$ we have

$$F_s : \sum b_{ik} r_i r_k = 0,$$

with $b_{ik} = \sum_{j,l} a_{ijkl} s_j s_l$. The conic degenerates to two lines if and only if the discriminant $|b_{ik}| = \det(b_{ik}) = 0$, and $|b_{ik}|$ is a homogeneous cubic polynomial in $\{b_{ik}\}$, hence (for fixed a_{ijkl}) is a homogeneous sextic polynomial in s_0, s_1, s_2. Therefore, Δ_π is a sextic plane curve in $\mathbb{P}^2$. It follows from known formulae [Is, §14.5] that then the third Betti number $b_3(F)$ is equal to 18. "In general", Δ_π is non-singular (for generic choice of a_{ijkl}). Therefore, by [B, Théoreme 2.1 (with $n = 1$)] the level 3 Hodge structure of F is of the form

$$H^3(F) = H^{2,1}(F) \oplus H^{1,2}(F),$$

so that the intermediate Jacobian $J(F) = H^{2,1}(F)/H^3(F, \mathbb{Z})$ is a normally polarized abelian variety of dimension $(1/2)b_3(F) = 1$. We know [V: 2.7, 2.8; B: 6.23; Di] the following:

Proposition. *A sufficiently general homogeneous sextic polynomial* $\Sigma(s) = \Sigma(s_0, s_1, s_2)$ *can be expressed as a symmetric determinant*

$$\det(b_{ik}(s)), \quad b_{ik}(s) = b_{ki}(s),$$

where $b_{ik}(s)$ *are quadratic forms in* $s = (s_0, s_1, s_2)$.

Moreover, given a sufficiently general sextic $\Sigma(s)$, *one may reconstruct uniquely the Fano 3-fold as a fibering by conics with the curve* $\triangle_\pi : \Sigma(s) = 0$ *as the base locus of its singular fibers.*

Thus it is that a general hyperplane section of the Fano 3-fold is a $K - 3$ surface S in $\mathbb{P}^7$, realized both as a sextic double cover of $\mathbb{P}^2$ and as a $K - 3$ surface of genus 7 whose general hyperplane section in $\mathbb{P}^7$ is a canonically imbedded curve of genus 7 in $\mathbb{P}^6$. Each type of $K - 3$ surface by itself has 19 moduli, but if we consider the family of $K - 3$ surfaces obtained as the general quadric section of $\mathbb{P}^2 \times \mathbb{P}^2$ in $\mathbb{P}^8$, then, as S. Tregub observed to me at the Yaroslavl' conference, and explained in greater detail in a later written communication, this implies that the number of moduli of $m(F^H)$ of this kind of F^H is equal to 18.

More importantly, Tregub has explained in the longer written communication [T] how to single out a subfamily of such F^H having 9 moduli. This is important because we should like, if possible, to link this set-up with the second symmetric hermitian space H_3 and Hermitian modular functions, for the reasons suggested earlier.

The 9-dimensional family described to me by Tregub is the family of $K - 3$ surfaces in our family which may be described as follows. Let $\mathcal{F}$ be the 18-dimensional family of $K - 3$ surfaces F^H described above, let $\mathcal{E}$ be the family of all $K - 3$ surfaces having a fixed-point free involution, which are therefore the 2-fold covers of Enriques surfaces. Put $\mathcal{F}_\mathcal{E} = \mathcal{F} \cap \mathcal{E}$. Then he shows that $\mathcal{F}_\mathcal{E}$ has a component $\mathcal{F}_{\mathcal{E},O} = M$, say, such that dim $M = 9$ and the Enriques surfaces corresponding to $K - 3$ surfaces from the family M are exactly those which contain a smooth rational curve (cf.[N]). In fact it would appear that the family M is exactly the family of Reye congruences described by Cossec [Co].

Now I present some highly speculative considerations and should like to raise the question whether there is some reasonable connection between this family M and some Zariski-open subset of a quotient of H_3 by some arithmetic group. At first glance this might seem unlikely. In fact, M is a family of $K - 3$ surfaces and should be (more than likely is) parametrized by a Zariski-open subset of an arithmetic quotient of a symmetric tube domain of type IV and of dimension 9. However, there are examples of pairs of quite different symmetric domains, say D_1 and D_2 on which arithmetic groups Γ_1 and Γ_2 operate such that if $(D/\Gamma)^\sim$ is a suitable smooth toroidal compactification of D/Γ, then there are normal crossing divisors $\triangle_1$ and $\triangle_2$ on $(D_1/\Gamma_1)^\sim$ and on $(D_2/\Gamma_2)^\sim$ such that $(D_1/\Gamma_1)^\sim$ is isomorphic to $(D_2/\Gamma_2)^\sim$, both equal to V, say, such that

$$V - \triangle_1 = D_1/\Gamma_1 \quad \text{and} \quad V - \triangle_2 = D_2/\Gamma_2. \tag{1}$$

Such examples are given by B. Hunt in [H] and by Hunt and S. Weintraub in [H-W].

Moreover, there is a natural geometric interpretation of both terms in (1) and the geometric relation is *not random*. In our case it seems necessary to see if there is some connection of this nature between a 9-dimensional domain of type IV, most naturally connected with certain $K - 3$ surfaces, and the domain H_3, most naturally connected with abelian varieties A *of dimension 6* such that $\mathrm{End}_Q(A)$ contains an imaginary quadratic number field, hence possibly with certain curves of *genus six*.

The point, it seems to me, is to look at the Prym varieties and the so-called Prym mapping which have been extensively studied in the articles [Co], [V], [Be], [D-5] and [Tyu]. For example, in [Co, p.750] it is stated that if $R(W)$ is the Reye congruence of a "good" web in $\mathbb{P}^5$, then its generic hyperplane section is a Prym-canonical curve of *genus six*. Of course, all of this is too vague to draw definite conclusions at this point. But to me, at least, it is quite suggestive and tantalizing.

References

[Ba1]	BAILY, W.L.,JR., *An exceptional arithmetic group and its Eisenstein series*, Ann. Math. 91(1970), 512-549.
[Ba2]	BAILY, W.L.,JR., *Theorems on the Finite Generation of Algebras of Modular Forms*, Amer. J. Math. 104(1982), 645-682.
[Ba3]	BAILY, W.L.,JR., *On the Theory of Hilbert Modular Functions I, Arithmetic Groups and Eisenstein Series*, J. Alg. 90(1984), 567-605.
[Ba4]	BAILY, W.L.,JR., *Reciprocity Laws for Special Values of Hilbert Modular Functions*, Proceedings of KIT Mathematics Workshop, Korea Institute of Technology, Taejon, Korea, 1986.
[Ba5]	BAILY, W.L.,JR., *On the proof of the reciprocity law for arithmetic Siegel modular functions*, Proc. Indian Acad. Sci., Math. Sci., 97(1987) 21-30 (1988).
[Be]	BEAUVILLE, A., *Variétés de Prym et Jacobiennes Intermédiaires*, Ann. Éc. Norm. Sup., 10(1977).
[Bo]	BOROVOI, M.V., *Langlands' Conjecture Concerning Conjugation of Connected Shimura Varieties*, Sel. Math. Sov. vol.3, No.1(1983/4).
[CF]	CHAI, C-L., FALTINGS, G., *Degeneration of Abelian Varieties*, Springer, 1990.
[Co]	COSSEC, F.R., *Reye Congruences*, Trans.AMS 280(1983), 737-751.
[DI]	DIXON, A.C., *Note on the Reduction of a Ternary Quantic to a Symmetrical Determinant*, Proc. Cambridge Phil. Soc., 11(1902), 350-351.
[D-S]	DONAGI, R. and SMITH, R.C., *The Structure of the Prym Map*, Acta. Math., v.146.
[H]	HUNT, B., *A Siegel Modular 3-fold that is a Picard Modular 3-fold*, pre-print, Göttingen, 1988(or Comp. Math. 76(1990), 203-242).
[H-W]	HUNT, B., and WEINTRAUB, S.H., *Janus-like Algebraic Varieties*, Preprint No. 194, Universität Kaiserslautern, 1991.
[Is]	ISKOVSKIKH, V.A., *Lectures on Three-Dimensional Algebraic Manifolds, Fano Manifolds*, Matematika, Moscow Univ., 1988 (Russian).

[Ka1]   KAREL, M., *Eisenstein Series and Fields of Definition*, Comp. Math. 37(1978), 121-169. Revised and Corrected Version of same to appear in Abh. Math. Sem. Hamburg.

[Ka2]   KAREL, M., *Special Values of Hilbert Modular Functions*, Revista Matemática Iberoamericana, 2(1986), 367-380.

[Ki]   KIM, H., *On a Modular Form on the Exceptional Domain of E_7*, Thesis, University of Chicago 1992.

[LaS]   LANG, S., *Complex Multiplication*, Springer, 1983.

[LaR]   LANGLANDS, R., *Ein Märchen, Automorphic Representations, etc.*, Proc. Sympos. Pure Math. v.33, Amer. Math. Soc., 1979.

[LaZ]   LAZARSFELD, R., and VAN DE VEN, A., *Topics in the Geometry of Projective Space, Recent Work of F.L.Zak with an addendum by F.L.Zak*, Birkhäuser, 1984.

[Mi]   MILNE, J., *The Action of an Automorphism of $\mathbb{C}$ on a Shimura Variety and its Special Points*, Progress in Math. 35, Birkhäuser 1983.

[N]   NAMIKAWA, Y., *Periods of Enriques Surfaces*, Math. Ann. 270(1985), 201-222.

[Sh1]   SHIMURA, G., *On arithmetic automorphic functions*, Proc. ICM 1970.

[Sh2]   SHIMURA, G., *Arithmetic Theory of Automorphic Functions*, Publ. Math. Soc. Jap. 11, Princeton University Press, 1971.

[Tr]   TREGUB, S., Written communication dated Sept. 1992.

[Tyu]   TYURIN, A.N., *The Geometry of the Poincaré Theta-Divisor of a Prym Variety*, Izvestia Mat., 39, No.5(1979) (Russian).

[V]   VERRA, A., *The Prym Map has Degree Two on Plane Sextics*, 1992.

On the Dimension of the Adjoint Linear System for Quadric Fibrations

M.C.Beltrametti, G.M.Besana and A.J.Sommese

Introduction. Let $L^\wedge$ be a very ample line bundle on a smooth, n-dimensional, projective manifold $X^\wedge$, i.e. assume that $L^\wedge \approx i^* \mathcal{O}_{\mathbb{P}^N}(1)$ for some embedding $i : X^\wedge \longrightarrow \mathbb{P}^N$. In [S1] it is shown that for such pairs, $(X^\wedge, L^\wedge)$, the Kodaira dimension of $K_{X^\wedge} \otimes L^{\wedge n-2}$ is ≥ 0, i.e. there exists some positive integer, t, such that $h^0\big((K_{X^\wedge} \otimes L^{\wedge n-2})^t\big) \geq 1$, except for a short list of degenerate examples. It is moreover shown that except for this short list there is a morphism $r : X^\wedge \longrightarrow X$ expressing $X^\wedge$ as the blow-up of a projective manifold X at a finite set B, and such that:

1. $K_{X^\wedge} \otimes L^{\wedge n-1} \approx r^*(K_X \otimes L^{n-1})$ where $L := (r_* L^\wedge)^{**}$ is an ample line bundle and $K_X \otimes L^{n-1}$ is ample;

2. $K_X \otimes L^{n-2}$ is nef, i.e. $(K_X \otimes L^{n-2}) \cdot C \geq 0$ for every effective curve $C \subset X$.

The hope (see [BS3] for a discussion of the problem and strong evidence) is that except for a few examples, $K_X \otimes L^{n-2}$ is not just nef, but spanned at all points by global sections.

For the case when the Kodaira dimension of $K_X \otimes L^{n-2}$ is 2, the first and the last authors classified the possible singular fibers of the map given by $(K_X \otimes L^{n-2})^N$ for large N. Using these results, the second author showed in [Bs1, Bs2] that the map given by $(K_X \otimes L^{n-2})^N$ for large N factors through a map with equal dimensional fibers, $\varphi : X \longrightarrow Y$, onto a smooth surface, Y, such that $K_X \otimes L^{n-2} \approx \varphi^*(K_Y \otimes H)$, where H is an ample line bundle. Using these new results the second author went on in [Bs1, Bs2] to show that $K_Y \otimes H$, and hence $K_X \otimes L^{n-2}$, is spanned except when certain invariants are small.

In this article we use the technique introduced in [BS3] and the results of the second author to show that $h^0(K_{X^\wedge} \otimes L^{\wedge n-2}) \geq 2$, when the Kodaira dimension of $K_X \otimes L^{n-2}$ is 2 with 3 possible exceptions (see Theorems (2.2), (3.3)).

The first author would like to thank the organizers of Yaroslavl' Conference, August 1992, for their kind hospitality. The second author wants to thank the CNR, Consiglio Nazionale delle Ricerche, for its support. The third author would also like to thank the Max-Planck-Institut in Bonn for its support during the final stages of this paper writing, and the National Science Foundation (NSF Grant DMS 89-21702) for their support.

0. Background material

We work over the complex numbers $\mathbb{C}$. Throughout the paper we deal with smooth, projective varieties, V. We denote by $\mathcal{O}_V$ the structure sheaf of V and by K_V the canonical bundle. For any coherent sheaf $\mathcal{F}$ on V, $h^i(\mathcal{F})$ denotes the complex dimension of $H^i(V, \mathcal{F})$.

Let L be a line bundle on V. L is said to be *numerically effective* (*nef*, for short) if $L \cdot C \geq 0$ for all effective curves C on V and in this case L is said to be *big* if $c_1(L)^n > 0$ where $c_1(L)$ is the first Chern class of L and $n = \dim V$.

(0.1) The notations used in this paper are standard from algebraic geometry. Let us only fix the following.

$\approx$ (respectively $\sim$), the linear (respectively numerical) equivalence of line bundles;

$\chi(L) = \sum_i (-1)^i h^i(L)$, the Euler characteristic of a line bundle L;

$|L|$, the complete linear system associated with a line bundle L on a variety V.

We denote by $\Gamma(L)$ the space of the global sections of L and we say that L is *spanned* if it is spanned at all points of V by $\Gamma(L)$;

$e(V) = c_n(V)$, the topological Euler characteristic of V, for V smooth, where $c_n(V)$ is the n-th Chern class of the tangent bundle of V. If V is a surface, $e(V) = 12\chi(\mathcal{O}_V) - K_V \cdot K_V$.

Line bundles and divisors are used with little (or no) distinction. Hence we shall freely switch from the multiplicative to the additive notation and viceversa.

(0.2) For a line bundle L on a variety V of dimension n the *sectional genus*, $g(L) = g(V, L)$, of (V, L) is defined by $2g(L) - 2 = \left(K_V + (n-1)L\right) \cdot L^{n-1}$.

(0.3) Reduction (see e.g. [S1], (0.5), [SV], (0.3) and [BS2], §3). Let $(X^\wedge, L^\wedge)$ be a smooth n-dimensional projective variety polarized with a very ample line bundle $L^\wedge$. A smooth polarized variety (X, L) is called a *(first) reduction* of $(X^\wedge, L^\wedge)$ if there is a morphism $r : X^\wedge \longrightarrow X$ expressing $X^\wedge$ as the blowing up of X at a finite set of points, B, such that $L := (r_* L^\wedge)^{**}$ is ample, and $L^\wedge \approx \pi^* L - [r^{-1}(B)]$ or, equivalently, $K_{X^\wedge} + (n-1)L^\wedge \approx r^*\left(K_X + (n-1)L\right)$.

Furthermore, except for four special cases in dimension 2 (see [SV]), $K_X + (n-1)L$ is very ample and the reduction is unique up to isomorphism.

Note also that there is a one to one correspondence between smooth divisors of $|L|$ which contain the set B and smooth divisors of $|L^\wedge|$.

Furthermore, well known results from adjunction theory (see [S1], [BS2]) state that, except for an explicit list of special polarized varieties $(X^\wedge, L^\wedge)$, such a reduction, (X, L), exists. Moreover, for $n \geq 3$, $K_X + (n-2)L$ is nef and big except for an explicit list of special polarized varieties (X, L). In this paper we deal with one of the special classes (X, L), of dimension $n = 3$, where $K_X + L$ is nef but not big (see (0.7) below).

(0.4) Pluridegrees. Let $(X^\wedge, L^\wedge), (X, L)$ be as in (0.3) with $n = 3$. Define the *pluridegrees*, for $j = 0, 1, 2, 3$,

$$d_j^\wedge = (K_{X^\wedge} + L^\wedge)^j \cdot L^{\wedge 3-j}, \quad d_j := (K_X + L)^j \cdot L^{3-j}.$$

If γ denotes the number of points blown up under $r : X^\wedge \longrightarrow X$, the invariants $d_j^\wedge, d_j$ are related by

$$d_0^\wedge = d_0 - \gamma; \quad d_1^\wedge = d_1 + \gamma; \quad d_2^\wedge = d_2 - \gamma; \quad d_3^\wedge = d_3 + \gamma.$$

We put $d := d_0, d^\wedge := d_0^\wedge$. Moreover, if $K_X + L$ is nef, by the generalized Hodge index theorem (see e.g. [BBS], (0.15)) one has

$$d_1^2 \geq dd_2; \quad d_2^2 \geq d_1 d_3. \tag{0.4.1}$$

(0.5) Double point formula. We need the following (see also [BBS], (2.11.4)).

(0.5.1) Theorem. *Let $(X^\wedge, L^\wedge)$ be a smooth projective 3-fold, polarized with a very ample line bundle $L^\wedge$. Assume that $|L^\wedge|$ embeds $X^\wedge$ in $\mathbb{P}^N$ with $N \geq 6$. Let $d_j^\wedge, j = 0, 1, 2, 3$, be the pluridegrees of $(X^\wedge, L^\wedge)$ as in (0.4). Let $S^\wedge$ be a smooth element of $|L^\wedge|$. Then*

$$e(X^\wedge) - 48\chi(\mathcal{O}_{X^\wedge}) + 84\chi(\mathcal{O}_{S^\wedge}) - 11d_2^\wedge - 17d_1^\wedge - d_3^\wedge + d^\wedge(d^\wedge - 20) \geq 0.$$

Proof. It is simply a particular case of the general formula (I, 37), Section D, p. 313 of [K]. It should be noted that the virtual normal bundle, ν, in that formula is defined in our situation by the exact sequence

$$0 \longrightarrow \mathcal{T}_{X^\wedge} \longrightarrow p^*\mathcal{T}_{\mathbb{P}^6} \longrightarrow \nu \longrightarrow 0$$

where $p : X^\wedge \longrightarrow \mathbb{P}^6$ is the restriction to $X^\wedge$ of the projection from a general $\mathbb{P}^{N-7}$ if $N > 6$ and $\nu = \mathcal{N}_{X^\wedge}^{\mathbb{P}^6}$, the usual normal bundle, if $N = 6$. Q.E.D.

(0.6) Tsuji inequality (see [S2], §1, [T], §5). Let $(X^\wedge, L^\wedge), (X, L)$ be as in (0.3) with $n = 3$ and let S be a smooth element of $|L|$. Then we have

$$(K_X + L)^3 + \frac{8}{3}K_S \cdot L_S \leq 32\big(2h^0(K_X + L) - \chi(\mathcal{O}_S)\big)$$

or

$$h^0(K_X + L) \geq \frac{d_3}{64} + \frac{d_1}{24} + \frac{\chi(\mathcal{O}_S)}{2}.$$

(0.7) Conic fibrations. Let $(V, \mathcal{A})$ be a smooth n-dimensional variety polarized with an ample line bundle $\mathcal{A}$. We say that $(V, \mathcal{A})$ is a *quadric fibration* if there exists a surjective morphism $\varphi : V \longrightarrow Y$ with connected fibers onto a variety Y of dimension m such that $K_V + (n - m)\mathcal{A} \approx \varphi^*\mathcal{H}$ for an ample line bundle $\mathcal{H}$ on Y. Note that quadric fibrations are called quadric bundles in [Bs1], [Bs2]. We also say that $(V, \mathcal{A})$ is a *conic fibration* if $n - m = 1$.

(0.7.1) Theorem ([Bs1], (5.4.1), (6.1.3), [Bs2], (1.3), (1.4)). *Let $(X^\wedge, L^\wedge), (X, L)$, and $r : X^\wedge \longrightarrow X$ be as in (0.3) with $n = 3$. Assume that (X, L) is a conic fibration over a normal surface Y with structural morphism $\varphi : X \longrightarrow Y$ and with $K_X + L \approx \varphi^*\mathcal{H}$ for an ample line bundle $\mathcal{H}$ on Y. Let $\Sigma \subset Y$ be the set of points over which there are divisorial fibers. Let $\pi : Z \longrightarrow Y$ be the blowing up of Y at Σ. Then*

1) Y *is a smooth surface;*

2) *there is an ample line bundle H on Y such that $\mathcal{H} = K_Y + H$;*

3) *there exists a flat map $\psi : X \longrightarrow Z$ such that the following diagram commutes*

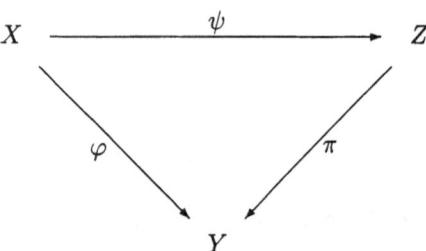

4) *there exists an ample line bundle $\mathcal{L}$ on Z such that (Y, H) is the first reduction of $(Z, \mathcal{L})$;*

5) *if $\mathcal{L} \cdot \mathcal{L} \geq 5$ then $K_X + L$ is spanned by its global sections.*

(0.7.2) *Let $(X, L), \varphi : X \longrightarrow Y, \mathcal{H}, H$ be as in (0.7.1). First, note that*

$$d_3 = (K_X + L)^3 = 0$$

since $K_X + L \approx \varphi^*\mathcal{H}$ is nef but not big. Let S be a smooth element of $|L|$ and let p be the restriction of φ to S. Then we have $K_S \approx p^*\mathcal{H}$ so that K_S is nef and big, i.e. S is a minimal surface of general type. Note also that p is a generically 2 to 1 morphism. Hence in particular

$$d_2 = K_S \cdot K_S = 2\mathcal{H} \cdot \mathcal{H} \geq 2 \quad and \quad d_1 = K_S \cdot L_S > 0.$$

We have

$$p_g(S) = p_g(Y) + h^0(K_X + L). \tag{0.7.2.1}$$

Indeed by the exact sequence

$$0 \longrightarrow K_X \longrightarrow K_X \otimes L \longrightarrow K_S \longrightarrow 0$$

we get

$$h^0(K_X) - h^0(K_X + L) + h^0(K_S) - h^1(K_X) = 0.$$

Since $h^0(K_X) = 0, h^1(K_X) = h^2(\mathcal{O}_X)$ by Serre duality, and $h^2(\mathcal{O}_X) = h^2(\mathcal{O}_Y)$ by Leray, we have the result. $\square$

For any further background material we refer to [Bs1], [Bs2], [BS2] and [BS3].

1. Preliminary results

In this section we reduce to the essential case which will be discussed in the following section and we prove an useful consequence of the double point formula (0.5).

(1.1) Proposition. *Let* $(X^\wedge, L^\wedge)$ *be a smooth connected threefold polarized with a very ample line bundle* $L^\wedge$. *Let* (X, L) *be the first reduction of* $(X^\wedge, L^\wedge)$ *and assume that* (X, L) *is a conic fibration. Let* S *be a smooth element in* $|L|$ *and let* $(Z, \mathcal{L})$ *be as in* (0.7.1). *Then* $h^0(K_{X^\wedge} + L^\wedge) = h^0(K_X + L) \geq 1$. *Furthermore* $h^0(K_{X^\wedge} + L^\wedge) \geq 2$ *unless* $\chi(\mathcal{O}_S) = 1$ *and* $\mathcal{L} \cdot \mathcal{L} \leq 4$.

Proof. Let L_S be the restriction of L to S. Recall that $d_1 = K_S \cdot L_S > 0$ (see (0.7.2)). Then Tsuji inequality (0.6) gives $h^0(K_X + L) \geq 1$ and furthermore $h^0(K_X + L) \geq 2$ as soon as $\chi(\mathcal{O}_S) > 1$.

If $\mathcal{L} \cdot \mathcal{L} \geq 5$ then $K_X + L$ is spanned by its global sections by (0.7.1), 5) and hence $h^0(K_X + L) \geq 3$.

Finally the equality $h^0(K_{X^\wedge} + L^\wedge) = h^0(K_X + L)$ is proved in [S2], (0.3.1). Q.E.D.

Thus from now on we can fix the following

(1.2) Assumptions. Notation as in (1.1). We can assume

$$h^0(K_X + L) = 1; \quad \chi(\mathcal{O}_S) = 1; \quad \mathcal{L} \cdot \mathcal{L} \leq 4. \qquad (1.2.1)$$

Then in particular the exact sequence

$$0 \longrightarrow \check{K}_X \longrightarrow K_X \otimes L \longrightarrow K_S \longrightarrow 0$$

gives $h^0(K_X + L) = \chi(K_X) + \chi(K_S) = \chi(\mathcal{O}_S) - \chi(\mathcal{O}_X)$, whence

$$\chi(\mathcal{O}_X) = \chi(\mathcal{O}_{X^\wedge}) = 0 \qquad (1.2.2)$$

as well as, since $h^0(K_X) = 0$,

$$p_g(S) = q(S) > 0. \qquad (1.2.3)$$

Therefore $q(X) > 0$, so we can also assume by the Barth-Lefschetz theorem that $|L^\wedge|$ embeds $X^\wedge$ in $\mathbb{P}^N$ with

$$N \geq 6. \qquad (1.2.4)$$

We may also assume

$$d_1 \geq 5. \qquad (1.2.5)$$

Indeed, let $S^\wedge$ be a smooth element in $|L^\wedge|$. Since $S^\wedge$ is a surface of general type (see (0.7.2)) we have from [LS], (0.6) that $d^\wedge = \deg(S^\wedge) > 8$. Hence $d := L^3 > 8$ and therefore $d_1^2 \geq d d_2$ yields $d_1 \geq 5$.

In the case of conic fibrations we derive from (0.5) the following useful numerical bound (compare with [BS3]).

(1.3) Theorem. *Let* $(X^\wedge, L^\wedge)$ *be a smooth projective threefold polarized with a very ample line bundle* $L^\wedge$. *Assume that* $(X^\wedge, L^\wedge)$ *has a conic fibration* (X, L) *as first reduction and let* γ *be the number of points blown up under* $r : X^\wedge \longrightarrow X$. *Let* $d^\wedge := L^{\wedge 3}$ *be the degree of* $(X^\wedge, L^\wedge)$ *and let* d_1, d_2 *be as in* (0.4). *Let* $S^\wedge$ *be a smooth element in* $|L^\wedge|$. *Suppose that* $\chi(\mathcal{O}_{S^\wedge}) = 1$, $\chi(\mathcal{O}_{X^\wedge}) = 0$ *and that* $|L^\wedge|$ *embeds* $X^\wedge$ *in* $\mathbb{P}^N$ *with* $N \geq 6$ *(cfr.* (1.2))*. Thus*

$$106 \geq (20 - d^\wedge)d^\wedge + 12d_2 + 17d_1 + 5\gamma.$$

Proof. Let $h^1(\mathcal{O}_{X^\wedge}) = a$. Then, since $\chi(\mathcal{O}_X) = \chi(\mathcal{O}_{X^\wedge}) = h^3(\mathcal{O}_{X^\wedge}) = 0$, we have $h^2(\mathcal{O}_{X^\wedge}) = a - 1$. Therefore the $h^{p,q} := h^{p,q}(X^\wedge) = h^q(\Omega^p_{X^\wedge})$ cohomology table for $X^\wedge$ looks like (recall that $h^{p,q} = h^{q,p}$ and the Serre duality $h^{p,q} = h^{3-p,3-q}$)

0	$a-1$	a	1
$a-1$	c	b	a
a	$h^{1,1} := b$	$h^{1,2} := c$	$a-1$
1	a	$a-1$	0

Note also that $h^{3,2} \leq h^{2,1}$ (see [ShS], (2.73), p. 47), so we can assume $c = a + \varepsilon$ for some non negative integer ε. Let $b_j := b_j(X^\wedge) = \sum_{j=p+q} h^{p,q}$ be the j-th Betti number of $X^\wedge$. Then

$$e(X^\wedge) := 1 - b_1 + b_2 - b_3 + b_4 - b_5 + 1 = 2b - 2a - 2\varepsilon - 2. \qquad (1.3.1)$$

Let S be a smooth element in $|L|$ corresponding to $S^\wedge$. Recall that $h^1(\mathcal{O}_{S^\wedge}) = h^1(\mathcal{O}_{X^\wedge}) = a$. Hence in particular $h^2(\mathcal{O}_{S^\wedge})(= h^2(\mathcal{O}_S)) = a$ since $\chi(\mathcal{O}_{S^\wedge}) = \chi(\mathcal{O}_S) = 1$. Therefore the $h^{p,q}$ cohomology table for $S^\wedge$ becomes

a	a	1
a	h	a
1	a	a

where $h := h^{1,1}(S^\wedge)$. Then

$$e(S^\wedge) := 1 - b_1(S^\wedge) + b_2(S^\wedge) - b_3(S^\wedge) + 1 = h - 2a + 2.$$

By the Lefschetz Theorem on hyperplane sections (see [GH], p. 157) one has $h^{1,1}(X^\wedge) = b \leq h$. Thus, by (1.3.1),

$$e(X^\wedge) \leq 2h - 2a - 2\varepsilon - 2 = 2(h - 2a + 2) + 2a - 2\varepsilon - 6 = 2e(S^\wedge) + 2a - 2\varepsilon - 6.$$

Therefore, by using the Noether inequality $d_2 = K_S \cdot K_S \geq 2(p_g(S) - 2) = 2a - 4$, we get

$$e(X^\wedge) \leq 2e(S^\wedge) + d_2 - 2. \qquad (1.3.2)$$

Now,

$$e(S^\wedge) = 12\chi(\mathcal{O}_S) - K_{S^\wedge} \cdot K_{S^\wedge} = 12 - (K_{S^\wedge} \cdot K_S - \gamma) = 12 - d_2 + \gamma.$$

Hence, by (1.3.2),

$$e(X^\wedge) \leq 22 - d_2 + 2\gamma. \qquad (1.3.3)$$

Since $\chi(\mathcal{O}_{X^\wedge}) = 0$, $\chi(\mathcal{O}_{S^\wedge}) = 1$, the double point formula (0.5) gives

$$e(X^\wedge) + 84 \geq 11d_2^\wedge + 17d_1^\wedge + d_3^\wedge + (20 - d^\wedge)d^\wedge = (20 - d^\wedge)d^\wedge + 11d_2 + 17d_1 + 7\gamma. \qquad (1.3.4)$$

By combining (1.3.3) and (1.3.4) we get the result. Q.E.D.

2. On the dimension of the adjoint bundle

In this section we prove the main results of the paper.

(2.1) Theorem. *Let $(X^\wedge, L^\wedge)$ be a smooth connected threefold polarized with a very ample line bundle $L^\wedge$. Let (X, L) be the reduction of $(X^\wedge, L^\wedge)$ and assume that (X, L) is a conic fibration. Let S be a smooth element of $|L|$. Then $h^0(K_{X^\wedge} + L^\wedge) \geq 2$ unless $d_2 = K_S \cdot K_S \in \{2, 4\}$.*

Proof. From (0.7.2) we know that $d_2 \geq 2$ and from (1.2.5) we can assume $d_1 \geq 5$. Recall that from (1.1) (see also (1.2)) we can also assume $h^0(K_X + L) = \chi(\mathcal{O}_S) = 1$, $\chi(\mathcal{O}_X) = \chi(\mathcal{O}_{X^\wedge}) = 0$.

Note that from Tsuji inequality (0.6) we obtain the bound $h^0(K_X + L) \geq 2$ as soon as $d_1 \geq 13$.

Assume $d_2 \geq 8$. Then from the inequality of (1.3) we find

$$106 \geq (20 - d^\wedge)d^\wedge + 96 + 85 = (20 - d^\wedge)d^\wedge + 181.$$

Hence $d^\wedge \leq 20$ is clearly not possible. Let $d^\wedge = 21$. Then $106 \geq -21 + 181$, again a contradiction. Thus $d \geq d^\wedge \geq 22$ so that $d_1^2 \geq dd_2$ gives $d_1^2 \geq 176$, or $d_1 \geq 14$ so we are done.

Let $d_2 = 6$. Therefore (1.3) leads to

$$106 \geq (20 - d^\wedge)d^\wedge + 72 + 85 = (20 - d^\wedge)d^\wedge + 157.$$

Again, if $d^\wedge \leq 22$, we clearly find a contradiction.

Let $d^\wedge = 23$. Then $d_1^2 \geq dd_2 \geq d^\wedge d_2 \geq 138$ gives $d_1 \geq 12$. If $d_1 \geq 13$ we are done. If $d_1 = 12$ the inequality in (1.3) yields the numerical contradiction $106 \geq -69 + 72 + 204$.

Let $d^\wedge = 24$. Then $d_1^2 \geq dd_2 \geq d^\wedge d_2 \geq 144$ gives $d_1 \geq 12$. If $d_1 \geq 13$ we are done. If $d_1 = 12$ the inequality in (1.3) yields the numerical contradiction $106 \geq -96 + 72 + 204$.

Thus $d^\wedge \geq 25$. Then $d_1^2 \geq d^\wedge d_2 \geq 150$ implies $d_1 \geq 13$ and we are done. Q.E.D.

We proceed further by considering the two remaining cases $d_2 = 2$, 4 of (2.1).

(2.2) Theorem. *Let $(X^\wedge, L^\wedge)$ be a smooth connected threefold polarized with a very ample line bundle $L^\wedge$. Let (X, L) be the first reduction of $(X^\wedge, L^\wedge)$ and assume that (X, L) is a conic fibration. Let $\varphi : X \longrightarrow Y$ be the structural morphism. Let S be a smooth element of $|L|$. Let H be the ample line bundle on Y such that $K_X + L \approx \varphi^*(K_Y + H)$. Then $h^0(K_{X^\wedge} + L^\wedge) \geq 2$ unless (Y, H) is a minimal surface of sectional genus $g(Y, H) = 2$, $p_g(S) = q(S) = 1$ and either*

(2.2.1)$d_2 = K_S \cdot K_S = 2$, Y is a $\mathbb{P}^1$ bundle, $p : Y \longrightarrow B$, over a smooth elliptic curve, B, of invariant $e = -1$, $H = 3E - f$, where E, f denote a section of minimal self-intersection $E^2 = 1$ and a fiber of p;

(2.2.2)$d_2 = K_S \cdot K_S = 4$, Y is an abelian surface, $H \cdot H = 2$, $h^0(H) = 1$ (described in [BLP], (2.7));

(2.2.3)$d_2 = K_S \cdot K_S = 4$, Y *is a hyperelliptic surface,* $H \cdot H = 2$, $h^0(H) = 1$ (*described in [BLP], (2.7)*).

Furthermore in each case φ has no divisorial fibres.

Proof. We use (0.7.1). Look at the commutative diagram

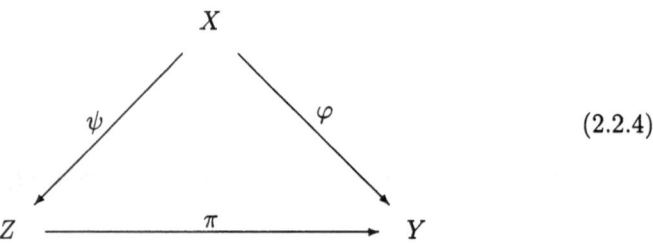

$$(2.2.4)$$

Recall that there exists an ample line bundle $\mathcal{L}$ on Z such that $K_X + L \approx \psi^*(K_Z + \mathcal{L})$ and $K_X + L$ is spanned if $\mathcal{L} \cdot \mathcal{L} \geq 5$. Hence we can assume $\mathcal{L} \cdot \mathcal{L} \leq 4$ since otherwise $h^0(K_X + L) \geq 3$. Recall also that (Y, H) is the reduction of $(Z, \mathcal{L})$ under π and π is the blowing up of Y at the points $y \in Y$ such that $\varphi^{-1}(y)$ is a divisor. Note that $K_Z + \mathcal{L}$ is nef and big since $K_Z + \mathcal{L} \approx \pi^*(K_Y + H)$ and $K_Y + H$ is ample. In view of (1.1) and (1.2) we may also assume

$$h^0(K_X + L) = 1, \quad \chi(\mathcal{O}_X) = 0, \quad \chi(\mathcal{O}_S) = 1, \quad p_g(S) = q(S) > 0. \qquad (2.2.5)$$

Then, since $\chi(\mathcal{O}_X) = \chi(\mathcal{O}_Z)$ we also have

$$\chi(\mathcal{O}_Z) = 0. \qquad (2.2.6)$$

Note that $h^0(K_X + L) = h^0(K_Z + \mathcal{L})$. By the Riemann-Roch Theorem on Z for $K_Z + \mathcal{L}$ we have

$$1 = h^0(K_Z + \mathcal{L}) = (\mathcal{L} \cdot \mathcal{L} + K_Z \cdot \mathcal{L})/2,$$

or $K_Z \cdot \mathcal{L} + \mathcal{L} \cdot \mathcal{L} = 2$, i.e. by the genus formula (0.2),

$$g(Z, \mathcal{L})(= g(Y, H)) = 2. \qquad (2.2.7)$$

From (0.7.2) we have

$$d_2 = K_S \cdot K_S = 2(K_Y + H)^2 = 2(K_Z + \mathcal{L})^2. \qquad (2.2.8)$$

Case $d_2 = 2$. By (2.2.8) we have $(K_Z + \mathcal{L})^2 = 1$. The genus formula gives $K_Z \cdot \mathcal{L} + \mathcal{L} \cdot \mathcal{L} = 2$. Then we get

$$K_Z \cdot K_Z + K_Z \cdot \mathcal{L} = -1. \qquad (2.2.9)$$

Assume that $h^0(mK_Z) > 0$ for some $m > 0$ and let $D \in |mK_Z|$ be an effective divisor. Then, by (2.2.9),

$$(K_Z + \mathcal{L}) \cdot D = m(K_Z \cdot K_Z + \mathcal{L} \cdot K_Z) = -m < 0.$$

This contradicts the nefness of $K_Z + \mathcal{L}$. Thus we conclude that $\kappa(Z) < 0$. Hence in particular $p_g(Z) = 0$. Since $p_g(Z) = p_g(Y)$ we have from (0.7.2.1) and (2.2.5) that $p_g(S) = q(S) = 1$.

From $\mathcal{L} \cdot \mathcal{L} \leq 4$ and $(K_Z + \mathcal{L})^2 = 1$ we find $K_Z \cdot K_Z + 2K_Z \cdot \mathcal{L} + 4 \geq K_Z \cdot K_Z + 2K_Z \cdot \mathcal{L} + \mathcal{L} \cdot \mathcal{L} = 1$, or

$$K_Z \cdot K_Z + 2K_Z \cdot \mathcal{L} \geq -3. \qquad (2.2.10)$$

Since $\mathcal{L} \cdot \mathcal{L} \geq 1$ by ampleness, we also have $1 = K_Z \cdot K_Z + 2K_Z \cdot \mathcal{L} + \mathcal{L} \cdot \mathcal{L} \geq K_Z \cdot K_Z + 2K_Z \cdot \mathcal{L} + 1$, or

$$K_Z \cdot K_Z + 2K_Z \cdot \mathcal{L} \leq 0. \qquad (2.2.11)$$

Since $p_g(Z) = \chi(\mathcal{O}_Z) = 0$ one has $q(Z) = 1$. Then [Be], VI, applies to give

$$K_Z \cdot K_Z \leq 0. \qquad (2.2.12)$$

By combining (2.2.9), (2.2.10), (2.2.11) and (2.2.12) we find the following possibilities

i) $K_Z \cdot K_Z = -2$, $\quad K_Z \cdot \mathcal{L} = \mathcal{L} \cdot \mathcal{L} = 1$;

ii) $K_Z \cdot K_Z = -1$, $\quad K_Z \cdot \mathcal{L} = 0$, $\quad \mathcal{L} \cdot \mathcal{L} = 2$, or

iii) $K_Z \cdot K_Z = 0$, $\quad K_Z \cdot \mathcal{L} = -1$, $\quad \mathcal{L} \cdot \mathcal{L} = 3$.

From the results of [BS1] (see in particular the table (*) [1] in the introduction of [BS1]) we see that only case iii) can occur. In this case Z is a $\mathbb{P}^1$ bundle over an elliptic curve, of invariant $e = -1$, $\mathcal{L} = 3E - f$, where E, f denote a section of minimal self-intersection, $E^2 = -1$, and a fiber. In particular, since Z is minimal, we have $(Z, \mathcal{L}) \cong (Y, H)$ and φ has no divisorial fibers. We are in case (2.2.1).

Case $d_2 = 4$. By (2.2.8) we have $(K_Z + \mathcal{L})^2 = (K_Y + H)^2 = 2$. Since $K_Y \cdot H + H \cdot H = 2$ by the genus formula, we get

$$K_Y \cdot K_Y + K_Y \cdot H = 0. \qquad (2.2.13)$$

Assume $K_Y \not\sim \mathcal{O}_Y$. Assume that $h^0(mK_Y) > 0$ for some $m > 0$ and let $D \in |mK_Y|$ be an effective divisor. Then, by (2.2.13),

$$(K_Y + H) \cdot D = m(K_Y \cdot K_Y + K_Y \cdot H) = 0.$$

This contradicts the ampleness of $K_Y + H$. Thus we conclude that $\kappa(Y) < 0$ and hence $\kappa(Z) < 0$. Since $p_g(Y) = 0$ we conclude from (0.7.2.1) and (2.2.5) that $p_g(S) = q(S) = 1$.

From $(K_Z + \mathcal{L})^2 = 2$, using $1 \leq \mathcal{L} \cdot \mathcal{L} \leq 4$, we find

$$K_Z \cdot K_Z + 2K_Z \cdot \mathcal{L} \geq -2 \qquad (2.2.14)$$

and

$$K_Z \cdot K_Z + 2K_Z \cdot \mathcal{L} \leq 1. \qquad (2.2.15)$$

From the genus formula $K_Z \cdot \mathcal{L} + \mathcal{L} \cdot \mathcal{L} = 2$ and $(K_Z + \mathcal{L})^2 = 2$ we get

$$K_Z \cdot K_Z + K_Z \cdot \mathcal{L} = 0. \qquad (2.2.16)$$

---

[1] (*) *Note:* the remark # 1 in that table is stated incorrectly. Indeed, as the proofs in the paper show, "S' not Abelian" should be "S' not Abelian if S has at least a rational singularity".

Note that $p_g(Z) = \chi(\mathcal{O}_Z) = 0$ implies $q(Z) = 1$, so the inequality (2.2.12) is still true. Therefore by combining (2.2.14), (2.2.15) and (2.2.16), we find the following possibilities:

i)$K_Z \cdot K_Z = K_Z \cdot \mathcal{L} = 0, \quad \mathcal{L} \cdot \mathcal{L} = 2$; or

ii)$K_Z \cdot K_Z = -K_Z \cdot \mathcal{L} = -1, \quad \mathcal{L} \cdot \mathcal{L} = 1.$

Recall that $q(Z) = 1$. Then [BLP], (3.3) applies to rule out case i) (see also the table in the introduction of [BS1]) and to say that Z is the blowing up, $\eta : Z \longrightarrow Z_0$, at a point of a $\mathbb{P}^1$ bundle Z_0 of invariant e such that $\mathcal{L} \approx \eta^* \mathcal{L}_0 - 2\mathcal{E}$, $\mathcal{E}$ exceptional divisor, and either $e = -1, \ \mathcal{L}_0 \sim 5E - 2f$, or $e = 0, \ \mathcal{L}_0 \sim 5E + f$, where E, f denote a section of minimal self-intersection $E^2 = -e$ and a fiber of η (note that there is a misprint in the table of [BLP], (3.3). Indeed "$5C_0 - f$" should be "$5C_0 - 2f$" as the argument in §3 shows: see in particular b), p.205). Let us compute $h^0(\mathcal{L}_0)$. Note that, in the first case, $\mathcal{L}_0 \sim K_{Z_0} + 7E - 3f$ and $7E - 3f$ is ample (see [Hr], p. 382). Then by the Kodaira vanishing and the Riemann-Roch theorem we have

$$h^0(\mathcal{L}_0) = \chi(\mathcal{L}_0) = ((5E - 2f)^2 - (-2E + f) \cdot (5E - f))/2 = 4.$$

The same argument gives us $h^0(\mathcal{L}_0) = 6$ in the second case.

Let $\mathcal{I}$ be the ideal sheaf of the point, p_0, blown up under η. Then by the projection formula, $\eta_* \mathcal{L} \approx \mathcal{L}_0 \otimes \mathcal{I}^2$. Since $h^0(\mathcal{L}_0) \geq 4$ we can find a non identically zero section of $h^0(\mathcal{L}_0)$ vanishing at p_0 with multiplicity 2. This section gives rise to a non zero section of $h^0(\eta_* \mathcal{L}) = h^0(\mathcal{L})$. On the other hand, since $\mathcal{L} \cdot \mathcal{L} = 1$, [Bs2], (7.13) (or [Bs1], (6.2.7)) applies to say that $h^0(\mathcal{L}) = 0$. Therefore we find a contradiction.

Thus we can assume $K_Y \sim \mathcal{O}_Y$, so that $\kappa(Y) = \kappa(Z) = 0$. Since $(K_Y + H)^2 = 2$ we have in this case $H \cdot H = 2$. Recall that $\chi(\mathcal{O}_Y) = 0$ by (2.1.6). Then we conclude that Y is either an abelian or a hyperelliptic surface and [BLP], (2.7.1) applies to describe the pair (Y, H). In particular $h^0(H) = 1$ (see also [BS1], p.181). To conclude the proof we have to show that $(Z, \mathcal{L}) \cong (Y, H)$ in this case. Assume otherwise. Thus in particular, since (Y, H) is the first reduction of $(Z, \mathcal{L})$, $K_Z + \mathcal{L}$ is nef and big but not ample. From the results of [BS1] (see in particular the table in the introduction) we see that this is not possible.

3. Extension to higher dimension

A standard inductive argument (see [Bs1], Ch. 7) can be used to lift the results of §2 to quadric fibrations of dimension $n \geq 4$ over a surface.

(3.1) **Assumption-Notation.** Let $(X^\wedge, L^\wedge)$ be a smooth connected variety of dimension $n \geq 4$ polarized with a very ample line bundle $L^\wedge$. Let $(X, L), r : X^\wedge \longrightarrow X$, be the reduction of $(X^\wedge, L^\wedge)$. Assume that (X, L) is a quadric fibration over a surface Y. Let $\varphi : X \longrightarrow Y$ be the structural morphism and let $\mathcal{H}$ be an ample divisor on Y such that $K_X + (n - 2)L \approx \varphi^* \mathcal{H}$. We denote by X_{n-1} a general smooth element of $|L|$. Similarly, for all $i = 3, \ldots, n-1$, X_i denotes a general smooth element of $|L_{i+1}|$, where L_j denotes the restriction of L to X_j. Put $X_n := X, \ L_n := L$.

(3.2) Theorem ([Bs1], Ch.7, [Bs2], §8). *Assumptions and notation as in* (3.1). *We have*

(3.2.1) Y *is smooth;*

(3.2.2) *there exists an ample line bundle H on Y such that $\mathcal{H} = K_Y + H$;*

(3.2.3) *there exists a i-dimensional smooth polarized pair $(X_i^\wedge, L_i^\wedge)$, $L_i^\wedge$ very ample, such that (X_i, L_i) is the first reduction of $(X_i^\wedge, L_i^\wedge)$;*

(3.2.4) (X_i, L_i) *is a quadric fibration over Y with structural morphism $\varphi \big|_{X_i}$.*

(3.3) Theorem. *Assumption and notation as in* (3.1), (3.2). *Then $h^0(K_X + (n-2)L) \geq 2$ unless (Y, H) is described as in* (2.2).

Proof. For all $i = 3, \ldots, n-1$, consider the exact sequence

$$0 \longrightarrow \mathcal{O}_{X_{i+1}}(-X_i) \longrightarrow \mathcal{O}_{X_{i+1}} \longrightarrow \mathcal{O}_{X_i} \longrightarrow 0,$$

tensor it by $K_{X_{i+1}} \otimes L_{i+1}^{\otimes(i-1)}$ and take cohomology. By the Kodaira Vanishing Theorem and adjunction formula, we see that the restriction map

$$H^0(K_{X_{i+1}} \otimes L_{i+1}^{\otimes(i-1)}) \longrightarrow H^0(K_{X_i} \otimes L_i^{\otimes(i-2)})$$

is surjective. If (Y, H) is not as in (2.2.1), (2.2.2), (2.2.3), we know by combining Theorems (2.2) and (3.2) that $h^0(K_{X_3} + L_3) \geq 2$. By induction on i we are done. Q.E.D.

References

[Be]   A.BEAUVILLE. *Surfaces algébriques complexes*, Astérisque 54, 1978.

[BBS]   M.C.BELTRAMETTI, A.BIANCOFIORE, A.J.SOMMESE. *"Projective n-folds of log-general type, I"*, Trans. A.M.S., v.314, n. 2(1989), 825-849.

[BLP]   M.C.BELTRAMETTI, A.LANTERI, M.PALLESCHI. *"Algebraic surfaces containing an ample divisor of arithmetic genus two"*, Arkiv för matematik, 25(1987), 189-210.

[BS1]   M.C.BELTRAMETTI, A.J.SOMMESE. *"On generically polarized Gorenstein surfaces of sectional genus two"*, J. reine angew. Math., 386(1988), 172-186.

[BS2]   M.C.BELTRAMETTI, A.J.SOMMESE. *"On the adjunction theoretic classification of polarized varieties"*, J. reine angew. Math., 427(1992), 157-192.

[BS3]   M.C.BELTRAMETTI, A.J.SOMMESE. *"On the dimension of the adjoint linear system for threefolds"*, preprint 1993.

[Bs1]   G.M.BESANA. *"The geometry of conic bundles arising in adjunction theory"*, Ph. D. Thesis, University of Notre Dame, April 1992.

[Bs2]   G.M.BESANA. *"On the geometry of conic bundles arising in adjunction theory"*, to appear in Math. Nachr.

[GH]   P.GRIFFITS, J.HARRIS. *Principles of Algebraic Geometry*, John Wiley & Sons, 1978.

[Hr]   R.HARTSHORNE. *Algebraic Geometry*, GTM 52, Springer-Verlag, 1977.

[K]   S.KLEIMAN. *"The enumerative theory of singularities"*, Real and Complex Singularities, Oslo 1976(P.HOLME, editor), Sijthoff and Noordhoff.

[LS]   E.L.LIVORNI, A.J.SOMMESE. *"Threefolds of non negative Kodaira dimension with sec-
          tional genus less than or equal to 15"*, Annali Scuola Normale Superiore, Serie IV, vol.
          XIII, n.4(1986), 537-558.

[ShS]   B.SHIFFMAN, A.J.SOMMESE. *Vanishing Theorems on Complex Manifolds,*
          Progress in Math., Birkhäuser, 1985.

[S1]   A.J.SOMMESE. *"On the adjunction theoretic structure of projective varieties"*, Complex
          Analysis and Algebraic Geometry, Proceedings Goettingen 1985, Lecture Notes Math.,
          1194 (1986), 175-213, Springer-Verlag.

[S2]   A.J.SOMMESE. *"On the nonemptiness of the adjoint linear system of a hyperplane section
          of a threefold"*, J. reine angew. Math., 402(1989), 211-220; "Erratum", J.reine angew. Math.,
          411(1990), 122-123.

[SV]   A.J.SOMMESE, A.VAN DE VEN. *"On the adjunction mapping"*, Math. Ann., 278(1987),
          593-603.

[T]   H.TSUJI. *"Stability of tangent bundles of minimal algebraic varieties"*, Topology 27
          (1988), 429-442.

On the Stability of M_E

David C. Butler

Introduction

This brief note outlines some of the author's results appearing in [5]. Suppose E is a vector bundle over a smooth irreducible projective curve C of genus g, and assume that global sections generate E. The natural evaluation map gives rise to a sequence of vector bundles:

$$0 \to M_E \to H^0(C, E) \otimes \mathcal{O}_C \to E \to 0.$$

When is the kernel M_E semistable (or stable)?

Theorem 1. *Let E be a semistable vector bundle over C with $\mu(E) \geq 2g$. Then M_E is semistable, and $\mu(M_E) = \frac{-\mu(E)}{\mu(E)-g} \geq -2$.*

Remark. If E is stable and $\mu(E) > 2g$, then M_E is stable.

In characteristic 0, Theorem 1 gives a vanishing theorem because tensor products preserve semistability. And this provides a surjectivity theorem on the tensor map of global sections for vector bundles E and F.

Corollary 2. *(characteristic 0) Let E and F be semistable vector bundles over a smooth irreducible projective curve C of genus g. If $\mu(E) \geq 2g$ and $\mu(F) > 2g$, surjectivity holds for the tensor map:*

$$\tau : H^0(C, E) \otimes H^0(C, F) \to H^0(C, E \otimes F).$$

The result holds in positive characteristic when either E or F is a line bundle or if $g(C) \leq 1$. Otherwise the problem remains open for positive characteristic.

When E and F are line bundles L_1 and L_2, Corollary 2 recovers a theorem of Mumford [18]. And setting $L_1 = L_2 = L$ shows any line bundle L with $\deg(L) \geq 2g + 1$ embeds C as a projectively normal variety (in which case L is said to be *normally generated*). This was first proved by Castelnuovo[6], then Mattuck, [15], and finally Mumford [18].

It seems likely the "$2g + 1$ Theorem" should generalize somehow to higher dimensions. Towards this end, Mukai observes that if A is an ample line bundle over C and B is nef, the "$2g + 1$ Theorem " says $L = \omega_C + 3A + B$ is normally generated. Reider shows that for any smooth surface S, $L = \omega_S + 4A + B$ is very ample [22]. Is L also normally generated? The results in this paper do not answer that question in such generality, but provide some results under the assumption that S is a ruled surface. Details appear in [4]. The idea is to use vector bundles over a curve to study line bundles over a ruled surface. A simple generalization of Corollary 2 then implies the following result.

Corollary 3.*(Characteristic 0) Let S be a ruled surface. If A is an ample line bundle over S and B is nef, then $L = K_S \otimes A^5 \otimes B$ is normally generated.*

In a different direction, Bertram, Ein, and Lazarsfeld solve the problem under the stronger hypothesis that A is *very* ample. Working over a smooth variety X of dim $= n$ and assuming that A is very ample, they show that with the exception of $(X, A, B) = (\mathbb{P}^n, \mathcal{O}_{\mathbb{P}^n}(1), \mathcal{O}_{\mathbb{P}^n})$ a line $L = K_X + nA + B$ is normally generated [4] and [7]. Andreatta, Ballico, and Sommese [1] and [2], prove this independently.

The results of this note will appear in the authors forthcoming paper [5]. Details of proofs (as well as omitted proofs) and related results can be found there. We wish to thank David Gieseker, Mark Green, and particularly our advisor Robert Lazarsfeld without whom none of this would have been possible.

Conventions and Notation

C is a smooth irreducible projective curve over $k = \bar{k}$

We assume $char\,(k) = 0$, unless stated otherwise.

The genus $g(C) = g$

E and F are algebraic vector bundles over C.

The slope of a vector bundle E is $\mu(E) = \deg(E)/rk\,(E)$.

E is (semi)stable if for all proper subbundles $S \subsetneq E, \mu(S) < \mu(E)$, $(\leq)$. Equivalently, E is (semi)stable if for all proper quotient bundles $E \to Q \to 0$, $\mu(S) > \mu(E)$, $(\geq)$.

Applications of Theorem 1

We need some basic facts about semistability. The first four are simple but are included for reference, the fifth is a deep but widely known theorem.

Proposition 4. *Let E and F be semistable vector bundles over C.*

If $\mu(E) < 0$, then $h^0(C, E) = 0$. $\hspace{6cm}$ (1)

If $\mu(E) > 2g - 2$, then $h^1(C, E) = 0$. $\hspace{5cm}$ (2)

If $\mu(E) > 2g - 1$, then E is generated by global sections. $\hspace{2.5cm}$ (3)

If $\mu(E) > 2g$, then $\mathcal{O}_{\mathbb{P}(E)}(1)$ is very ample. $\hspace{3.8cm}$ (4)

$E \otimes F$ is semistable if char $(k) = 0$. $\hspace{5cm}$ (5)

Proof of Proposition 4. For (1), if $h^0(C, E) > 0$ then $\mathcal{O}_C \hookrightarrow E$, and hence $\mu(E) \geq 0$. Part (2) follows from (1) and Serre duality. Parts (3) and (4) follow from (2) and the fact that for $p \in C$, $\mu(E(-p)) = \mu(E) - 1$. Finally, (5) is a deep theorem [10], [3], [20], [19], [12] [9], (see also [14] for higher dimensions and [17] for an elementary proof). □

Remark. Proposition 4 part 5 fails in positive characteristic when $g \geq 2$ (see [8] for examples).

Proof of Corollary 2. By Proposition 4, global sections generate E. So consider the sequence,

$$0 \to M_E \to H^0(C, E) \otimes \mathcal{O}_C \to E \to 0,$$

tensor by F and take global sections. By Theorem 1 M_E is semistable with $\mu(M_E) \geq -2$. So $M_E \otimes F$ is semistable by Proposition 4, and $\mu(M_E \otimes F) = \mu(M_E) + \mu(F) > 2g - 2$. Hence $h^1(C, M_E \otimes F) = 0$ by Proposition 4. □

Sadly enough, some vector bundles are not semistable; however, all vector bundles have a unique HN Filtration in terms of semistable bundles [11]. This fact forces the results given here to generalize for unstable bundles. One considers a slightly different invariant of a vector bundle which seems to originate with Mehta [16].

Definition. If E is a vector bundle over C then $\mu^-(E) = \min\{\mu(Q)|E \to Q \to 0\}$.

One may similarly define μ^+ for $S \subset E$. One can then generalize proposition 4 as follows.

Proposition 4*.*Let E and F be vector bundles over C.*

- If $\mu^+(E) < 0$, then $h^0(C, E) = 0$.
- If $\mu^-(E) > 2g - 2$, then $h^1(C, E) = 0$.
- If $\mu^-(E) > 2g - 1$, then E is generated by global sections.
- If $\mu^-(E) > 2g$, then $\mathcal{O}_{\mathbb{P}(E)}(1)$ is very ample.
- $\mu^-(E \otimes F) = \mu^-(E)^- + \mu^-(F)$ if char $(k) = 0$.

Parts 1 through 4 are trivial. Part 5 can be proven by tensoring the HN Filtration for E by the HN Filtration of F. Now the other results generalize.

Theorem 1*. *Let E be a vector bundle over C with $\mu^-(E) \geq 2g$. Then $\mu^-(M_E) = \frac{-\mu^-(E)}{\mu^-(E) - g} \geq -2$.*

Corollary 2*.(*characteristic 0*) *Let E and F be vector bundles over C. If* $\mu^-(E) \geq 2g$ *and* $\mu^-(F) > 2g$, *surjectivity holds for the tensor map:*

$$\tau : H^0(C, E) \otimes H^0(C, F) \to H^0(C, E \otimes F).$$

Corollary 2* can be restated in terms of line bundles over projective bundles.

Corollary 5.(*Characteristic 0*) *Let E be a vector bundle over C. And let* $X = \mathbb{P}(E)$. *If L is a line bundle over X, ample on the fibres and* $\mu^-(\pi_* L) > 2g$, *then L is normally generated.*

Corollary 3 follows from Corollary 5.

Proof of Main Theorem

Proof of Theorem 2. That $\mu(M_E) = \frac{-\mu(E)}{\mu(E)-g} \geq -2$ follows from a simple calculation:

$$\mu(M_E) \qquad = \frac{-\deg(M_E)}{rk\,(M_E)}$$
$$= \frac{-\deg(E)}{h^0(C,E) - rk\,(E)}$$
$$= \frac{-\deg(E)}{\deg(E) - rk\,(E)g}$$
$$\mu(M_E) \qquad = \frac{-\mu(E)}{\mu(E)-g}.(*)*$$

To show M_E is semistable, it suffices to show that if $N \subsetneq M_E$ is stable and of maximal slope, then $\mu(N) \leq \mu(M_E)$. So consider the following diagram:

$$
\begin{array}{ccccccccc}
& & 0 & & 0 & & & & \\
& & \downarrow & & \downarrow & & & & \\
0 & \longrightarrow & N & \longrightarrow & V \otimes \mathcal{O}_C & \longrightarrow & G & \longrightarrow & 0 \\
& & \downarrow & & \downarrow & & \downarrow{\scriptstyle \alpha} & & \\
0 & \longrightarrow & M_E & \longrightarrow & H^0(C, E) \otimes \mathcal{O}_C & \longrightarrow & E & \longrightarrow & 0.
\end{array}
$$

V is taken to be the minimal such vector space. G is then a vector bundle with no trivial summands. The map α must be nonzero by a simple diagram chase. The proof now comes down to a few simple steps. First show that $\mu(G) \leq \mu(E)$. Then if $h^1(C, G) = 0$, the formula (*) shows $\mu(N) \leq \mu(M_E)$. Finally, if $h^1(C, G) > 0$, the stability of M_{ω_C} and the stability of N shows that $\mu(N) \leq -2 \leq \mu(M_E)$.

To show $\mu(G) \leq \mu(E)$, let S be the image of α. $V \subseteq H^0(C, S)$, therefore, $N \subseteq M_S$. If $\deg(G) > \deg(S)$, then $-\deg(G) < -\deg(S)$ or $\deg(N) < \deg(M_S) < 0$, and hence $\mu(N) < \mu(M_S)$. But this is impossible because N is of maximal slope. Therefore $\deg(G) \leq \deg(S)$, and so $\mu(G) \leq \mu(S) \leq \mu(E)$.

If $h^1(C, G) = 0$, then (*) implies that $\mu(N) \leq \frac{-\mu(G)}{\mu(G)-g} \leq \frac{-\mu(E)}{\mu(E)-g} = \mu(M_E)$. Now assume $h^1(C, G) > 0$. The classification of vector bundles over rational and elliptic curves shows there are no special vector bundles with no trivial components that are generated by global sections [3] and [10]. So $g \geq 2$. Since $h^1 > 0$, there is a nonzero map $G \to \omega_C$ such that the image of $V \subseteq H^0(C, G)$ generates ω_C. This gives rise to a nonzero map $N \to M_{\omega_C}$. Since $\mu(M_{\omega_C}) = -2$ and M_{ω_C} is semistable [21], the stability of N implies $\mu(N) \leq -2$. $\square$

References

[1] M. Andreatta, E. Ballico and A. Sommese On the projective normality of the adjunction bundles II

[2] M. Andreatta and A. Sommese On the projective normality of the adjunction bundles I

[3] M. F. Atiyah 414 – 452 Vector bundles over an elliptic curve VII(1957) 27 Proc. Lond. Math. Soc. (3)

[4] A. Bertram, L. Ein and R. Lazarsfeld Vanishing theorems, a theorem of Severi, and the equations defining projective varieties. (to appear)

[5] D. Butler Normal generation of vector bundles over a curve (to appear)J. of Diff. Geo.

[6] G. Castelnuovo Sui Multipli di uni serie di gruppi di punti appartenente ad una curva algebrica RendĊircṀatṖalermo 7 (1892) 99 – 119

[7] L. Ein, R. Lazarsfeld A theorem on the syzygies of smooth projective varieties of arbitrary dimension

[8] D. Gieseker Stable vector bundles and the Frobenius morphism Ann. École Norm. Sup 6 (1973) 95 – 101

[9] On a theorem of Bogomolov on Chern classes of stable bundles Amer. Jour. of Mathematics 101 (1979) 77 – 85

[10] A. Grothendieck Sur la classification des fibres holomorphes sur la sphère de Riemann AmerĴ. Math·79 (1957) 121 – 38

[11] G. Harder and M. S. Narasimhan On the cohomology groups of moduli space of vector bundles on curves Math. Ann. 212 (1975) 215 – 248

[12] R. Hartshorne Ample vector bundles on curves Nagoya Math. J. 43 (1971) 73 – 90

[13] R. Lazarsfeld A sampling of vector bundle techniques in the study of linear series In: Lectures on Riemann Surfaces World Scientific Press, 1989 500 – 559.

[14] M. Maruyama The theorem of Grauer-Mulich-Spindler Math. Ann. 1981 255 317 – 333

[15] A. Mattuck Symmetric products and Jacobians Am. J. Math. 83 (1961) 189 – 206

[16] V.B. Mehta On some restriction theorems for semistable vector bundles In: Invariant theory (Montecantini 1982). Lecture Notes in Math 966 145 – 153 Springer

[17] Y. Miyaoka The Chern class and Kodaira dimension of a minimal variety In: Algebraic Geometry – Sendai 1985, Adv. Studies in Pure Math 10 449 – 476

[18] D. Mumford Varieties defined by quadratic equations Corso CIME in: Questions on algebraic varieties, Rome (1970) 30 – 100

[19] M. S. Narasimhan and C. S. Seshadri Stable and unitary vector bundles on a compact Riemann surface Ann. of Math 82 (1965) 213 – 224

[20] T. Oda Vector bundles on an elliptic curve Nagoya Math. J. 43 (1971) 41 – 72

[21] A. Paranjape and S. Ramanan On the canonical ring of an algebraic curve In: Algebraic Geometry and Commutative Algebra (in honor of M. Nagata) (1988) Kinokaniya

[22] I. Reider Vector bundles of rank 2 and linear systems on an algebraic surface Ann. of Math. 127 (1988) 309 – 316

On a Class of Del Pezzo Fiber Spaces

Harry D'Souza

Abstract. Prymians and Intermediate Jacobians of threefolds enable us to distinguish between rational and non-rational varieties. Hence they are important invariants from that standpoint alone. We show that the class of varieties under consideration has a Chow group of codimension two cycles, $A^2(X)$. Furthermore, by analyzing the singular fibers of the threefold we are able to show that the Prym variety of this threefold is isomorphic to the Intermediate Jacobian.

Introduction

In the study of projective 3-folds the class with $\kappa(X) = -\infty$ forms an intriguing set, in that we have many anamolies. The class of threefolds whose hyperplane sections are elliptic surfaces belong to this group. As shown in [D 1], these are Del Pezzo fiber spaces. Using the results of Kanev [Ka], we show the existence of a curve E in X, which determines a Prym-Tyurin variety. This Prym-Tyurin variety has a natural polarization Ξ via which it is isomorphic to the Intermediate Jacobian of X, $J(X)$ with its natural polarization Θ. Moreover, we show that via the cylinder morphism ϕ, that $\phi\big(H_1(E, \mathbb{Z})\big)$ has finite index in $H_3(X, \mathbb{Z})$.

We also study the relationship between the Intermediate Jacobians and the Prym variety. We analyze the singular fibers of this fibration and show by using the cylinder homomorphism of [C-G] or [T], that the Prym variety is isomorphic to the Intermediate Jacobian of the threefold. The result of Kanev in [Ka], on the existence of a unique Principal Polarization plays an important role in the paper. The discussion of the Chow group of codimension 2-cycles $A^2(X)$, is a result of discussions in [B-M] and [Mu], where the theorem of isomorphism of the torsion cycles yields the nature of $A^2(X)$.

0. Some notation and background material

Throughout this paper X will be an *irreducible complex projective three-dimensional manifold*, and L an ample and spanned (i.e. generated by global sections) line bundle over X.

(0.1) Let L be a line bundle over X. We say that L is *big* if $c_1(L)^n > 0$. We say that L is *nef* if $c_1(L) \cdot [C] \geq 0$, for all effective curves C on X. We say that L is semi-ample if there exists an $m > 0$, such that $\mathrm{Bs}|mL|$, the base locus of $|mL|$, is empty.

(0.2) Definition. (*reduction and minimal pair*) Let (X', L') be a polarized manifold. A *reduction* of (X', L') is a *polarized manifold* (X, L) such that:

a) there exists a morphism $\pi : X' \longrightarrow X$ expressing X' as X with a finite set F in X blown up,

b) $L = \pi^*(L') \otimes [\pi^{-1}(F)]^{-1}$ or equivalently $K_X \otimes L^{n-1} = \pi^*(K_{X'} \otimes L'^{n-1})$,

c) if in addition to (a) and (b), $\pi(L') = L = [S]$, where S is a minimal model of S', where $L = [S]$, then (X, L) is called a *minimal pair*.

(0.3) Definition. (*locally ample, spanned*) Let L be a line bundle over a smooth variety X, and $p : X \longrightarrow Y$ be a morphism. Then L is said to be locally ample (spanned by global sections) if p restricted to Δ_y, where Δ_y is a neighbourhood of $y \in Y$, is ample (spanned by global sections) for every $y \in Y$.

The following theorem provides the structure for the minimal pair for dim $X = n$.

(0.4) Theorem. *Let L' be an ample and spanned line bundle on an n-dimensional complex projective manifold X' with $n \geq 3$. Assume that the intersection of $(n-2)$ generic members of $|L'|$ is a smooth elliptic surface S' of Kodaira dimension $\kappa(S') = 1$. Assume moreover that $\kappa(A') \geq 0$, where $A' \in |K_{X'} \otimes L'^{n-2}|$, then there exists a minimal reduction (X, L) of (X', L') such that there exists a morphism $p_n : X \longrightarrow C$, where C is a smooth curve, and $K_X \otimes L^{n-2} = p_n^*(M)$, for some line bundle M over C. Moreover L is ample and locally spanned with respect to p_n.*

Proof. See [D 2; (1.1), (1.2) and (1.3)]. $\square$

(0.5) Definition. (*log(X, S)*) Let S be a smooth ample divisor on a smooth 3-fold X. Then the *logarithmic Kodaira dimension* $\log(X, S) = \mathrm{tr_c} \bigoplus_{n \geq 0} \Gamma\left(X, (K_X \otimes L)^n\right) - 1$. Note that $\log(X, S)$ is a *birational invariant*.

(0.6) Remark. (Iitaka) $\log(X, S) < 0 \iff$ *either* $\kappa(S) < 0$ *or* X is a $\mathbb{P}^1$-bundle over a smooth surface $S^\sim$ with S as a mermorphic (non-holomorphic) section.

(0.7) By $\mathbb{F}_r$ with $r \geq 0$, we denote the $r\underline{th}$ Hirzebruch surface, i.e. the unique $\mathbb{P}^1$ bundle $\pi : \mathbb{F}_r \longrightarrow \mathbb{P}^1$ with a section E satisfying $E \cdot E = -r$. For $r \geq 1$, we let $\mathbb{F}_r^\sim$ denote the normal surface obtained by contracting E, and let $\pi_1 : \mathbb{F}_r \longrightarrow \mathbb{F}_r^\sim$ be the contraction map. Given a line bundle L on $\mathbb{F}_r^\sim$, the pullback of L to $\mathbb{F}_r$ is of the form $([E] \otimes [f]^r)^a$ for some integer a, where f is a fiber of π, and E (unique) such that $E \cdot E = -r$.

If π denotes the blow down map of E i.e. $\pi : \mathbb{F}_r \longrightarrow \tilde{\mathbb{F}_r}$. Suppose L_1 and L_2 are two line bundles on $\tilde{\mathbb{F}_r}$, where $\pi^*(L_k) = ([E] \otimes [f]^r)^a k$ then $L_i \cdot L_k = a_i a_k r$.

1. Preliminary results

The following definition can be found in [T] or [C-G].

(1.0) Definition. (*Cylinder homomorphism*) Let X be a nonsingular, irreducible threefold. Suppose we have the following situation:p is a *surjective* morphism,C a nonsingular (not necessarily irreducible, but having smooth components) curve,F a nonsingular surface, and η a morphism such that the restriction of η to a *generic* fiber of p is an *embedding*.

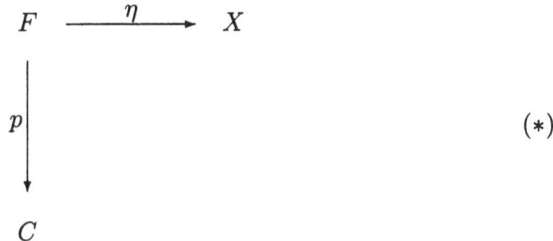

$$(*)$$

Let $c \in C$, then ℓ_c denotes the *algebraic* 1-cycle $\eta_* p^*(c)$. From the above diagram (*) we get the *cylinder homomorphism* $\Phi =: \eta_* p^* : H_1(C, \mathbb{Z}) \longrightarrow H_3(X, \mathbb{Z})$.

Before proceeding further, we need the definition of incidence correspondence.

(1.1) Definition. (*NR-condition*) Let p, η, C, F and X be as in (1.0), *suppose* furthermore the following condition (**NR**) is satisfied:

For any $c \in C$, there exists *only a finite number* of points $c' \in C$, such that
(**NR**) $\text{Supp}(\ell_c) \cap \text{Supp}(\ell_{c'}) \neq \emptyset$. There exists an *open dense subset* $U \subset C$, such that η restricted to $p^{-1}(U)$ is *non-ramified*.

The following discussion, also noted in [Ka], shows that (1.1) is natural.

Let ζ be the *closed* subset of F, given by $\zeta = \{x : \eta(x) \in \ell_{p(x)} \cap \ell_c, \text{ for some } c \neq p(x)\}$. Let ζ_i denote the union of the i-dimensional *components* of ζ. By (**NR**), ζ_1 contains no fibers of $p : F \longrightarrow C$. Let $n : \tilde{\zeta_1} \longrightarrow \zeta_1$ denote the desingularization of ζ_1. By identifying $\zeta_1 - \text{Sing}(\zeta_1)$ with $n^{-1}(\zeta_1 - \text{Sing}(\zeta_1))$, we see that the restriction of p and η on ζ_1 induces morphisms $\pi : \zeta_1 \longrightarrow C$, and $\tau : \zeta_1 \longrightarrow X$ respectively. Then there exists an open dense subset $U_0 \subset U(\text{see (\textbf{NR})})$ satisfying the following properties $p^{-1}(U_0)$ no *zero-dimensional* component of ζ, and for any $c \in U_0$, the fiber $p^*(c)$ is a *nonsingular irreducible* curve, and the restriction of η to $p^*(c)$ is an embedding. If $V_0 = \pi^{-1}(U_0)$, then $n(V_0) \subset \zeta_1 - \text{Sing}(\zeta_1)$ and π is étale on V_0. The *curve* $\Gamma_0 = \tau(V_0)$ is *nonsingular*, and the map $\eta : V_0 \longrightarrow \Gamma_0$ is étale.

Let $Z = \eta(F)$, then by **(NR)** it follows that $\mathrm{Sing}(Z) = \Gamma$. Then by [H], we can desingularize X by successive blow-ups of X along smooth points or irreducible nonsingular curves, to get:

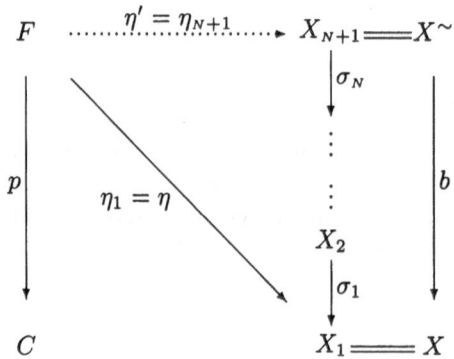

where $\eta_2 = (\sigma_1)^{-1} \circ \eta, \ldots, \eta_{N+1} = (\sigma_N)^{-1} \circ (\sigma_{N-1})^{-1} \ldots (\sigma_1)^{-1} \circ \eta$ are rational maps, $\eta = \eta_1, \ldots, \eta' = \eta_{N+1}$. Let $b_j = \sigma_j \circ \ldots \circ \sigma_N$ and $b_1 = b$. By [H], we can choose the blow-ups in such a way that for any $j \in \{1, \ldots, N\}$ the center of σ_j is sitting inside $\mathrm{Sing}(\eta_j(F))$. There is a Zariski open set $C_0 \subset U_0$, such that, if $F_0 = p^{-1}(C_0)$, then for any $j \in \{1, \ldots, N\}$, the map $\eta_j : F_0 \longrightarrow X_j$ is a *morphism*. Furthermore, if the center of σ_j is a point q, then $q \notin \eta_j(F_0)$.

The following structure theorem provides the basis for the study of the properties of the class of threefolds mentioned in the introduction.

(1.2) Theorem. *Let S be a smooth elliptic surface that is an ample divisor on a smooth threefold X. Assume moreover that (X, S) is a minimal pair with $\log(X, S) \geq 0$, and that S is not a $K - 3$ surface, then there exists a surjective morphism $p : X \longrightarrow C$, where C is a smooth curve, and $p_s : S \longrightarrow C$ is an elliptic fibration without multiple fibers and the generic fiber of p is a Del Pezzo surface.*

Proof. See [D 3]. $\square$

(1.3) Remarks. (i) $K_X + L = p^*(\mathcal{D})$ is *nef*, where $\mathcal{D}$ is an ample divisor in C, and if the genus of C is g, then $\deg \mathcal{D} = \chi(\mathcal{O}_S) + 2g - 2 > 0$.

(ii) $\kappa(X) = -\infty$ and so X is *uniruled* by Mori-Miyaoka [M-M].

(iii) If $g(C) \neq 0$, then X *cannot* be *unirational*.

(iv) $p_{(i)}(\mathcal{O}_X) = 0$ for $i > 0$.

(v) $\log(X, S) = 1$.

(1.4) Lemma. *Let (X, S) be as in (1.3). Let $p^{-1}(y) + X_y$, a Del Pezzo surface for a generic y, and ℓ a line in X_y (i.e. $\ell^2 = -1 = K_y \cdot \ell$). Then $Hilb_X$ (the Hilbert scheme parametrizing closed subschemes of X) is smooth, and of dimension one at the point corresponding to ℓ.*

Proof. Consider the standard normal bundle sequence

$$0 \longrightarrow N_{\ell/X_y} \longrightarrow N_{\ell/X} \longrightarrow N_{X_y/X|_\ell} \longrightarrow 0.$$

Since $N_{X_y/X} = \mathcal{O}_{X_y}$ and $\ell \cdot \ell = -1$ in X_y, we see that this reduced to (using $\ell \cong \mathbb{P}^1$), that

$$0 \longrightarrow \mathcal{O}_\ell(-1) \longrightarrow N_{\ell/X} \longrightarrow \mathcal{O}_\ell \longrightarrow 0.$$

And by considering the associated long exact sequence and that $H^0(\ell, \mathcal{O}_\ell(-1)) = 0 = H^1(\ell, \mathcal{O}_\ell) = H^1(\ell, \mathcal{O}_\ell(-1))$, we get $H^0(\ell, N_{\ell/X}) = 1$ and $H^1(\ell, N_{\ell/X}) = 0$ and

$$N_{\ell/X} = \mathcal{O}_\ell(-1) \oplus \mathcal{O}_\ell.$$

The rest follows immediately. $\square$

(1.5) Remark. As is well known [Ma], there are 1, 3, 6, 20, 16, 27, 56 and 240 lines on a Del Pezzo surface of degree (i.e. $d = K_y \cdot K_y$ on X_y) 8, 7, 6, 5, 4, 3, 2 and 1 respectively.

Let X_y be a *smooth* fiber of $p : X \longrightarrow C$, and $\ell_1, \ldots, \ell_d$ be the lines on it. Let E'_j be the maximal reduced subscheme of the irreducible component of $Hilb_X$ containing the point representing the line ℓ_j. Let $E' = \bigcup_{1 \leq i \leq d} E_i$ (in general the number of components are $\leq d$ since some of the E_i might coincide). Then by (1.3), there exists a *proper, flat* family of 1-dimensional fibers $p' : F' \longrightarrow E'$ and a morphism $\eta' : F' \longrightarrow X$ such that for any $e \in E', \eta'_e : F'_e \longrightarrow E'$ is a *closed immersion*, and $\eta'(F'_e)$ is the *subscheme* of X, *corresponding* to $e \in E'$. Let $n : E \longrightarrow E'$ be the *desingularization* of E' and $F =$ desingularization of $\{(F' \underset{E'}{\times} E)_{red}\}$. We thus have diagram of morphisms

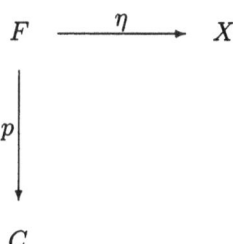

(1.6) Remarks. (i) The above family satisfies the (**NR**) condition (1.1).

(ii) $\ell_e = \eta_*\big(p^*(e)\big)$ is an *algebraic* 1-cycle in X. The *curves* E' and E are equipped with the morphisms $\pi' : E' \longrightarrow C$ and $\pi : E \longrightarrow C$ of *degree d*. If $y \in C$ is generic then $\pi^{-1}(y) = \{e_1, \ldots, e_d\}$.

(iii) Let $U \subset C$ be the *open* subset of points y for which X_y is *smooth*. Let $E_0 = \pi^{-1}(U)$, then by (1.1) the points of $\pi'^{-1}(U)$ are smooth, whence $E_0 = \pi'^{-1}(U)$, and so $p^{-1}(E_0) \cong p'^{-1}\pi'^{-1}(U)$, and for $y \in U$, if $\pi^{-1}(y) = \{e_1, \ldots, e_d\}$, then $\ell_1, \ldots, \ell_d$ are the *corresponding* lines on X_y.

(1.7) Definitions. (i) Let C be a non-singular curve. A correspondence of C is a divisor $T \in \mathrm{Div}(C \times C)$ satisfying the condition that if $z \in \mathrm{Div}(C)$ and $\pi_1^*(z)$ and T have *no common* irreducible components, then for $z \in J(C)$, we let $T(z) = \pi_{2*}\big(\pi_1^*(z) \cdot T\big)$ and for $\gamma \in H_1(C, \mathbb{Z})$, we define $T(\gamma) = \pi_{2*}\big(\pi_1^*(\gamma) \cdot T\big)$.

(ii) For any $k \geq 1$, let $D_k = $ *reduced effective divisor* on $E \times E$ such that $\mathrm{Supp}D_k = \{(x,x') \mid x \neq x' \text{ and } \pi(x) = \pi(x') \in U \text{ and } \ell_x \cdot \ell_{x'} = k \text{ on } X_{\pi(x)}\}$. Let $D = \sum_{k \geq 0} kD_k$, then D is called the *incidence correspondence*.

[*Note:* k above is the number of blow-ups required to desingularize the point $\pi(x)$. See((1.8); (i)) below.]

(1.8) Remarks. (i) $D_k = 0$ if $k \geq 4$ and $D_3 \neq 0 \iff K_y \cdot K_y = 1$, and $D_2 \neq 0 \iff K_y \cdot K_y = 1$ or 2. Hence $D = D_1$ if $K_y \cdot K_y \geq 3$; $D = D_1 + 2D_2$ if $K_y \cdot K_y = 2$; and if $K_y \cdot K_y = 1, D = D_1 + 2D_2 + 3D_3$.

(ii) D induces the endomorphisms $\iota : J(E) \longrightarrow J(E)$ and $\iota_* : H_1(E, \mathbb{Z}) \longrightarrow H_1(E, \mathbb{Z})$, where $\iota(z) = \pi_{2*}\big(\pi_1^*(z) \cdot D\big)$ and $J(E) = \underset{1 \leq i \leq s}{\times} J(E_i)$, where $s \leq d$ (see (1.7)).

(iii) $\deg_E D(e) = \sum_{1 \leq j \leq d} (\ell, \ell_j)_{X_e} + 1 = n$, where X_e is the fiber over $e \in E$, and ℓ is any line on X_e. By [Dem], $\sum_{1 \leq j \leq d} (\ell, \ell_j)_{X_e}$ is independent of ℓ. We note that $n = 2, 3, 5, 10, 29,$ 241 for $K_y \cdot K_y = 6, 5, 4, 3, 2, 1$ respectively.

(iv) If $W = \eta(F)$, then on $X, \ell \cdot W = (\ell, \sum_j \ell_j)_{X_e}$.

We can now state Kanev's generalization of Tyurin's lemma:

(1.9) Lemma. (Tyurin-Kanev) *Let p, η, C, F and X be as in* (1.0). *Using the notations of* (1.8); *if $E = \underset{1 \leq r \leq s}{\bigcup} E_r, \gamma \in H_1(E_r, \mathbb{Z}), \gamma' \in H_1(E, \mathbb{Z}), d_r = \ell_r \cdot W$ and $n_r = \deg_E D(e)$, where ℓ_r is the line in the fiber over a generic $e_r \in E_r$, then*

$$\Phi(\gamma) \cdot \Phi(\gamma') = (d_r - n_r)\gamma \cdot \gamma' + \iota_* \gamma \cdot \gamma'.$$

The left hand side is the intersection in X, and the right hand side, that on E.

Proof. (See [Ka; (1.4) for very intricate details]). The discussion following (1.1) helps us keep track of the intersections of the lines in X_e, and the results of [T, pg. 14] help prove the result. $\square$

(1.10) Proposition. *Let D be as in* (1.8), *then D is symmetric, i.e. $D = D^t$, and D is effective. If $1 \leq K_y \cdot K_y \leq 6$, then there exists an integer n, such that for any $e \in E, \deg_E D(e) = n$; and for $\gamma, \gamma' \in H_1(E, \mathbb{Z})$*

$$\Phi(\gamma) \cdot \Phi(\gamma') = (\iota_* - 1)\gamma \cdot \gamma',$$

where Φ is as in (1.0), *and the left hand side is the intersection in X, and the right hand side, that on E.*

Proof. (See [Ka]). By (1.8) (iii), the existence of n, with $\deg_E D(e) = n$ is clear. Let X_e be a fiber over a generic $e \in E$, then $W \cdot X_e = \sum_{1 \leq i \leq d} \ell_i$. So on $X, d = \ell \cdot W = [\ell \cdot (W \cdot X_e)]_{X_e} = (\ell, \sum_j \ell_j)_{X_e} = n - 1$ by (1.8) (iii). Now by (1.9), we see that the result follows using $d - n = -1$. $\square$

(1.11) Proposition. *Let $\iota : J(E) \longrightarrow J(E)$ be the endomorphism induced by $D \in \text{Div}(E \times E)$. If $1 \leq K_y \cdot K_y \leq 5$, then there exists $q \geq 2$, such that $(\iota - 1) \circ (\iota - q + 1) = 0$. If $K_y \cdot K_y = 6$, then $(\iota - 1) \circ (\iota + 1) \circ (\iota + 2) = 0$.*

Proof. See (1.12), also see [Ka]. $\square$

(1.12) Proposition. *Let C and E be as in* (1.1) *and* (1.10), *then $\pi : E \longrightarrow C$ is a finite morphism of degree d, where d is the number of lines on X_y. If $1 \leq K_y \cdot K_y \leq 6$, then there exist $q, m > 0$ such that for any $e \in E$:*

(a) $D\big(D(e)\big) + (q - 2)D(e) - (q - 1)(e) = m\pi^*\big(\pi(e)\big)$ for $1 \leq K_y \cdot K_y \leq 5$.

(b) $D\big(D(D(e))\big) + 2D\big(D(e)\big) - D(e) - 2e = 2\pi^*\big(\pi(e)\big)$ for $K_y \cdot K_y = 6$.

Proof. See [Ka]. The first part follows from (1.8, ii). For (a) and (b), we use the transitivity of the Weyl group W_r(see [Dem] or [Ma]) on the root system $R_r = \{\lambda : K_y \cdot \lambda = 0$ and $\lambda \cdot \lambda = -2$ on $X_y\}$ of type A_4, D_5, E_6, E_7, E_8 for $K_y \cdot K_y = 5, 4, 3, 2, 1$ respectively, and for $K_y \cdot K_y = 6$, we can verify directly, and verify case by case. Note that $q = 3, 4, 6, 12$ and 60, and $m = 1, 2, 5, 20$ and 300 respectively. $\square$

At this point we introduce some more notation! Let Σ ($\subset C$) be a *finite* subset of points over which the fibers of $p : X \longrightarrow C$ are *singular*. Let R be an *irreducible* component of some singular fiber. Let $R^\sim$ be its desingularization and $e_{R^\sim}$ be the composite of the two maps $R^\sim \longrightarrow R \subset X$, then $(e_{R^\sim})_*\big(H_3(R^\sim, \mathbb{Z})\big)$ as a subgroup of $H_3(X, \mathbb{Z})$ is *independent of the desingularization $R^\sim$.*

(1.13) Theorem. (Kanev) *Let $p : X \longrightarrow C$ be as in* (1.2), *and E in* (1.8). *If $1 \leq K_y \cdot K_y \leq 7$, then*

$$H_3(X, \mathbb{Q}) = \Phi\big(H_1(E, \mathbb{Q})\big) + \sum_{y \in \Sigma, \, R \subset X_y} (e_{R^\sim})_*\big(H_3(R^\sim, \mathbb{Q})\big).$$

Proof. See [Ka] for details. To sketch the idea of the proof, it hinges on the fact that if $\pi : X^\sim \longrightarrow X$ is a proper modification of X, such that it is a desingularization of (see (1.8(iv))) $W \bigcup_{c \in C} X_c$, where X_c is a singular fiber of p, then $\text{Ker}\big(\pi_* : H_3(X^\sim, \mathbb{Q}) \longrightarrow H_3(X, \mathbb{Q})\big) = im\big(i_* : \bigoplus_{1 \leq e \leq k} H_3(Y_e, \mathbb{Q}) \longrightarrow H_3(X^\sim, \mathbb{Q})\big)$, where $\{Y_e\}_{1 \leq e \leq k}$ are the exceptional divisors of the modification, and by the results of Deligne [Del], we have degeneration of the spectral sequence associated with $i' : X'^\sim \longrightarrow X^\sim$, where $X'^\sim = X^\sim - S^\sim$ and $S^\sim$ is the union of the strict transforms of W. The rest follows (tediously!) from the Leray spectral sequence, Lefschetz duality and the Kunneth formula. $\square$

2. Consequences of the possible components of singular fibers of $p : X \longrightarrow C$

Let F be a general fiber of $p : X \longrightarrow C$. We know that F is a Del Pezzo surface and that L is locally very ample with respect to p (see (0.3)). Hence we see that $3 \leq \deg L_F \leq 9$.

Since the degree is preserved by flat maps it follows that for any possible singular fiber Γ of p, $3 \leq \deg L_\Gamma \leq 9$. Let S be a generic element of $|L|$ and let $\gamma = \Gamma \cap S$. Note that Γ is a possible singular fiber of $p : S \longrightarrow C$. Moreover S is an elliptic surface with no multiple fibers. Hence the possible fibers of γ are I_1, I_2, I_b with $3 \leq b \leq 9$, II, III, IV and I_b^* with $b = 0$, 1([11], (6.2)).

(2.1) Lemma. *Let $\Gamma = \sum_\alpha \Gamma^\alpha$ denote a possible reducible fiber of p. Let $\gamma = \Gamma \cap S$ where S is a general member of $|L|$. If γ is of type I_2, I_b with $3 \leq b \leq 9$, III, IV and I_b^* with $b = 0$, 1; then either $\Gamma^\alpha \cong \mathbb{F}_r$ for $r \geq 0$, or $\mathbb{F}_r^\sim$ with $r \geq 1$.*

Proof. Since γ is a sum of smooth rational curves of self-intersection -2, by ([Ko], (6.2)) and moreover since $\gamma^\alpha (= \Gamma^\alpha \cap S)$ is ample on Γ^α, by ([So], (0.6.1)) it follows that $\Gamma^\alpha \cong \mathbb{F}_r$ with $r \geq 0$, or $\mathbb{F}_r^\sim$ with $r \geq 1$. See (0.7). $\square$

(2.2) Remark. In order to prove that Γ is irreducible and reduced it suffices to prove that Γ is irreducible. This is because the fibers of p are reduced and L is ample in X.

(2.3) Theorem. *If $p : X \longrightarrow C$ is as in (1.2) and E as in (1.6), with $1 \leq K_y \cdot K_y \leq 7$, then*

$$H_3(X, \mathbb{Q}) = \Phi\big(H_1(E, \mathbb{Q})\big), \quad i.e. \quad \Phi \quad is \; surjective.$$

Proof. Since the reducible fibers have either $\mathbb{F}_r$ with $r \geq 0$, or $\mathbb{F}_r^\sim$ with $r \geq 1$ as its components, we see that [see (1.13) and the note before it] $(e_{R^\sim})_* \big(H_3(R^\sim, \mathbb{Z})\big) = 0$ for every irreducible component $R^\sim$. Hence the result follows from (1.13). $\square$

(2.4) Corollary. *If $p : X \longrightarrow C$ is as in (1.2) and $1 \leq K_y \cdot K_y \leq 7$, then $\Phi\big(H_1(E, \mathbb{Z})\big)$ has finite index in $H_3(X, \mathbb{Z})$.*

Proof. This is immediate from (2.3). $\square$

3. Prym-Tyurin varieties, Intermediate Jacobians and Abel-Jacobi maps

(3.1) Definition. *(Prym-Tyurin variety)* The abelian subvariety of $J(C)$ defined by $P(C, \iota)$, where $P(C, \iota) = (1 - \iota)J(C)$ is called the *Prym-Tyurin* variety.

(3.2) Theorem. (Bloch-Murre) *If $P = P(C, \iota), J = J(C), B = (\iota + q - 1)J$ and $J_q =$ points of order q in J, then $B + P = J$; $B \cap P \subset J_q$; $B = \big(\mathrm{Ker}(\iota - 1)\big)^{\cdot}$; $P = \big(\mathrm{Ker}(\iota + q - 1)\big)^{\cdot}$, where $G^{\cdot}$ denotes the connected component of G containing $0 \in G$.*

Proof. See [B-M]. $\square$

(3.3) Remarks. If $p : X \longrightarrow C$ is as in (1.2), then since the generic fiber is a Del Pezzo surface, by ((1.3); iv), $p_i(\mathcal{O}_X) = 0$ for $i > 0$; by the standard Leray spectral sequence $H^3(X, \mathcal{O}_X) = 0$. So by Serre duality $H^{3,0}(X) = 0$. Hence by Lefschetz decomposition $H^3(X, \mathbb{C}) = H^{2,1}(X) \oplus H^{1,2}(X)$.

(3.4) Definitions. (i) The *Intermediate Jacobian* $J(X) = H^{2,1}(X)^*/H_3(X,\mathbb{Z})$. Note that $J(X)$ is a complex torus.

(ii) The *Principal polarization* θ on $J(X)$, is one whose associated bilinear Riemann form E_θ is given by $E_\theta(\delta,\delta') = \delta \cdot \delta'$ where $\delta,\delta' \in H_3(X,\mathbb{Z})$ mod torsion.

(iii) The *Abel-Jacobi homomorphism* $\phi : J(E) \longrightarrow J(X)$ associated to the family of curves is given as follows: let $z \in J(E)$ be represented by a divisor $\sum n_j e_j$ of degree 0 on *each irreducible component of E*, and such that e_j is an element of E_0 (see (1.8)). If γ is a 1-chain on E_0 such that $\partial\gamma = \sum n_j e_j$ and $\Gamma = \eta p^{-1}(\gamma)$ the corresponding 3-chain in X, then $\phi(z) = \int_\Gamma \omega$ mod $H_3(X,\mathbb{Z})$, where $\omega \in H^{2,1}(X)$.

(3.5) Remark. (i) $\phi_* : H_1(E,\mathbb{Z}) \longrightarrow H_3(X,\mathbb{Z})$ coincides with the cylinder homomorphism Φ modulo torsion.

(ii) In (i) we therefore have $^t\Phi : H_3(X,\mathbb{Z}) \longrightarrow H_1(E,\mathbb{Z})$. And indeed if ϕ is as in (3.4) (iii), and $\phi^\wedge$ the dual map, then $^t\Phi = -(\phi^\wedge)_*$ and $\phi^\wedge\phi = 1 - \iota$.

(3.6) Proposition. *Let $p : X \longrightarrow C$ is as in (1.2). If $\phi : J(C) \longrightarrow J(X)$ is as in (3.4(iii)) then* $(\mathrm{Ker}\phi)^. = (\iota + q - 1)J(C)$. *In particular, $\phi \circ (\iota + q - 1) = 0$.*

Proof. See [Ka]; follows immediately from (3.2), where we note that $B = (\iota + q - 1)J$, and by (2.3) and (1.8). $\square$

(3.7) Theorem. *Let $p : X \longrightarrow C$ is as in (1.2), and ϕ as in (3.4(iii)). If $1 \le K_y \cdot K_y \le 5$, then the restriction to P of the Abel-Jacobi homomorphism, $\phi_p : P(C,\iota) \longrightarrow J(X)$ is an isogeny.*

Proof. This is a direct consequence of (2.4) and (3.5 (i)). $\square$

(3.8) Definition. We call Θ, the *canonical principal polarization* on $J(E)$, where

$$J(E) = H^{1,0}(E)^*/H_1(E,\mathbb{Z})$$

with the associated Riemann form $E_\Theta(\gamma,\gamma') = -(\gamma \cdot \gamma')$, where $\gamma,\gamma' \in H_1(E,\mathbb{Z})$.

(3.9) Theorem. *Let P be as in (3.1), and assume that the incidence correspondence D (see (1.8)) is without fixed points (i.e. $(\mathrm{Supp}D) \cap \Delta = \emptyset$, where $\Delta = $ diagonal on $E \times E$), then there exists a principal polarization Ξ, on P, such that [see (3.8) too]*

$$\Theta \cdot P \approx q\Xi,$$

where $\approx$ denotes algebraic equivalence.

Proof. See [B-M] or [Ka]. $\square$

(3.10) Remark. Let $\Lambda_p = \mathrm{Ker}(\iota + q - 1)$, then Λ_p is a lattice of P in $H_1(E,\mathbb{Z})$. The *principal polarization* Ξ is given by the associated bilinear form $E_\Xi(\beta,\beta') = -q^{-1}(\beta,\beta')$.

(3.11) Lemma. *With the notations of (3.2), we have $P_q = (\iota - 1)J_q$.*

Proof. By [B-M] (see [Ka]), $\Theta \cdot P \approx q\Xi$ is equivalent to $\mathrm{Ker}(\iota - 1) = (\iota + q - 1)J$. The rest follows from the q-divisibility of J. $\square$

(3.12) Lemma. *Let $p : X \longrightarrow C$ is as in (1.2). With the notations of (3.2) and (3.4), there exists a homomorphism $\psi : P \longrightarrow J(X)$ such that $\phi_p = q\psi$, where ϕ_p is the restriction of ϕ to P.*

Proof. (See [Ka]). If $p \in P_q, p = (\iota - 1)p' = (\iota + q - 1)j$, where $j \in J_q$. Now simply note that $\phi(j) = 0$, hence we see that $P_q \subset \text{Ker}\phi_p$, which proves the result. $\square$

(3.13) Theorem. *Let $p : X \longrightarrow C$ is as in (1.2), and ϕ as in (3.4(iii)). If $1 \leq K_y \cdot K_y \leq 5$ and the incidence correspondence $D(\text{see }(3.9))$ is without fixed points, then (P, Ξ) is isomorphic to a direct summand of $(J(X), \theta)$.*

Proof. (See [Ka]). By (3.10) and (3.11), all that remains to show is that $\psi^* E_\theta \cong E_\Xi$. So let $\beta \in \text{Ker}(\iota + q - 1)$, then

$$q^2 E_\theta (\psi_*(\beta), \psi_*(\beta')) = (q\psi_*(\beta), q\psi_*(\beta')) = (\phi_*(\beta), \phi_*(\beta')) \quad \text{by} \quad (3.11)$$

$$= ((\iota_* - 1)\beta, \beta') \quad \text{by} \quad (1.8)$$

$$= -(q\beta, \beta') \quad \text{by} \quad (\iota + q - 1)\beta = 0$$

$$= q^2 E_\Xi(\beta, \beta'). \qquad \square$$

4. Chow groups of codimension two cycles

(4.1) Definitions. (i) Let Z_1 and Z_2 be two 1-cycles, then Z_1 and Z_2 are said to be *algebraically equivalent* denoted $Z_1 \approx Z_2$, if there exists an *irreducible* curve T and any two points t_1 and t_2 in T, and a *cycle* Z on $X \times T$ such that $Z \cdot (X \times \{t_i\}) = Z_i$ for $i = 1, 2$. The intersection of Z, which takes place in the Chow ring of X, i.e. $A(X)$, is such that $Z \cdot (X \times \{t\})$ is always an element of $A^{n-1}(X)$.

(ii) For *rational equivalence* we replace T by $\mathbb{P}^1$, and traditionally, the points t_1, t_2 by 0, 1 respectively.

(iii) The *codimension two* cycles $A^2(X)$ is the group of algebraic 1-cycles on X, *algebraically equivalent to zero* modulo the cycles *rationally equivalent to zero*.

(4.2) Remark. Let $w : A^2(X) \longrightarrow J(X)$ be the Abel-Jacobi homomorphism (3.4 (iii)). The family of lines in (1.8) defines homomorphism

$$\rho^a = \eta_* \circ p^* : J(E) \longrightarrow A^2(X) \quad \text{and} \quad {}^t\rho^a = p_* \circ \eta^* : A^2(X) \longrightarrow J(E).$$

We also note that $w \circ \rho^a = \phi$ and ${}^t\phi \circ w = {}^t\rho^a$.

(4.3) Definitions. (i) Let A be an abelian variety. A *homomorphism* $\phi : A^2(X) \longrightarrow A$ is said to be *regular* if for *every* algebraic family $\{Z(t)\}_{t \in T}$ on X, where T is a parametrizing variety and $Z \in CH^2(T \times X)$, the composite map $\phi \circ w_z : T \longrightarrow A^2(X) \longrightarrow A$, is a *morphism* from $T \longrightarrow A$. Where $w_z : T \longrightarrow A^2(X)$ is given by $w_z(t) = Z(t) - Z(t_0)$ for some fixed $t_0 \in T$.

(ii) A *regular morphism* $\phi_0 : A^2(X) \longrightarrow A_0$ is said to be *universal*, if for any other morphism $\phi : A^2(X) \longrightarrow A$, we can find a (unique) map $f : A_0 \longrightarrow A$, such that $\phi = f \circ \phi_0$.

(iii) A_0 in (ii) is said to be the *algebraic representative* of $A^2(X)$.

(4.4) Remark. By a result of Murre [Mu], this homomorphism ϕ in (4.3 (i)) is regular, universal homomorphism. Moreover for any prime number l, the restriction of ϕ on the l-torsion points, $\phi_l : A^2(X)_{l-tors} \longrightarrow J(X)_{l-tors}$ is an isomorphism.

(4.5) Theorem. *Let* $p : X \longrightarrow C$ *is as in* (1.2), *and* ψ *as in* (3.10), (3.12). *If* $1 \le K_y \cdot K_y \le 5$ *and the incidence correspondence* D(*see* (3.9)) *is without fixed points. Let* (P, Ξ) (*see* (3.1), (3.2)) *be a principal polarized abelian variety, then there exists a regular universal isomorphism* $^t\rho^a : A^2(X) \longrightarrow P$ *and an isomorphism* $\psi^a : P \longrightarrow A^2(X)$, *such that* $q\psi^a = \rho^a \big|_P$ *and* $^t\rho^a = -(\psi^a)^{-1}$.

Proof. By the remark in (4.2) $w \circ \rho^a = \phi$ and $^t\phi \circ w = {}^t\rho^a$. Indeed by [B-M], w in (4.2) is an isomorphism because the generic fiber of p is a Del Pezzo, and hence rational, surface. Hence w is an isogeny, and by (4.4), w is an isomorphism. $\square$

Acknowledgements: even though the work on this material started before it, the appearance of [Ka] changed the course of this article dramatically, and simplified matters considerably. Hence my indebtness to Kanev's work is deep. I would like to thank Igor Dolgachev for bringing Kanev's work to my attention.

It is a pleasure to thank A.Tikhomirov for organizing a wonderful conference and special thanks to Slava Nikulin for his kind hospitality.

References

[B-M]   S.BLOCH and J.P.MURRE. *On the Chow group of certain type of Fano threefolds*, Compositio Math., 39, 47-105(1979).

[C-G]   H.CLEMENS and P.GRIFFITHS. *Intermediate Jacobian of the cubic threefold*, Ann. Math. 95, 281-356(1972).

[D 1]   H.D'SOUZA. *Threefolds whose hyperplane sections are elliptic surfaces*, Pac. J. Math. 134, 57-78(1988).

[D 2]   H.D'SOUZA. *A general result on local spannedness*, C. R. Math. de l'Academie des Sciences, Royal Soc. Canada, 11, 143-148(1989).

[D 3]   H.D'SOUZA. *A complete structure theorem for a class of Del Pezzo fiber spaces* (preprint).

[Del]   P.DELIGNE. *Théorie de Hodge II*, Publ. Math. I.H.E.S., 40, 5-57(1972).

[Dem]   M.DEMAZURE. *Surfaces de Del Pezzo*, SLN 777, 21-69(1980).

[H]   H.HIRONAKA. *Resolution of singularities of an algebraic variety over a field of characteristic zero I*, Ann. Math. 79, 109-180(1964).

[Ka]   V.KANEV. *Intermediate Jacobians and Chow groups of threefolds with a pencil of Del Pezzo surfaces*, Ann. di Mat. Pura ed Applicata (IV) 13-48(1989).

[Ko]   K.KODAIRA. *On compact analytic surfaces II*, Ann. Math. 77, 563-623(1963).

[Ma]   Y.MANIN. *Cubic forms*, North-Holland Publishing Co., Amsterdam (1973).

[M-M]   Y.MIYAOKA and S.MORI. *A numerical criterion for uniruledness*, Ann. Math. 124, 65-69(1986).

[Mu]   J.P.MURRE. *Un résultat en théorie des cycles algebrique de codimension deux*, C. R. Acad. Sc. Paris, 296, 981-984(1983).

[So]   A.J.SOMMESE. *On the configuration of −2 rational curves on the hyperplane sections of elliptic threefolds*, Classification of algebraic and analytic manifolds, Progr. Math. 39, 465-497(1980).

[T]   A.N.TYURIN. *Five lectures on three dimensional varieties*, Uspehi Nauk USSR, 29, 3-50(1972).

The Decomposition, Inertia and Ramification Groups in Birational Geometry

M.H.Gizatullin

Let X be an irreducible scheme, let $\mathrm{Bir}(X)$ be its group of birational automorphisms, let G be a subgroup of $\mathrm{Bir}(X)$. If $g \in G$, then $\mathrm{dom}(g)$ denotes the domain of definition of the map g, g^* denotes the corresponding automorphism of the total ring of fractions on X. Let Y be an irreducible reduced subscheme of X, let p_Y be the generic point of Y, let A_Y be the local ring of p_Y, let m_Y be the maximal ideal of A_Y.

The decomposition group of Y in G is

$$G_Y = \{g \in G \mid p_Y \in \mathrm{dom}(g) \cap \mathrm{dom}(g^{-1}), \quad g(p_Y) = p_Y\}.$$

In other words, G_Y is the stabilizer of p_Y in G. There is a natural homomorphism

$$r : G_Y \longrightarrow \mathrm{Bir}(Y). \tag{1}$$

The group G_Y operates on the local ring A_Y, G_Y preserves all the powers of the maximal ideal m_Y, therefore G_Y operates on A_Y / m_Y^{i+1}, $i \geq 0$. The i-th ramification group of Y in G is

$$G_{iY} = \{g \in G_Y \mid g \text{ operates trivially on } A_Y / m_Y^{i+1}\}.$$

In other words, G_{iY} consists of those elements of G_Y that operate trivially on the i-th infinitesimal neighbourhood of Y in X. Especially, G_{0Y} is the kernel of the homomorphism (1). The group G_{0Y} is called the inertia group of Y in G.

Example 1. This example is the origin of the terminology used in the paper. If A is a Dedekind domain, K is its field of fractions, L/K is a finite Galois extension, B is the integral closure of A in L, $X = \mathrm{Spec}\,B$, Y is a subscheme of X defined by a maximal ideal P of B, $G = \mathrm{Gal}(L/K)$, then G_Y, G_{0Y}, G_{iY} are respectively the decomposition, inertia and ramification groups of P in L/K in the usual sense (see [1], chap. 5, §10).

Example 2. This example indicates that a notion of congruence subgroup is a special case of the notion of the ramification group. Let X be $\mathrm{Proj}\,\mathbb{Z}[z_0, z_1]$, i.e. X is the projective line over $\mathbb{Z}$, let Y be a curve on X defined by a prime number p (see picture in Mumford's fifth lecture [2]). The group $\mathrm{Bir}(X)$ is isomorphic to $\mathrm{PGL}(2, \mathbb{Q})$, the group $\mathrm{Bir}(X)_Y$ consists of linear fractional transformations with integer coefficients and determinant coprime to p. If $G = \mathrm{PSL}(2, \mathbb{Z})$, $G \subset \mathrm{Bir}(X)$, then $G_Y = G$, G_{iY} is the congruence subgroup $\Gamma(p^{i+1})$.

Let us return to the general situation.

Theorem 1. *For each $i \geq 0$ the group G_{iY} is a normal subgroup of G_Y.*

Proof. G_{iY} is the kernel of a natural homomorphism

$$G_Y \longrightarrow \mathrm{Aut}(A_Y / m_Y^{i+1}).$$

Theorem 2. $(G_{iY}, G_{jY}) \subset G_{(i+j)Y}$, *where $i \geq 0$, $j \geq 0$, (A, B) is a subgroup generated by $a^{-1}b^{-1}ab$, $a \in A$, $b \in B$.*

Proof. If $a \in G_{iY}$, $b \in G_{jY}$, $x \in A_Y$, then

$$a^*(x) = x + \sum_l \left(\prod_{k=0}^{i} y_{lk} \right), \qquad b^*(x) = x + \sum_n \left(\prod_{m=0}^{j} z_{nm} \right),$$

where y_{lk}, z_{nm} are elements of m_Y. Further

$$b^*a^*(x) \in \left(x + \sum_n \left(\prod_{m=0}^{j} z_{nm} \right) + \sum_l \left(\prod_{k=0}^{i} (y_{lk} + m_Y^{j+1}) \right) \right),$$

$$a^*b^*(x) \in \left(x + \sum_l \left(\prod_{k=0}^{i} y_{lk} \right) + \sum_n \left(\prod_{m=0}^{j} (z_{nm} + m_Y^{i+1}) \right) \right),$$

hence

$$b^*a^*(x) - a^*b^*(x) \in m_Y^{i+j+1}, \qquad (bab^{-1}a^{-1})^*(x) - x \in m_Y^{i+j+1}.$$

Q.E.D.

Theorem 1 and example 2 motivate the following terminology. We shall say that a triple (X, Y, G_Y) has the positive solution of the congruence subgroups problem (briefly, (X, Y, G_Y) has p.s.c.s.p.), if each nontrivial normal subgroup of G_Y contains some nontrivial ramification subgroup G_{iY}. In the opposite case we shall say that this triple has the negative solution of that problem (briefly, (X, Y, G_Y) has p.s.c.s.p.). The congruence subgroup problem for some triples is connected with the unsolved problem of the simplisity of the Cremona group $\mathrm{Bir}(\mathbb{P}_2)$.

Theorem 3. *Let $\mathbb{P}_2$ be the projective plane over an algebraically closed field k, let P be a closed point on the plane, let $G = \mathrm{Bir}(\mathbb{P}_2)$ be the Cremona group. If the triple $(\mathbb{P}_2, P, G_P)$ has p.s.c.s.p., then the group G is simple. In other words, if G is not simple, then there exists a normal subgroup $H \subset G_P$ such that $\{e\} \neq (H \cap G_{iP}) \neq G_{iP}$ for each $i \geq 0$.*

Proof. Let H be a nontrivial normal subgroup of G.

Lemma 1. $H \cap G_P \neq \{e\}$.

Proof. Let g be a nontrivial element of H, $P_0 \in \mathrm{dom}(g) \cap \mathrm{dom}(g^{-1})$. A replacement of g by a suitable conjugate element makes $P_0 = P$. Suppose that $g(P) = Q$, $Q \neq P$ (if $Q = P$, then $g \in H \cap G_P$),

$$M = \{h \in \mathrm{Aut}(\mathbb{P}_2) \mid h(P) = Q, h(Q) = P\}.$$

The set M is a connected 4-dimensional subvariety of $\mathrm{Aut}(\mathbb{P}_2)$. There exists $h \in M$ such that $hgh^{-1} \neq g^{-1}$. Indeed, if g is a projective transformation, then the existence of h is evident, if g has a fundamental point, then a suitable h moves aside fundamental points of g apart ones of g^{-1}. Thus the transformation $hgh^{-1}g$ is nontrivial and belongs to $H \cap G_P$. Q.E.D.

If the triple $(\mathbb{P}_2, P, G_P)$ has p.s.c.s.p., then $H \cap G_P \supset G_{iP}$ for some natural i. Let x, y be the affine coordinates on the projective plane, let P be the origin. The transformation g, defined by the formulae

$$x' = x, \quad y' = y + x^{i+1} \tag{2}$$

belongs to G_{iP}, hence $g \in H$. It is sufficient to prove that the normal closure $\langle g \rangle$ of g in the group G is the whole G (normal closure of a subset is the smallest normal subgroup containing this subset). The transformation (2) preserves the pencil of lines $x = \mathrm{const}$, therefore it is sufficient to prove the following lemma.

Lemma 2. *If a nontrivial birational transformation of the projective plane preserves a pencil of lines, then the normal closure of this transformation in the Cremona group is the whole Cremona group.*

Proof. *Case 1.* Let g be a projective transformation. Since the projective group $\mathrm{Aut}(\mathbb{P}_2)$ over an algebraically closed field is simple, we have $\langle g \rangle \supset \mathrm{Aut}(\mathbb{P}_2)$, therefore the involution h, defined by $x' = 1-x$, $y' = 1-y$, belongs to g. If s is the standard quadratic transformation $x' = 1/x$, $y' = 1/y$, then $s^2 = e$, $(hs)^3 = e$, hence $s = (hs)h(hs)^{-1} \in \langle g \rangle$. The set $\mathrm{Aut}(\mathbb{P}_2) \cup \{s\}$ generates G, therefore $\langle g \rangle = G$.

Case 2. Let g be an arbitrary nontrivial element of the group J of all the birational transformations preserving the pencil $x = \mathrm{const}$. Elements of J are defined by the formulae of the following form

$$x' = (px + q)/(rx + s), \quad y' = \big(a(x)y + b(x)\big)/\big(c(x)y + d(x)\big), \tag{3}$$

where p, q, r, $s \in k$, a, b, c, $d \in k[x]$, $ps - qr \neq 0$, $ad - bc \neq 0$. As well as in the proof of lemma 1, there is a transformation h of the form (3) with $p = s = 1$, $q = r = 0$, $\{a, b, c, d\} \subset k$ such that $ghg^{-1}h^{-1} \neq e$. Thus the transformation $g_0 = ghg^{-1}h^{-1}$ is nontrivial, g_0 is of the form (3) with $p = s = 1$, $q = r = 0$, i.e. g_0 is an element of the group

$$\mathrm{Aut}\big(k(x,y)/k(x)\big) = \mathrm{PGL}(2, k(x)).$$

Each nontrivial normal subgroup of the last group contains $\mathrm{PSL}(2, k(x))$, hence $\langle g \rangle$ contains a nontrivial projective transformation, therefore, as it was established in the first case, $\langle g \rangle = G$. Lemma 2 and theorem 2 are proved.

Corollary 1. *Let L be a line on the projective plane over an algebraically closed field, let G be the Cremona group. If the triple $(\mathbb{P}_2, L, G_L)$ has p.s.c.s.p., then the group G is simple.*

Proof. Let H be a nontrivial normal subgroup of G, let x, y be the affine coordinates on $\mathbb{P}_2$. Suppose that $x = 0$ is the equation of L, P is the origin, f is the Cremona transformation defined by $x' = x$, $y' = xy$. By lemma 1 there exists $g_0 \in H \cap G_P$, $g_0 \neq e$. The transformation $g_1 = f^{-1}g_0 f$ belongs to G_L, therefore $H \cap G_L \neq \{e\}$, hence this intersection contains some subgroup G_{iL}. The transformation g of the form (2) belongs to G_{iL}, hence $g \in H$. From lemma 2 it follows that $\langle g \rangle = G$. Q.E.D.

Corollary 2. *Let Y be a plane rational curve which admits a birational straightening, i.e. there is a Cremona transformation f such that the generic point p_Y of Y belongs to* $\text{dom}(f) \cap \text{dom}(f^{-1})$ *and $f(p_Y)$ is the generic point of some line L. If the triple $(\mathbb{P}_2, Y, G_Y)$ has p.s.c.s.p., then G is simple.*

Proof. If f is the straightening, H is a normal subgroup of G, then according to the proof of corollary 1 $H \cap G_L \neq \{e\}$, hence $H \cap f^{-1}G_L f \neq \{e\}$, i.e. $H \cap G_Y \neq \{e\}$. If H contains G_{iY}, then H contains $G_{iL} = f G_{iY} f^{-1}$. By corollary 1 $H = G$. Q.E.D.

Remark on lemma 2. The methods used in the proofs of lemmas 1, 2 lead to the following result. If a Cremona transformation g transforms some pencil of lines into a pencil of curves of degree d, $d \leq 4$, then $\langle g \rangle = G$. Especially, if $\deg(g) \leq 7$, then $\langle g \rangle = G$. We omit the proof.

Example 3, or more precisely, a construction of some elements of the decomposition group. Let $\mathbb{P}_n$ be the n-dimensional projective space over a field k of characteristic different from 2, $G = \text{Bir}(\mathbb{P}_n)$, let Y be a reduced irreducible hypersurface in $\mathbb{P}_n$, $\deg Y = d$, $d \geq 2$, let P be a k-rational point of $\mathbb{P}_n$ such that the multiplicity of P on Y is equal to $d - 2$. We shall construct two involutory transformations $T_P \in G_Y$ and $R_P \in G_{0Y}$. Both of these involutions preserve the general lines through P. Let L be such a line, $L \cap Y = \{P, A, B\}$. The involutions T_P and R_P are defined by the following conditions $T_P(A) = B$, $(T_P \mid L)(P) = P$, $R_P(A) = A$, $R_P(B) = B$. Especially, if $d = 2$, then P does not belong to the quadric Y, $T_P \in \text{Aut}(\mathbb{P}_n)$, R_P is the quadratic inversion with the centre P and fixed quadric Y. If $d = 3$, then Y is a cubic hypersurface, the restriction $t_P = T_P \mid Y$ (i.e. t_P is the image of T_P by the homomorphism (1)) is the involution used by Yu.I. Manin in [3]. If $x = (x_1, ..., x_n)$ are the affine coordinates in $\mathbb{P}_n$, $P = (0, ..., 0)$, $f = 0$ is the equation of Y, where $f = F_{d-2}(x) + F_{d-1}(x) + F_d(x)$, $F_i \in k[x]$, F_i is a homogeneous polynomial of degree i, then the above transformations are defined by the following formulae

$$T_P : \quad x_i' = -x_i F_{d-2}/(F_{d-2} + F_{d-1}),$$

$$R_P : \quad x_i' = -x_i(F_{d-1} + 2F_{d-2})/(F_{d-1} + 2F_d),$$

$$i = 1, ..., n,$$

$$i = 1, ..., n,$$

or

$$R_P : \quad x_i' = x_i - \left(2x_i/\left(2F_d + F_{d-1}\right)\right)f,$$

$$i = 1, ..., n.$$

Note that the condition $\operatorname{char} k \neq 2$ is essential only for the construction of R_P.

Theorem 4. *There are the following relations between the above transformations T_P, R_P and $W \in \operatorname{Aut}(\mathbb{P}_n, Y)$ (i.e. W is a projective k-automorphism preserving Y)*

$$R_P^2 = e, \quad T_P R_P = R_P T_P, \quad W R_P W^{-1} = R_{W(P)}, \tag{4}$$

$$T_P^2 = e, \quad W T_P W^{-1} = T_{W(P)}. \tag{5}$$

Moreover if A, B, C are three collinear points of the multiplisity $d - 2$ on Y, then

$$(T_A T_B T_C)^2 = e. \tag{6}$$

Proof. The relations (4), (5) are evident consequences of the construction of T_P, R_P. Let L be the line containing A, B, C. Since the family of planes through L is invariant with respect to T_A, T_B, T_C, then it is sufficient to prove that the restriction of $T_A T_B T_C$ on the general plane of this family is involutory. Thus we shall deal with the case when $n = 2$, Y is a curve (maybe reducible) of degree d, A, B, C are three collinear points of the multiplicity $d - 2$ on Y. If $d \geq 4$, then $L \subset Y$ and either

(i) $Y = Z + (d - 3)L$, where Z is a reduced curve of the degree 3, L is not a component of Z, the points A, B, C are simple on Z, or

(ii) $Y = Z + (d - 2)L$, where Z is a reduced conic, the points A, B, C don't lie on the conic.

Let x, y be the affine coordinates such that $y = 0$ is the equation of L, let $h = 0$ be the equation of Z, let Z_t be the general member of the pencil $h + ty^3 = 0$ (resp., $h + ty^2 = 0$) in the case (i) (resp., (ii)). The involutions $T_A^{(t)}$, $T_B^{(t)}$, $T_C^{(t)}$ constructed with the help of Z_t (instead of Z) are independent of the parameter t, i.e. they coincide with T_A, T_B, T_C respectively. It is obvious that $\left((T_A T_B T_C) \mid Z_t\right)^2 = e$, therefore (6) is true. Q.E.D.

We shall touch upon the question of the surjectivity of the restriction map $r : G_Y \longrightarrow \operatorname{Bir}(Y)$ mentioned in (1). In the classical situation of the example 1 the map r is epimorphism (see [4], ch. 5, n°2, Th.2, (ii)).

Theorem 5. *Let $\mathbb{P}_3$ be the projective space over a perfect field k, $G = \operatorname{Bir}(\mathbb{P}_3)$, let $Y \subset \mathbb{P}_3$ be a minimal smooth cubic surface defined over k. Then the map r is epimorphism, moreover the exact sequence*

$$\{e\} \longrightarrow G_{0Y} \longrightarrow G_Y \longrightarrow \operatorname{Bir}(Y) \longrightarrow \{e\}$$

splits, i.e. there exists a homomorphism $s : \operatorname{Bir}(Y) \longrightarrow G_Y$ such that $rs = id$.

Proof. Let L/k be a quadratic extension of k, A and B be two $\operatorname{Gal}(L/k)$-conjugate points of Y, $C \in Y(k)$ be a rational point collinear with A and B. Then the transformation $S_{AB} = T_A T_B T_C$ is defined over k. Indeed $\operatorname{Gal}(L/k)$-conjugate to S_{AB} is $T_B T_C T_A$ which coincides with S_{AB} by (6), hence $S_{AB} \in G$. The transformations $t_P = T_P \mid Y = r(T_P) \in \operatorname{Bir}(Y)$, $s_{AB} = S_{AB} \mid Y = r(S_{AB}) \in \operatorname{Bir}(Y)$ (where $P \in Y(k)$, $A, B \in Y(L)$ for some quadratic extension L/k, A and B are $\operatorname{Gal}(L/k)$-conjugate) together with the set $\operatorname{Aut}(\mathbb{P}_3, Y)$ generate the group $\operatorname{Bir}(Y)$ (see [3], ch. 5). Therefore r is epimorphic. Defining relations between the mentioned generators are the following

$$t_P^2 = e, \quad s_{AB}^2 = e, \quad (t_P t_Q t_R)^2 = e \text{ for each collinear triple } P, Q, R,$$

$$wt_P w^{-1} = t_{w(P)}, \quad ws_{AB}w^{-1} = s_{w(A)w(B)}, \text{ where } w \in \text{Aut}(\mathbb{P}_3, Y) \mid Y$$

(see [3]). The analogous relations are true for the transformations T_P, S_{AB} (see (5), (6)), therefore the definition of the required map s by $s(t_P) = T_P$, $s(s_{AB}) = S_{AB}$ is correct. Q.E.D.

Theorem 6. *If Y is a plane smooth cubic curve over a perfect field k, $Y(k) \neq \emptyset$, $G = \text{Bir}(\mathbb{P}_2/k)$, then the map (1) is surjective.*

Proof. If $Y(k) \neq \emptyset$, then the group $\text{Bir}(Y)$ is generated by the set $\text{Aut}(\mathbb{P}_2, Y) \mid Y$ and the reflections $t_P = T_P \mid Y$, where $P \in Y(k)$. Q.E.D.

The next point is the nontriviality of the higher ramification groups.

Theorem 7. *If Y is an irreducible reduced k-hypersurface in the projective space $\mathbb{P}_n$ over a field k of characteristic different from two, the set $M(Y)$ of points $P \in \mathbb{P}_n(k)$ with $\text{mult}_P(Y) = d - 2$ is infinite, then all the groups G_{iY} (where $G = \text{Bir}(\mathbb{P}_n/k)$) are different from $\{e\}$.*

Proof. Let $P \in M(Y)$, R_P be the transformation constructed in example 3, $f = 0$ be an affine equation of Y.

Lemma 1. $R_P^*(f) \equiv -f (\text{mod } m_Y^2)$.

Proof. Let $x = (x_1, \ldots, x_n)$ be the affine coordinates in $\mathbb{P}_n$, $P = (0, \ldots, 0)$, $f = F_{d-2} + F_{d-1} + F_d$, where $F_m = F_m(x)$ is a homogeneous polynomial of degree m. If $D = (F_{d-1} + 2F_{d-2})/(2F_d + F_{d-1})$, $E = 2/(2F_d + F_{d-1})$, then $D = -1 + Ef$, $R_P^*(x_i) = -x_i D$, $i = 1, \ldots, n$ (see example 3). Therefore

$$
\begin{aligned}
R_P^*(f) &= F_{d-2}(-xD) + F_{d-1}(-xD) + F_d(-xD) = (-D)^{d-2}(F_{d-2} - DF_{d-1} + D^2 F_d) \\
&= (-D)^{d-2}(F_{d-2} + F_{d-1} + F_d - Ef(F_{d-1} + 2F_d) + E^2 f^2 F_d) \\
&= (-D)^{d-2}(-f + E^2 f^2 F_d) \equiv -f(\text{mod } m_Y^2).
\end{aligned}
$$

Q.E.D.

Lemma 2. *If P, $Q \in M(Y)$, $P \neq Q$, then $R_P R_Q \neq e$, $(R_P R_Q)^*(f) \equiv f(\text{mod } m_Y^2)$.*

Proof. The points P, Q are among the fundamental points of the transformation $R_P R_Q$, hence this transformation is nonprojective. The congruence of the lemma follows from the preceding lemma. Q.E.D.

Lemma 3. *If P, Q, U, V are four different points of $M(Y)$, $a = R_P R_Q$, $b = R_U R_V$, $g = (a, b)$, i.e. $g = a^{-1} b^{-1} ab$, then $g \neq e$, $g \in G_{1Y}$.*

Proof. These four points belong to the set of fundamental points of g, hence $g \neq e$. If z is an element of the local ring A_Y, then $a^*(z) \equiv z + sf(\text{mod } m_Y^2)$, $b^*(z) \equiv z + tf(\text{mod } m_Y^2)$ because of a, $b \in G_{0Y}$. By lemma 2 $b^* a^*(z) \equiv z + (s+t)f(\text{mod } m_Y^2)$, $a^* b^*(z) = z + (s+t)f(\text{mod } m_Y^2)$, hence $b^* a^*(z) \equiv a^* b^*(z)(\text{mod } m_Y^2)$, $(a, b)^*(z) \equiv z(\text{mod } m_Y^2)$, i.e. (a, b) is an element of G_{1Y}. Q.E.D.

Let $(P, Q, U, V), (P_1, Q_1, U_1, V_1), (P_2, Q_2, U_2, V_2), \ldots$ be disjoint quadruples of points of $M(Y)$, let $g, g_1, g_2, \ldots$ be the corresponding commutators constructed in lemma 3. Then these commutators belong to G_{1Y}, $(g, g_1) \in G_{2Y}$ by theorem 2, $(g, g_1) \neq e$ by the reason pointed in the proofs of lemmas 2, 3, $((g, g_1), g_2)$ is a nontrivial element of G_{3Y}, etc. Q.E.D.

References

[1]   O. ZARISKI, P. SAMUEL. *Commutative algebra*, vol.1, Van Nostrand, 1958.

[2]   D. MUMFORD. *Lectures on curves on an algebraic surface*, Princeton, 1966.

[3]   Yu. I. MANIN. *Cubic forms*, Moscow, 1972 (Engl. transl.: North Holland, Amsterdam, 1974, 2-nd edition 1986).

[4]   N. BOURBAKI. *Eléments de Mathématique: Algebre commutative*, chap. 1-7, Paris, Hermann, 1961-1965.

Helix Theory and Nonsymmetrical Bilinear Forms

A.L.Gorodentsev

In this paper we will discuss the connections between the description of the combinatorical structure of the set of exceptional vector bundles and some natural questions about the unimodular (nonsymmetrical) bilinear forms. For investigation of the set of exceptional vector bundles (such bundles E that $\mathrm{Ext}^i(E,E) = 0$ for $i \neq 0$ and $\dim \mathrm{Ext}^0(E,E) = 1$) there was developed five years ago in [GR], [G1], [G2] the theory of helices. In section 1 we give a review of the main results of this theory. In section 2 we formulate some questions about classes of exceptional vector bundles in Grothendieck group K_0 in terms of a lattice with nonsymmetrical nondegenerate integer bilinear form. Arithmetical problems of helix theory are equivalent to the description of the orbits of isometries and orbits of the natural action of braid group on semiorthogonal bases of a lattice. In section 3 we speak about the general properties of nonsymmetrical forms and their isometries and then give some examples in section 4.

1. Survey of helix theory

1.1. Helices on $\mathbb{P}_n$. Let (A, B) be a pair of coherent sheaves on $\mathbb{P}_n$. Consider the canonical map

$$lcan : \mathrm{Hom}(A,B) \otimes A \longrightarrow B$$

(tensor product of vector space $\mathrm{Hom}(A,B)$ and sheaf A may be considered as $\dim \mathrm{Hom}(A,B)$-times direct sum $A \oplus \ldots \oplus A$). If $lcan$ is surjective, then we call its kernel *left mutation of B in pair* (A, B) and denote it by $L_A B$. In this case we have the exact triple of sheaves:

$$0 \longrightarrow L_A B \longrightarrow \mathrm{Hom}(A,B) \otimes A \longrightarrow B \longrightarrow 0.$$

Similarly, if partial dualisation of $lcan$:

$$rcan : A \longrightarrow \mathrm{Hom}(A,B)^* \otimes B$$

is injective, then its cokernel is called *right mutation of A in pair* (A, B) and denoted by $R_B A$. We have

$$0 \longrightarrow A \longrightarrow \mathrm{Hom}(A, B)^* \otimes B \longrightarrow R_B A \longrightarrow 0.$$

There is a remarkable infinite collection of sheaves on $\mathbb{P}_n$: $\sigma_0 = \{\mathcal{O}(i)\}$, $i \in \mathbb{Z}$. Let us denote $\mathcal{O}(i)$ by E_i and consider the following properties of this collection:

(H1) All E_i are *exceptional*. This means that $\mathrm{Ext}^i(E_i, E_i) = 0$ for $i \neq 0$ and $\dim \mathrm{Hom}(E_i, E_i) = 1$.

(H2) All subcollections of the form $(E_i, E_{i+1}, \ldots, E_{i+n})$ are *semiorthogonal* with respect to the functor $\mathrm{Ext}^{\cdot}$. This means that $\mathrm{Ext}^i(E_\alpha, E_\beta) = 0$ for all i if $\beta + n \geq \alpha > \beta$.

(H2') If $\alpha < \beta$, then $\mathrm{Ext}^i(E_\alpha, E_\beta) = 0$ for $i \neq 0$.

(H3) σ_0 is $(n+1)$-periodical in the sense that all consecutive mutations

$$L^{(k)} E_i = L_{E_{i-k}} L_{E_{i-k+1}} \cdots L_{E_{i-2}} L_{E_{i-1}} E_i,$$

$$R^{(k)} E_i = R_{E_{i+k}} R_{E_{i+k-1}} \cdots R_{E_{i+2}} R_{E_{i+1}} E_i,$$

are well defined for any $E_i \in \sigma_0$ and $1 \leq k \leq n$ and

$$L^{(n)} E_i = E_{i-n-1}; \quad R^{(n)} E_i = E_{i+n+1}.$$

(H4) σ_0 is $(n+1)$-periodical in the sense that for any E_i

$$E_{i+n+1} = E_i \otimes \omega_{\mathbb{P}_n}^*.$$

(H5) All subcollections $(E_i, E_{i+1}, \ldots, E_{i+n})$ are bases of the category of the coherent sheaves on $\mathbb{P}_n$ in the sense that for any coherent sheaf F there exists a functorial with respect to F spectral sequence $E_r^{p,q}$ with limit

$$E_\infty^n = \begin{cases} F & \text{for } n = 0, \\ 0 & \text{for } n \neq 0, \end{cases}$$

and $E_1^{p,q} \neq 0$ only for $-n \leq p \leq 0$, $0 \leq q \leq n$ and equals (for these p, q):

$$E_1^{p,q} = \mathrm{Ext}^q(E_{i-p}, F) \otimes L^{(-p)} E_{i-p}.$$

It is very simple to verify the properties **(H1)**, **(H2)**, **(H2')**. The property **(H4)** is evident. The property **(H5)** is a well-known Beilinson theorem. To explain **(H3)** we consider external powers of Euler exact triple on $\mathbb{P}_n = P(V)$:

$$0 \longrightarrow \mathcal{O} \longrightarrow V \otimes \mathcal{O}(1) \longrightarrow T \longrightarrow 0,$$
$$0 \longrightarrow T \longrightarrow \wedge^2 V \otimes \mathcal{O}(2) \longrightarrow \wedge^2 T \longrightarrow 0,$$
$$0 \longrightarrow \wedge^2 T \longrightarrow \wedge^3 V \otimes \mathcal{O}(3) \longrightarrow \wedge^3 T \longrightarrow 0,$$
$$\cdots \cdots \cdots$$
$$0 \longrightarrow \wedge^{n-1} T \longrightarrow \wedge^n V \otimes \mathcal{O}(n) \longrightarrow \wedge^n T = \mathcal{O}(n+1) \longrightarrow 0.$$

Since $\wedge^k V$ may be canonically identified with

$$\mathrm{Hom}(\mathcal{O}(k), \wedge^k T) = \mathrm{Hom}\left(\wedge^{k-1} T, \mathcal{O}(k)\right)^*$$

the triple

$$0 \longrightarrow \wedge^{k-1} T \longrightarrow \wedge^k V \otimes \mathcal{O}(k) \longrightarrow \wedge^k T \longrightarrow 0$$

naturally coincides with the left mutation triple for the pair $(\mathcal{O}(k), \wedge^k T)$ and coincides with the right mutation triple for the pair $(\wedge^{k-1} T, \mathcal{O}(k))$. Hence, all consecutive mutations $R^{(k)}\mathcal{O}$ and $L^{(n-k)}\mathcal{O}(n+1)$ are well defined and equal to $\wedge^k T$ for all $k = 1, 2, \ldots, n$. For other details see [G1].

1.1.1. Definition. An infinite collection $\sigma = (E_i)$, $i \in \mathbb{Z}$ of coherent sheaves on $\mathbb{P}_n$ is called a *helix on* $\mathbb{P}_n$ if it satisfies the conditions **(H1)**, **(H2)**, **(H2')**, **(H3)** and **(H4)** above.

1.1.2. Definition. *Left mutation* $L_\alpha\sigma$ *of the helix* σ on $\mathbb{P}_n$ (where α is a fixed class in $\mathbb{Z}/(n+1)\mathbb{Z}$) is the collection obtained from σ by changing all pairs (E_i, E_{i+1}) with $i \equiv \alpha(\mathrm{mod}\ n+1)$ for pairs $(L_{E_i} E_{i+1}, E_i)$. Similarly, *righ mutation* R_α changes all pairs (E_i, E_{i+1}) with $i \equiv \alpha(\mathrm{mod}\ n+1)$ for pairs $(E_{i+1}, R_{E_{i+1}} E_i)$.

1.1.3. Theorem. a) *Left and right mutations of a helix are helices for all* $\alpha \in \mathbb{Z}/(n+1)\mathbb{Z}$.

b) *Left and right mutations* L_α, R_α *are inverse to each other and satisfy the braid group relations:*

$$R_\alpha L_\alpha = L_\alpha R_\alpha = 1,$$
$$L_\alpha L_{\alpha+1} L_\alpha = L_{\alpha+1} L_\alpha L_{\alpha+1},$$
$$L_\alpha L_\beta = L_\beta L_\alpha \quad \text{if} \alpha \neq \beta \pm 1.$$

c) *All helices on* $\mathbb{P}_n$ *satisfy the property* **(H5)** *above.*

This theorem has been proved in [GR] and [G1] starting with more strong conditions in the definition of a helix. The present version may be deduced from [G2], [G3] and [B1]. Using the theorem we can construct infinite set of helices and semiorthogonal bases of category of coherent sheaves. This set has very interesting combinatorical structure, because admits the action of the braid group.

The conditions of surjectivity of $lcan$ and injectivity of $rcan$ in definition of mutations are very awkward. There are manifolds M with ample anticanonical class and exceptional collections on M, which satisfy all properties **(H1)**-**(H5)** except **(H2')** and **(H3)**, because $\mathrm{Hom}(A, B)$ may be equal to zero or $lcan$ may be not surjective. To provide the set of exceptional sheaves on such manifolds with combinatorical structure as above we need more general technique.

1.2. Helix theory in triangulated categories. Let T be a triangulated category (see [GM]) in which $\bigoplus_i \mathrm{Hom}^i(X, Y)$ are finite-dimensional vector spaces over k. Object E of T is called *exceptional* if $\mathrm{Hom}^i(E, E) = 0$ for $i \neq 0$ and $\mathrm{Hom}^0(E, E) = k$. As above, for any two objects A, B we may consider the canonical maps:

$$lcan : \mathrm{Hom}^{\cdot}(A, B) \otimes A \longrightarrow B,$$
$$rcan : A \longrightarrow \mathrm{Hom}^{\cdot}(A, B)^* \otimes B,$$

where by definition

$$\mathrm{Hom}^{\cdot}(A, B) \otimes A = \bigoplus_i \mathrm{Hom}^i(A, B) \otimes A[-i],$$
$$\mathrm{Hom}^{\cdot}(A, B)^* \otimes B = \bigoplus_i \mathrm{Hom}^{-i}(A, B)^* \otimes B[-i],$$

(dualisation of graded vector space changes indexes for opposite ones). *Left mutation $L_A B$* and *right mutation $R_B A$* are, by definition, the objects inserted into distinguished triangles

$$L_A B \longrightarrow \mathrm{Hom}^{\cdot}(A, B) \otimes A \xrightarrow{lcan} B \longrightarrow L_A B[1],$$
$$R_B A[-1] \longrightarrow A \xrightarrow{rcan} \mathrm{Hom}^{\cdot}(A, B)^* \otimes B \longrightarrow R_B A.$$

Although distinguished triangle is not canonically defined by morphism, the following result has been proved in [G2], [G3]:

1.2.1. Theorem. *If E is exceptional, then left and right mutations by E:*

$$L_E : T \longrightarrow T : X \longmapsto L_E X,$$
$$R_E : T \longrightarrow T : X \longmapsto R_E X,$$

are well defined functors. Moreover, if we denote by $^{\perp}E$ and $E^{\perp}$ left and right orthogonals to E in T:

$$^{\perp}E = \{X \in T \mid \mathrm{Hom}^{\cdot}(X, E) = 0\},$$
$$E^{\perp} = \{X \in T \mid \mathrm{Hom}^{\cdot}(E, X) = 0\},$$

and restrict L_E onto $^{\perp}E$ and R_E onto $E^{\perp}$, we get the equivalences of these subcategories, which are inverse to each other:

$$^{\perp}E \underset{R_E}{\overset{L_E}{\underset{\sim}{\rightleftarrows}}} E^{\perp}.$$

1.2.2. Definitions. Ordered collection of objects $\varepsilon = (E_1, \dots, E_n) \subset T$ is called *semiorthogonal* or *exceptional* if all E_i are exceptional and $\mathrm{Hom}^{\cdot}(E_\alpha, E_\beta) = 0$ for $\alpha > \beta$. *Left mutation L_i* (where $i = 1, 2, \dots, n - 1$) changes the pair (E_i, E_{i+1}) for the pair $(L_{E_i} E_{i+1}, E_i)$. *Right mutation R_i* changes it for $(E_{i+1}, R_{E_i} E_{i+1})$.

1.2.3. Theorem. *Mutations L_i and R_i preserve the exceptionality of collections and satisfy the relations of braid group B_n from 1.1.3 (b).*

1.2.4. Theorem 1.2.3 has been proved in [G2]. For other general results about exceptional collections see [G2], [G3], [B1], [B2], [BK]. Here we are interested in *categories T generated by exceptional collection* $(E_1, \dots, E_n)$. In this case the next results follow from general theory:

(a) There exists Serre functor $F : T \longrightarrow T$ such that

$$\mathrm{Hom}^{\cdot}(X,Y) = \mathrm{Hom}^{\cdot}(Y,FX)^* \quad \text{for any } X,Y.$$

Moreover, $FX = L_{E_1} L_{E_2} \dots L_{E_n} X$. In particular, $FE_n = L_{E_1} L_{E_2} \dots L_{E_n}$. F is automatically an autoequivalence of T.

(b) Infinite extension $\sigma = (E_i)$, $i \in \mathbb{Z}$ of basic collection $(E_1, \dots, E_n)$ by the rules:

$$E_{i+n+1} = F^{-1} E_i; \quad E_{i-n-1} = FE_i$$

satisfies all properties (H1)-(H5) except (H2'). We call such infinite collections σ helices in T.

(c) Any strictly full inclusion $T \hookrightarrow D$ in other triangulated category has left and right adjoint functors.

1.3. Applications to exceptional sheaves. To apply previous technique to the exceptional sheaves, let us consider a variety M such that the derived category of coherent sheaves $D^b(Sh\,M)$ is generated by the exceptional collection of sheaves. In this case by 1.2.4.(b) we have helix theory in $D^b(Sh\,M)$ and we can constuct an infinite set of exceptional collections in $D^b(Sh\,M)$ by mutations of basic collection of exceptional sheaves. The problem is that these mutations may be not sheaves, but complexes of sheaves. It takes place, for example, on ruled surface $\mathbb{F}_2$. Next, two results solve this problem in two special cases.

1.3.1. Theorem. *If a helix $\sigma = (E_i)$, $i \in \mathbb{Z}$ in $D^b(Sh\,M)$ consists of sheaves, has a period $\dim M + 1$ and satisfies the condition (H2'), then all its mutations in $D^b(Sh\,M)$ do the same.*

1.3.2. Theorem. *If M is a surface with ample anticanonical divisor, then any exceptional object in $D^b(Sh\,M)$ has only one nonzero cohomology sheaf (and, hence, it is quasiisomorphic to this sheaf modulo action of translation functor).*

First of these theorems has been proved by A. Bondal in [B1], the second one was in special cases proved in [G2], [G3], [N]. Using methods of [G2], D.Orlov proved the general case in [O]. It was shown in [G2], [O] that from 1.3.2. follows

1.3.3. Corollary. *For any exceptional pair of sheaves (A, B) on Del Pezzo surface the long exact seqience of cohomology sheaves of the distinguished triangle*

$$L_A B \longrightarrow \mathrm{Hom}^{\cdot}(A,B) \otimes A \longrightarrow B \longrightarrow L_A B[1]$$

always coincides with one of the following exact triples:

$$0 \longrightarrow L_A B \longrightarrow \mathrm{Hom}(A,B) \otimes A \longrightarrow B \longrightarrow 0,$$
$$0 \longrightarrow \mathrm{Hom}(A,B) \otimes A \longrightarrow B \longrightarrow L_A B \longrightarrow 0,$$
$$0 \longrightarrow B \longrightarrow L_A B \longrightarrow \mathrm{Ext}^1(A,B) \otimes A \longrightarrow 0.$$

In particular, $\mathrm{Ext}^i(A,B) \neq 0$ either only for $i = 0$ or only for $i = 1$ and in the first case the canonical map lcan is either surjective or injective. The same is right for the right mutation of the pair (A, B).

1.4. Main problems of helix theory. A description of a set of exceptional vector sheaves on a given manifold M in terms of helix theory contains solutions of the following three problems.

Problem 1. Describe the orbits of the braid group action on a set of helices in $D^b(Sh\ M)$.

Problem 2. Investigate for each exceptional object, pair, triple,... in $D^b(Sh\ M)$ a set of helices which includes this object, pair, triple,... In particular, clear up, is it empty or not.

Problem 3. Describe the set of exceptional sheaves and orbits of exceptional sheaves in the set of all exceptional objects in $D^b(Sh\ M)$.

I now only two results solving the third problem - theorems 1.3.1 and 1.3.2 above. But I have

1.4.1. Conjecture. All exceptional objects in $D^b(Sh\ \mathbb{P}_n)$ are (modulo translations) exceptional vector bundles.

It is clear, that the solution of the second problem for a given collection $\varepsilon = (E_1, \dots, E_k)$ is equivalent to the solution of the first one for the subcategory $\varepsilon^\perp \subset D^b(Sh\ M)$. This subcategory in general case is not equivalent to $D^b(Sh\ M')$ for any M'. Hence, we need to consider the first problem for any triangulated category T instead of $D^b(Sh\ M)$. Some results on the first two problems may be found in [DL], [R1], [R2], [N]. Here we formulate two conjectures, corresponding to the theorems 1.3.1 and 1.3.2 above.

1.4.2. Conjecture. Each exceptional collection of sheaves on Del Pezzo surface is contained in some helix. Braid group acts transitively (modulo translations in derived category and twists by inversable line bundles) on the set of all helices and on the set of helices containing a given exceptional collection.

1.4.3. Conjecture. Let $T = D^b(A)$ be the derived category of abelian category A and let the cohomological dimension of A be equal to n. If T is generated by helix, which consists of the objects of A, has a period $n + 1$ and satisfies the condition **(H2')**, then the braid group acts transitively on a set of all helices.

One reason for these conjectures is that their arithmetical variants have been proved in many cases.

2. Arithmetical problems of helix theory

2.1. Mukai lattices. A *Mukai lattice* is a finitely generated free $\mathbb{Z}$-module V with unimodular bilinear form (which can be nonsymmetrical) $V \times V \longrightarrow \mathbb{Z} : (v, w) \longmapsto \langle v, w \rangle \in \mathbb{Z}$. A Gram matrix of this form in any given basis of V will be denoted by χ.

Basic example of Mukai lattice is Grothendieck group $K_0(M)$ of manifold M in the case when the derived category $D^b(Sh\ M)$ is generated by the finite exceptional collection. In any case there is (nonsymmetrical) bilinear form on $K_0(M)$ given by the formula

$$\langle E, F \rangle = \sum_{i=0}^{\dim M} (-1)^i \dim \operatorname{Ext}^i(E, F).$$

By the Serre duality $\langle E, F \rangle = (-1)^{\dim M} \langle F, E \otimes \omega_M \rangle$. Hence, this form is symmetrical only in the case when $\omega_M = \mathcal{O}_M$. If $K_0(M)$ is generated by the exceptional collection, then the Gram matrix χ of the form in this basis has a form

$$\begin{pmatrix} 1 & & & & \\ 0 & 1 & & * & \\ 0 & 0 & 1 & & \\ \vdots & \vdots & \vdots & \ddots & \\ 0 & 0 & 0 & \cdots & 1 \end{pmatrix}$$

In particular, $\det \chi = 1$.

2.2. Exceptional bases and braid group action. Basis $\varepsilon = \{e_0, e_1, \ldots, e_n\}$ of Mukai lattice V is called exceptional, if $\langle e_i, e_i \rangle = 1$ for all i and $\langle e_j, e_i \rangle = 0$ for $j > i$. If we permute two neighbouring vectors (e_i, e_{i+1}) of the exceptional basis, then new basis $(e_0, \ldots, e_{i-1}, e_{i+1}, e_i, e_{i+2}, \ldots, e_n)$ will not be exceptional, but it can be orthogonalized by Gram-Shmidt process. There are two ways to do this: "left way" and "right way". In the first case we replace e_{i+1} by $L_{e_i}(e_{i+1}) = e_{i+1} - \langle e_i, e_{i+1} \rangle e_i$, in the second case we replace e_i by $R_{e_{i+1}}(e_i) = e_i - \langle e_i, e_{i+1} \rangle e_{i+1}$. New collections will be exceptional and are called left mutation $L_i \varepsilon$ and right mutation $R_i \varepsilon$ of the previous collection ε. Direct calculations show that these mutations L_i and R_i $(i = 0, \ldots, n-1)$ satisfy the relation $L_i R_i = R_i L_i = id$ and standard relations for generators g_i of the braid group B_{n+1}. Hence, there is a natural action of the braid group on the set of exceptional bases of Mukai lattice.

2.3. Isometries. The action of B_{n+1} on the exceptional bases, as a rule, changes their Gram matrices. Besides B_{n+1}, the group of isometries of the form $\langle v, w \rangle$ acts on the set of exceptional bases. By definition, $\varphi : V \longrightarrow V$ is an isometrical operator, if $\langle \varphi v, \varphi w \rangle = \langle v, w \rangle$ for all $v, w \in V$. If $V = K_0(M)$, there are two important examples of isometries: translation functor $T : E \longmapsto E[1]$ in derived category acts on $K_0(M)$ by changing E by $-E$, and twists by line bundles $E \longmapsto E \otimes \mathcal{L}$ are isometries too.

2.4. Arithmetical problems. The problems of helix theory, mentioned above, correspond to the following questions on Mukai lattices:

Question 1. To describe orbits of braid group and group of isometries on the set of exceptional bases of a given Mukai lattice V (we suppose, that such bases exist in V).

Question 2. Can we extend a given exceptional collection $\{e_0, \ldots, e_k\}$ $(k < \dim V)$ to exceptional basis or not? In particular, has V exceptional basis or not?

Question 3. In the case $V = K_0(M)$ describe the exceptional collections in V, which correspond (modulo isometries of V) to exceptional collections in $D^b(Sh\, M)$.

D.Yu.Nogin has shown in [N] that all exceptional bases of $V = K_0(M)$ can be obtained from any one given by the action of the braid group and isometries in the cases when M is a ruled surface, $\mathbb{P}_3$, 3-dimensional quadric or Fano threefold with $\operatorname{Pic}(M) = \mathbb{Z}$. In all these cases (except the case of Fano threefold V_{22}, when the question is open) all exceptional bases of $K_0(M)$ arise (modulo isometries) from the exceptional bases of $D^b(Sh\, M)$.

Questions 2 and 3 almost have not been investigated. In fact, cohomological conditions $\mathrm{Ext}^i(E, E) = 0$ for $i \neq 0$ are much stronger than the arithmetical condition $\langle E, E \rangle = 1$. For example, if $V = K_0(\mathbb{P}_n)$, $n \geq 2$, then, as a rule, the given vector e with $\langle e, e \rangle = 1$ is not equal to the class of any exceptional object of $D^b(Sh\, \mathbb{P}_n)$ in V (modulo isometries of V). But in cases, which I have calculated, such vectors e can not be included in any exceptional basis of $K_0(\mathbb{P}_n)$. Moreover, from Nogin's results one can deduce, that for M, mentioned above, vector e with $\langle e, e \rangle = 1$ can be included into any exceptional basis of $K_0(M)$ only if it can be obtained from the class of the exceptional object of $D^b(Sh\, M)$ by the action of the group of isometries V. I have a conjecture, that it is true for $K_0(\mathbb{P}_n)$ for any $n \geq 2$.

2.5. Canonical operator. To give the unimodular bilinear form $\langle \cdot, \cdot \rangle : V \times V \longrightarrow \mathbb{Z}$ is the same as to give the pair of *correlations*

$$l : V \xrightarrow{\sim} V^* = \mathrm{Hom}_{\mathbb{Z}}(V, \mathbb{Z}) : \quad l(v) = \langle v|,$$
$$r : V \xrightarrow{\sim} V^* = \mathrm{Hom}_{\mathbb{Z}}(V, \mathbb{Z}) : \quad r(v) = |v\rangle,$$

such that they are isomorphisms of $\mathbb{Z}$-modules and dual each other (this means that l^* is a map between dual modules $(V^*)^* = V$ and V^* is equal to r). In this case the operator $\kappa = r^{-1}l$ is called the *canonical operator*. It is well defined by the condition

$$\langle v, w \rangle = \langle w, \kappa v \rangle \quad \text{for all} v, w \in V.$$

In the case $V = K_0(M)$, $\dim M = n$, κ can be obtained from the Serre duality:

$$\kappa(E) = (-1)^{\dim M} E \otimes \omega_M.$$

2.5.1. Proposition. *If* $\omega_M = \mathcal{O}(-D)$ *and* D *is effective divisor on* M, *then* $\kappa = (-1)^n \mathrm{Id} + \eta$, *where* η *is nilpotent.*

The proof follows from the exact triple

$$0 \longrightarrow E \otimes \omega_M \longrightarrow E \longrightarrow E \mid_D \longrightarrow 0,$$

(where E is any locally free sheaf on M) which shows, that the operator $\kappa - (-1)^n \mathrm{Id}$ equals to the operator of restriction to the divisor D (multiplied by $(-1)^n$), and restriction operator is nilpotent.

This proposition shows, that the forms $\langle \cdot, \cdot \rangle$ in cases $V = K_0(M)$ considered above are special. To explane this, we give some general results classifying nondegenerate bilinear forms over algebraically closed field k in the next section.

3. General results about bilinear forms

3.1. Notations. In this section we consider only *vector spaces* V *over a closed field* k. On this space we fix bilinear (nonsymmetrical) form

$$\langle \cdot, \cdot \rangle : V \times V \longrightarrow k$$

and denote by χ its Gram matrix. In particular, in coordinates we have

$$\langle v, w \rangle = v^t \cdot \chi \cdot w = w^t \cdot \chi^t \cdot v.$$

We always suppose, that $\det \chi \neq 0$. We denote by $\kappa : V \longrightarrow V$ the *canonical operator* (see above).

3.2. Canonical algebra. For a given $\varphi \in \mathrm{End}(V)$ there exist operators $\varphi^\vee$ and $^\vee\varphi$ such that for all $v, w \in V$

$$\langle ^\vee\varphi v, w \rangle = \langle v, \varphi w \rangle,$$
$$\langle \varphi v, w \rangle = \langle v, \varphi^\vee w \rangle.$$

In terms of matrices

$$^\vee\varphi = (\chi^t)^{-1}\varphi^t\chi^t,$$
$$\varphi^\vee = \chi^{-1}\varphi^t\chi.$$

$^\vee\varphi$ and $\varphi^\vee$ are called *left* and *right adjoint* to φ. The operator is called *reflexive*, if $\varphi^{\vee\vee} = \varphi$. It is equivalent to the properties $^{\vee\vee}\varphi = \varphi$ and $^\vee\varphi = \varphi^\vee$. The operator φ is reflexive if and only if $\varphi\kappa = \kappa\varphi$. Subalgebra $\mathcal{Z}(\kappa) \subset \mathrm{End}(V)$ of such operators is called the *canonical algebra*. The map $\varphi \longrightarrow \varphi^\vee$ is involutive antiautomorphism of $\mathcal{Z}(\kappa)$, hence $\mathcal{Z}(\kappa)$ has decomposition $\mathcal{Z}(\kappa) = \mathcal{Z}_+ \oplus \mathcal{Z}_-$ with $\mathcal{Z}_+ = \{\varphi \in \mathrm{End}(V) \mid \langle \varphi v, w \rangle = \langle v, \varphi w \rangle\}$ and $\mathcal{Z}_- = \{\varphi \in \mathrm{End}(V) \mid \langle \varphi v, w \rangle = -\langle v, \varphi w \rangle\}$. Operators of the first type are called *selfadjoint* (or SA-operators), and of the second type – *antiselfadjoint* (or AS-operators).

3.3. Isometries. The subgroup $\mathrm{Isom}(\chi) \subset \mathrm{GL}(V)$ by definition consists of all $\varphi \in \mathrm{GL}(V)$ such that

$$\langle v, w \rangle = \langle \varphi v, \varphi w \rangle \quad \text{for all } v, w \in V.$$

The important example of isometry is the canonical operator $\kappa : \langle v, w \rangle = \langle w, \kappa v \rangle = \langle \kappa v, \kappa w \rangle$. Evidently $\mathrm{Isom}(\chi) \subset \mathcal{Z}(\kappa)$.

3.3.1. Theorem. $Lie(\mathrm{Isom}\chi) = \mathcal{Z}_-$.

The proof is quite similar to the proof of the fact that Lie algebra of the group SO_n consists of antisymmetrical matrices.

3.3.2. To write the equations which define $\mathrm{Isom}(\chi)$ as a submanifold in the space $\mathcal{Z}(\kappa)$ it is convenient to fix in $\mathcal{Z}(\kappa)$ a special basis $\xi_0, \xi_1, \ldots, \xi_k, \eta_{k+1}, \ldots, \eta_m$ in which $\xi_0 = \mathrm{Id}$; $\xi_1, \ldots, \xi_k$ are SA-operators and $\eta_{k+1}, \ldots, \eta_m$ are AS-operators. Consider the bilinear forms $s_i(\varphi, \psi)$, $A_j(\varphi, \psi)$ on $\mathcal{Z}(\kappa)$, defined by the formula

$$\langle \varphi v, \psi w \rangle = \sum_{i=0}^{k} S_i(\varphi, \psi)\langle v, \xi_i w \rangle + \sum_{j=k+1}^{m} A_j(\varphi, \psi)\langle v, \eta_i w \rangle$$

for all $v, w \in V$; $\varphi, \psi \in \mathcal{Z}(\kappa)$. It is easy to show that all forms S_i are symmetrical and all forms A_j are antisymmetrical: $S_i(\varphi, \psi) = S_i(\psi, \varphi)$, $A_j(\varphi, \psi) = -A_j(\psi, \varphi)$. Evidently, $\varphi \in \mathrm{Isom}(\chi)$ if and only if $S_0(\psi, \psi) = 1$ and $S_i(\psi, \psi) = A_j(\psi, \psi) = 0$ for $i \neq 0$. Equations $A_j(\psi, \psi) = 0$ hold automaticaly. Hence, we obtain the quadratic equations

$$S_0(\psi, \psi) = 0, \qquad S_i(\psi, \psi) = 0 \qquad \text{for} \quad i = 1, \ldots, k,$$

which define $\text{Isom}(\chi)$ in $\mathcal{Z}(\kappa)$.

3.3.3. Let $\varphi \in \text{Isom}(\chi)$. For each λ such that $\text{Ker}(\varphi - \lambda\text{Id}) \neq 0$ consider the subspace $K_\lambda = \{v \in V \mid (\varphi - \lambda\text{Id})^n v = 0 \text{ for some } n\}$.

Theorem. *V splits in biorthogonal direct sum of subspaces K_{+1}, K_{-1} and $K_\lambda \oplus K_{\lambda^{-1}}$ for $\lambda \neq \pm 1$. Restrictions of the given form $\langle \cdot, \cdot \rangle$ on V to K_{+1}, K_{-1}, $K_\lambda \oplus K_{\lambda^{-1}}$ is nondegenerate, restrictions of $\langle \cdot, \cdot \rangle$ to K_λ and $K_{\lambda^{-1}}$ are zeroes, pairing $\langle K_\lambda, K_{\lambda^{-1}} \rangle$ defines the isomorphism $K_\lambda^* \simeq K_{\lambda^{-1}}$ and the nilpotente operators $(\varphi - \lambda\text{Id}) \mid_{K_\lambda}$ and $(\varphi - \lambda^{-1}\text{Id}) \mid_{K_{\lambda^{-1}}}$ have the same cyclic types.*

3.4. Canonical orthogonal decomposition of V**.** In the case of canonical operator κ the last theorem has more strong form:

3.4.1. Theorem. *Any finitely dimensional vector space V over algebraically closed field k with nondegenerate bilinear form $V \times V \longrightarrow k$ canonically splits in biorthogonal direct sum of spaces with nondegenerate bilinear forms of following types*

ε-type (where $\varepsilon = \pm 1$): space of ε-type has

$$\kappa = \varepsilon\text{Id} + \eta,$$

where η is nilpotent and must contain only even number of cycles of length k for any integer k such that $(-1)^k = \varepsilon$.

λ-type (where $\lambda \neq \pm 1$): space of λ-type has a form $V_+ \oplus V_-$, restrictions of the form to V_+ and to V_- are zeroes and pairing $\langle V_+, V_- \rangle$ identifies V_+^* with V_-. $\kappa \mid_{V_+} = \lambda\text{Id} + \eta_+$, $\kappa \mid_{V_-} = \lambda^{-1}\text{Id} + \eta_-$, where η_+ and η_- are any nilpotent operators of the same cyclic type.

3.4.2. Remark. This theorem shows, that the forms on $K_0(M)$, considered above, are very special. In fact, "general" form on V splits in sum of 2-dimensional subspaces of λ-type with distinguished λ.

3.5. Theorem. *Two forms $\langle \cdot, \cdot \rangle_1$ and $\langle \cdot, \cdot \rangle_2$ have the same canonical operator κ if and only if there is SA-operator φ (with respect to each form) such that*

$$\langle v, w \rangle_1 = \langle \varphi v, \varphi w \rangle_2.$$

If V is indecomposable in the direct biorthogonal sum, then for given $\langle \cdot, \cdot \rangle_1$ and $\langle \cdot, \cdot \rangle_2$ there are only two such SA-operators $\varphi_1 = -\varphi_2$.

3.6. Remarks about proofs. Invariants of matrices with respect to transformations $A \longmapsto CAC^t$ by $C \in \text{GL}(V)$ were investigated in terms of λ-matrices in classical period of invariant theory. Corresponding results are contained in [HP] t.1, II.IX. §§5,6. In particular, canonical forms of bilinear form and the fact, that the invariants of the *form A* with respect to the action $A \longmapsto CAC^t$ and the invariants of the *operator* $\kappa = A^{-1}A^t$ with respect to the action $\kappa \longmapsto C\kappa C^{-1}$ are the same, were known. I will soon publish the proofs of these results and results about calculations of the groups $\text{Isom}(\chi)$ and their invariants in my new paper. Here I give only two examples.

4. Two examples

4.1. $K_0(\mathbb{P}_n)$. Let $V = K_0(\mathbb{P}_n)$. It follows from 2.5.1 that the canonical operator $\kappa = (-1)^n \mathrm{Id} + \eta$, where $(-1)^n \eta$ is the operator of the restriction to the divisor $(n+1)H(H \subset \mathbb{P}_n$ is a hyperplane). Hence, η is nilpotent and contains only one cycle of length $(n+1) = \dim V$.

The canonical algebra $Z(\kappa) \simeq k[\eta]/\eta^{n+1}$ and $\dim Z(\kappa) = n + 1$. Since this algebra is commutative, the group $\mathrm{Isom}(\chi)^0$ is a *direct sum of 1-dimensional unipotent additive groups* (where $\mathrm{Isom}(\chi)^0$ is the connected component of Id). To compute it more precisely we must construct in $Z(\kappa)$ a basis consisting of AS and SA-operators.

The natural basis in $Z(\kappa)$ is $1, \eta, \dots, \eta^n$. It is easy to calculate, that

$$(\eta^k)^\vee = (-1)^k \sum_{\nu \geq 0} (-\varepsilon)^\nu \binom{k + \nu - 1}{k - 1} \eta^{k+\nu}, \quad \text{where} \varepsilon = (-1)^n.$$

Hence, the operators $\gamma_k = (-1)^k (\eta^k)^\vee + \eta^k$ form the basis in $Z(\kappa)$ too. These operators are SA for even $k = 0, 2, \dots$ and AS for $k = 1, 3, 5, \dots$ We obtain

4.1.1. Proposition. $\dim \mathrm{Isom}\,(K_0(\mathbb{P}_n)) = \left[\dfrac{n+1}{2}\right].$

To obtain the equations on the coefficients $a_0, a_1, \dots, a_n$ of the given $\varphi = a_0 + a_1\eta + \dots + a_n\eta^n \in \mathrm{Isom}(\chi)$, we consider the forms $Q_k(\varphi, \psi)$ such that for all $v, w \in K_0(\mathbb{P}_n)$

$$\langle \varphi v, \psi w \rangle = \sum_{k \geq 0} Q_k(\varphi, \psi) \langle v, \eta^k w \rangle.$$

These forms Q_k are nonsymmetrical (as in 3.5.2), but from 3.5.2 and the fact that the transformation $\eta^k \longmapsto \gamma_k$ is given by the triangle matrix it follows, that in the system of equations
$$Q_0(\varphi, \varphi) = 1; \quad Q_i(\varphi, \varphi) = 0 \quad \text{for} i \geq 1$$
(it is equivalent to $\langle \varphi v, \varphi w \rangle = \langle v, w \rangle$ and $\varphi \in \mathrm{Isom}(\chi)$) all equations $Q_i(\varphi, \varphi) = 0$ with $i = 1, 3, 5, \dots$ hold for any φ automatically.

We have for $\varphi = a_0 + a_1\eta + \dots + a_n\eta^n$ and $\psi = b_0 + b_1\eta + \dots + b_n\eta^n$ the following expressions for $Q_i(\varphi, \psi) = Q_i(a, b)$:

$Q_0 = a_0 b_0,$
$Q_1 = a_0 b_1 - a_1 b_0,$
$Q_2 = [a_0 b_2 - a_1 b_1 + a_2 b_0] + \varepsilon a_1 b_0,$
$Q_3 = [a_0 b_3 - a_1 b_2 + a_2 b_1 - a_3 b_0] + \varepsilon[a_1 b_1 - 2a_2 b_0] - a_1 b_0,$
$Q_4 = [a_0 b_4 - \dots + a_4 b_0] + \varepsilon[a_1 b_2 - 2a_2 b_1 + 3a_3 b_0] - [a_1 b_1 - 3a_2 b_0] + \varepsilon a_1 b_0, \dots.$

In general

$$Q_k(a, b) = P_k(a, b) + \sum_{m=1}^{k-1} (-\varepsilon)^m \sum_{\nu=0}^{k-m-1} (-1)^\nu \binom{k - \nu}{m} a_{k-m-\nu} b_\nu,$$

where

$$P_k(a, b) = a_0 b_k - a_1 b_{k-1} + \ldots = \sum_{\mu=0}^{k} (-1)^{\mu} a_{\mu} b_{k-\mu}.$$

Taking $b = a$ we get $Q_{2k+1}(a, a) = 0$ tautologically and see that $a_0 = \pm 1$; $a_1, a_3, a_5, \ldots$ are free parameters and a_{2k} is expressed by $a_{2k-1}, a_{2k-2}, a_{2k-3}, \ldots, a_0$. This proves

4.1.2. Proposition. Isom $(K_0(\mathbb{P}_n))$ *is rational and has two connected components.*

4.1.3. Example. Isom $(K_0(\mathbb{P}_2))$ consists of operators

$$\varphi_t = 1 + t\eta + \frac{1}{2}t(t-1)\eta^2,$$
$$\tilde{\varphi}_t = -1 + t\eta - \frac{1}{2}t(t+1)\eta^2 = -\varphi_{-t}.$$

Here t is additive parameter: $\varphi_t \varphi_s = \varphi_s \varphi_t = \varphi_{s+t}$. It is easy to see, that the integer isometries are generated by twists $E \longmapsto E \otimes \mathcal{O}(n)$(this is φ_n) and the translation in the derived category $E \longmapsto E[1]$ (class $E[-1]$ in K_0 is $-E$), this isometry is $\tilde{\varphi}_0$.

4.2. Classification of nondegenerate bilinear forms with antisymmetrical part of rank 2. Let V be a vector space over a closed field k and bilinear (nondegenerate) form $\langle \cdot, \cdot \rangle$ is represented as

$$\langle v, w \rangle = S(v, w) + A(v, w),$$

where S is symmetrical and A — antisymmetrical form and suppose, that $rk\, A = 2$. Then from 3.4.1 it follows that $V = U \oplus W$, where $\langle U, W \rangle = \langle W, U \rangle = 0$, U is any space with *symmetrical nondegenerate* form $\langle \cdot, \cdot \rangle |_S$ and W is one of the following:

i) dim $W = 2$ and W is the space of λ-type: $W = V_\lambda \oplus V_{\lambda^{-1}}$, dim $V_\lambda = $ dim $V_{\lambda^{-1}} = 1$ and the Gram matrix χ in this basis is

$$\chi = \begin{pmatrix} 0 & 1 \\ \lambda & 0 \end{pmatrix}$$

ii) dim $W = 2$ and W is Artin plane with

$$\chi = \begin{pmatrix} 0 & 1 \\ -1 & 0 \end{pmatrix}$$

iii) dim $W = 2$ and $W \simeq K_0(\mathbb{P}_1)$ with

$$\chi = \begin{pmatrix} 1 & 2 \\ 0 & 1 \end{pmatrix}$$

iv) dim $W = 3$ and $W \simeq K_0(\mathbb{P}_2)$ with

$$\chi = \begin{pmatrix} 1 & 3 & 3 \\ 0 & 1 & 3 \\ 0 & 0 & 1 \end{pmatrix}$$

v) dim $W = 4$ and in some basis the Gram matrix is

$$\chi = \begin{pmatrix} 0 & 1 & 0 & 1 \\ -1 & 0 & -1 & 0 \\ 0 & -1 & 0 & 0 \\ 1 & 0 & 0 & 0 \end{pmatrix}$$

The first case (i) takes place if and only if $\det(\kappa - \lambda \mathrm{Id}) = 0$, where $\kappa = \chi^{-1}\chi^t$ is the canonical operator. If all roots of the polynomial $\det(\kappa - t\mathrm{Id})$ are rational, then the classification above takes place over $k = \mathbb{Q}$.

References

[B1]   BONDAL A.I. *Representations of associative algebras and coherent sheaves.* Math. USSR. Isv. 53 (1989), N1, p.25-44 (Russian).

[B2]   BONDAL A.I. *Helices, Representations of Quivers and Koszul Algebras.* Lond. Math. Soc. L.N.S. 148 (1990), p.75-97.

[BK]   BONDAL A.I., KAPRANOV M.M. *Representable functors, Serre functors and mutations.* Math. USSR Isv. 53 (1989), N6, p.1183-1205 (Russian).

[DL]   DREZET J.-M., LE POTIER J. *Fibres stables et fibres exceptionnels sur* $\mathbb{P}_2(\mathbb{C})$. Ann. E.N.S., t.18 (1985), p.193-244.

[GM]   GELFAND S.I., MANIN YU.I. *Methods of homological algebra, I.* Moscow, "Nauka", 1989 (Russian).

[G1]   GORODENTSEV A.L. *Mutations of exceptional vector bundles on* $\mathbb{P}_n$. Math. USSR Isv. 52 (1988), N1, p.3-15 (Russian).

[G2]   GORODENTSEV A.L. *Exceptional vector bundles on surfaces with moveable anticanonical divisor.* Math. USSR. Isv. 52 (1988), N4, p.740-757 (Russian).

[G3]   GORODENTSEV A.L. *Exceptional objects and mutations in derived categories.* Lond. Math. Soc. L.N.S. 148 (1990), p.57-75.

[GR]   GORODENTSEV A.L., RUDAKOV A.N. *Exceptional vector bundles on projective spaces.* Duke Math. J. v.54 (1987), N1, p.115-130.

[HP]   HODGE W.V.D., PEDOE D. *Methods of Algebraic Geometry.* Cambridge, 1947, v.1.

[N]   NOGIN D.YU. *Helices of period 4 and equations of the Markov type.* Math. USSR Isv. 54 (1990), N4, p.862-878 (Russian).

[O]   ORLOV D.O. *Projective bundles, monoidal transforms and derived categories of coherent sheaves.* Math. USSR. Isv. 55 (1991) (Russian).

[R1]   RUDAKOV A.N. *The Markov numbers and exceptional bundles on* $\mathbb{P}_2$. Math. USSR Isv. 52 (1988), 99-112 (Russian).

[R2]   RUDAKOV A.N. *Exceptional bundles on a quadric.* Math. USSR Isv. 53 (1989), p.115-138 (Russian).

On the Unramified 2-covers of the Curves of Genus 3

P.I.Katsylo

0

The field $\mathbb{C}$ of complex numbers is taken as the ground field.

Let M_3 be the moduli variety of curves of genus 3, and $M_3^{[2]}$ be the moduli variety of curves of genus 3 with a fixed unramified 2-cover. There is the canonical morphism

$$\pi : M_3^{[2]} \longrightarrow M_3, \qquad \deg \pi = 63.$$

Theorem 0.1. $M_3^{[2]}$ *is rational.*

Let X be a curve of genus 3 and $\varphi : Y \longrightarrow X$ be an unramified 2-cover of X. We have $g(Y) = 5$ by Hurwitz formula. The curve Y has the canonical involution without fixed points.

Let Y be a curve of genus 5 and $\tau : Y \longrightarrow Y$ be an involution without fixed points. Then

$$X = Y//\{1_Y, \tau\}$$

is the curve of genus 3 by Hurwitz formula. The canonical morphism $\varphi : Y \longrightarrow X$ is the unramified 2-cover.

The previous consideration gives us the following form of theorem 0.1.

Theorem 0.1'. *Let M_5^τ be the moduli variety of curves of genus 5 with a fixed involution without fixed points. Then M_5^τ is rational.*

We shall prove theorem 0.1' in §1,2.

1

Let V be 5-dimensional linear space, PV be a projectivization of the space V, $(e_0, e_1, f_0, f_1, f_2)$ be a basis of V, $(x_0, x_1, y_0, y_1, y_2)$ be a dual basis of V^*,

$$\tau : V \longrightarrow V,$$
$$(x_0, x_1, y_0, y_1, y_2) \longmapsto (-x_0, -x_1, y_0, y_1, y_2),$$

be an involution. We have a decomposition

$$S^2 V^* = S_+ \oplus S_-, \quad where \quad \tau\big|_{S_+} = 1, \quad \tau\big|_{S_-} = -1,$$

$$S_+ = \langle x_0^2, x_0 x_1, x_1^2, y_0^2, y_0 y_1, y_0 y_2, y_1^2, y_1 y_2, y_2^2 \rangle,$$

$$S_- = \langle x_0 y_0, x_0 y_1, x_0 y_2, x_1 y_0, x_1 y_1, x_1 y_2 \rangle.$$

We set

$$X(N) = \{ \bar{v} \in PV \mid f(v) = 0 \quad for\ all \quad f \in N \},$$

for any 3-dimensional linear subspace N of $S^2 V^*$. If N is a 3-dimensional linear subspace of S_+ in general position then $X(N)$ is the canonical curve in PV with the fixed involution τ without fixed points. Let G be a centralizer of the element τ in $GL(V) \simeq GL_5$. We have

$$G = \left\{ \begin{pmatrix} g & 0 \\ 0 & h \end{pmatrix} \mid g \in GL_2, \quad h \in GL_3 \right\} \simeq GL_2 \times GL_3.$$

The canonical curves $X(N_1), X(N_2)$ in general position (where $N_1, N_2 \subset S_+$) with the fixed involution τ(without fixed points) are isomorphic iff $G \cdot N_1 = G \cdot N_2$.

Let $X \subset PV$ be a canonical curve with a fixed involution σ without fixed points. The involution σ defines uniquely (up to multiplication by ± 1) the involution of the space V (we denote it by the same letter σ). Let

$$V = V_+ \oplus V_-, \quad \sigma\big|_{V_+} = 1, \quad \sigma\big|_{V_-} = -1.$$

If $\dim V_+ \leq 1$ then $PV_- \cap X \neq \emptyset$ are fixed points of σ. Hence $\dim V_3 \geq 2$. Similarly $\dim V_- \geq 2$. Replacing σ by $-\sigma$ we can assume, that

$$\dim V_+ = 3, \quad \dim V_- = 2.$$

Let α be an automorphism of V such that $\alpha(V_-) = \langle e_0, e_1 \rangle, \alpha(V_+) = \langle f_0, f_1, f_2 \rangle$. Then $\alpha^{-1} \tau \alpha = \sigma$. The subspace

$$N\big(\alpha(X)\big) = \{ f \in S^2 V^* \mid f(v) = 0 \quad for\ all \quad \bar{v} \in \alpha(X) \}$$

is 3-dimensional. We have

$$\alpha(X) = X\Big(N\big(\alpha(X)\big) \Big),$$

and hence

$$\tau \cdot N\big(\alpha(X)\big) = N\big(\alpha(X)\big),$$

$$N\big(\alpha(X)\big) = N\big(\alpha(X)\big) \cap S_+ \oplus N\big(\alpha(X)\big) \cap S_-.$$

If $\dim N\big(\alpha(X)\big) \cap S_- \geq 1$, then $PV_+ \cap \alpha(X) \neq \emptyset$ is the set of fixed points of τ. Hence

$$N\big(\alpha(X)\big) \subset S_+.$$

So we obtain: a canonical curve $X \subset PV$ with a fixed involution σ without fixed points is isomorphic to a canonical curve $X(N)$ with the fixed involution τ without fixed points (where $N \subset S_+$ is 3-dimensional linear subspace).

The summary of our consideration is

Lemma 1.1. *The isomorphism of fields*

$$\mathbb{C}(M'_S) \approx \mathbb{C}\big(G(3, S_+)\big)^G \approx \mathbb{C}\big(G(6, S^*_+)\big)^G$$

holds.

2

Lemma 2.1. *The field*

$$\mathbb{C}\big(G(6, S^*_+)\big)^G$$

is rational.

Proof. We set

$$V(2) = S^2\langle x_0, x_1 \rangle,$$

$$V(0,1) = \langle y_0, y_1, y_2 \rangle,$$

$$V(2,0) = S^2\langle f_0, f_1, f_2 \rangle.$$

$$V(2)^* = S^2\langle e_0, e_1 \rangle,$$

$$V(0,3) = S^3\langle y_0, y_1, y_2 \rangle,$$

We have:

$$S^*_+ = V(2)^* \oplus V(2,0).$$

The group GL_2 acts canonically in the space $\langle e_0, e_1 \rangle$, the group GL_3 acts canonically in the space $\langle f_0, f_1, f_2 \rangle$. So, we have the canonical actions of the group $G = GL_2 \times GL_3$ in the spaces $V(2), V(2)^*, V(2,0), V(0,3), V(0,1), S^*_+, \ldots$ We have to prove the rationality of the field

$$\mathbb{C}\Big(G\big(6, V(2)^* \oplus V(2,0)\big)\Big)^{\mathrm{GL}_2 \times \mathrm{GL}_3}.$$

Let

$$\psi : V(2,0) \times V(0,3) \longrightarrow V(0,1),$$

$$\varphi : V(2)^* \times V(2) \longrightarrow \mathbb{C}$$

be contractions. Consider $\mathrm{GL}_2 \times \mathrm{GL}_3$-equivariant bilinear mapping

$$\gamma : [V(2)^* \oplus V(2,0)] \times [V(2) \otimes V(0,1) \oplus V(0,3)] \longrightarrow V(0,1),$$

$$((\alpha_1, \alpha_2), (\beta_1 \otimes \beta_2, \beta_3)) \longmapsto \varphi(\alpha_1, \beta_1)\beta_2 + \psi(\alpha_2, \beta_3).$$

Lemma 2.2. *There exist*

$$\pi_0 \in G\big(6, V(2)^* \oplus V(2,0)\big),$$

$$\rho_0 \in V(2) \otimes V(0,1) \oplus V(0,3)$$

such that

$$\pi_0 = \mathrm{Ker}\gamma(\,\cdot\,, \rho_0),$$

$$\mathbb{C}\rho_0 = \{v \in V(2) \otimes V(0,1) \oplus V(0,3) \mid \gamma(\pi_0, v) = 0\}.$$

Proof. We set

$$\pi_0 = V(2)^* \oplus \langle f_0^2 - 6f_1f_2, f_1^2 - 6f_2f_0, f_2^2 - 6f_0f_1 \rangle,$$

$$\rho_0 = (0, y_0^3 + y_1^3 + y_2^3 + y_0y_1y_2).$$

It is easy to verify that π_0, ρ_0 satisfy to lemma.

Put

$$\Gamma : P\big(V(2) \otimes V(0,1) \oplus V(0,3)\big) \longrightarrow G\big(6, V(2)^* \oplus V(2,0)\big),$$

$$\mathbb{C}\rho \longmapsto \mathrm{Ker}\gamma(\,\cdot\,, \rho).$$

Lemma 2.2 implies that Γ is a birational $\mathrm{GL}_2 \times \mathrm{GL}_3$-equivariant isomorphism. We have the isomorphism of fields

$$\mathbb{C}\Big(G\big(6, V(2)^* \oplus V(2,0)\big)\Big)^{\mathrm{GL}_2 \times \mathrm{GL}_3} \approx \mathbb{C}\Big(P\big(V(2) \otimes V(0,1) \oplus V(0,3)\big)\Big)^{\mathrm{GL}_2 \times \mathrm{GL}_3}$$

Let us prove the rationality of the field

$$\mathbb{C}\Big(P\big(V(2) \otimes V(0,1) \oplus V(0,3)\big)\Big)^{\mathrm{GL}_2 \times \mathrm{GL}_3} \approx \mathbb{C}\big(V(2) \otimes V(0,1) \oplus V(0,3)\big)^{\mathrm{GL}_2 \times \mathrm{GL}_3}.$$

The subvariety
$$Y = (x_0^2 \otimes y_0 + x_0 x_1 \otimes y_1 + x_2^2 \otimes y_2) \oplus V(0,3)$$
is $(\mathrm{GL}_2 \times \mathrm{GL}_3, H)$-section of the variety $V(2) \otimes V(0,1) \oplus V(0,3)$. The subgroup H is isomorphic to GL_2. We have the isomorphism of fields

$$\mathbb{C}\big(V(2) \otimes V(0,1) \oplus V(0,3)\big)^{\mathrm{GL}_2 \times \mathrm{GL}_3} \approx \mathbb{C}(Y)^H.$$

Last remark is that the field
$$\mathbb{C}(Y)^H \approx \mathbb{C}\big(V(0,3)\big)^H$$

is rational [1].

References

[1] P.I.KATSYLO. *Rationality of fields of invariants of reducible representations of the group* SL$_2$, Vestnik Moscov. Univ., Ser. 1, 1984, no. 5, p. 77-79.

Spatial Polygons and Stable Configurations of Points in the Projective Line

Alexander A. Klyachko

Introduction

Let $\widehat{C}_n(m)$ be a variety of spatial polygons $P = (a_1, a_2, \ldots, a_n)$ with the vector-side $a_i \in \mathbb{E}^3$ of a given length $m_i = |a_i|$, $i = \overline{1,n}$. The polygons are considered up to motion in Euclidean space $\mathbb{E}^3$.

On the other hand let us consider a configuration of n points $p_i \in \mathbb{P}^1$ in the projective line $\mathbb{P}^1$. Suppose that for each point p_i there is given a positive *weight* (or *multiplicity*) m_i. The configuration of weighted points is called *semistable* (resp. *stable*) if sum of the weights of equal points does not exceed (resp. less then) half the weight of all points. The group $\mathrm{PSL}_2(\mathbb{C})$ naturally acts in the space of configurations. The orbits space of stable configurations is a nonsingular algebraic variety. It will be denoted by $C_n(m)$, where $m = (m_1, m_2, \ldots, m_n)$ is the vector of weights.

In a similar way we denote by $\overline{C}_n(m)$ a categorical factor of the space of semistable configurations with respect to the action of $\mathrm{PSL}_2(\mathbb{C})$. If all the weights are equal to one we reduce the notations to $\overline{C}_n$ and C_n respectively.

The main observation of the paper is a natural complex structure on the variety $\widehat{C}_n(m)$ and its identification with a space of semistable configurations $\overline{C}_n(m)$. We make use of this interpretation to construct a cell decomposition of the space $\widehat{C}_n(m) = \overline{C}_n(m)$ and to find out its Betti numbers.

There are several sources of interest in the varieties $\overline{C}_n(m)$.

First of all they are closely related to the classical invariant theory of binary forms. Such a form of degree n up to proportionality is determined by its divisor of zeros in $\mathbb{P}^1$. Hence a coarse moduli space of semistable forms of degree n may be identified with the factor $\overline{C}_n/S_n$ with respect to a natural action of the symmetric group S_n by permutation of points.

This means that a projective coordinate ring of the variety $\overline{C}_n/S_n$ is the classical ring $S(2, n)$ of invariant of binary forms of degree n:

$$Proj(S(2, n)) = \overline{C}_n/S_n \tag{1}$$

An explicit description of this ring is a classical open problem. Springer [Spr 77] found out the Poincaré series of these rings for all n. Nevertheless its complete structure is known only for degrees less then 7 [Pop 77, Mum 65].

The varieties $\overline{C}_n(m)$ also naturally arise in the theory of vector bundles and torsion free sheaves [Kly 91, Kly 89]. Let us consider for example the moduli space $\overline{\mathcal{M}}(D)$ of stable sheaves of rank two and discriminant $D = c_1^2 - 4c_2$ on the projective plane $\mathbb{P}^2$. The variety $\overline{\mathcal{M}}(D)$ is projective and nonsingular for D odd and $D \equiv 4 \bmod 8$. The structure of the moduli space $\overline{\mathcal{M}}(D)$ in essential is determined by a fixed points space $\overline{\mathcal{M}}(D)^T$ with respect to a natural action of a maximal torus $T \subset \mathbb{P}GL_2$. For odd discriminant D all connected components of the space $\overline{\mathcal{M}}(D)^T$ are products of the projective lines [Kly 89]. In the case $D \equiv 4 \bmod 8$ the connected components are of the form $\overline{C}_n(m)$. So the structure of the varieties $\overline{C}_n(m)$ is a key for investigation of the moduli space $\overline{\mathcal{M}}(D)$.

1   Spatial polygons

Let $\widehat{C}_n(m)$ be a variety of spatial polygons $P = (a_1, a_2, \ldots, a_n)$ with the vector-side $a_i \in \mathbb{E}^3$ of a given length $m_i = |a_i|$, $i = \overline{1, n}$. The polygons are considered up to motion in Euclidean space $\mathbb{E}^3$. The sum of vector-sides is zero

$$a_1 + a_2 + \cdots + a_n = 0, \tag{2}$$

and their lengths $m = (m_1, m_2, \ldots, m_n)$ have to satisfy the polygon inequalities

$$m_i < m_1 + m_2 + \cdots + \widehat{m_i} + \cdots + m_n; \ i = \overline{1, n}. \tag{3}$$

1.1   Tangent spaces and singularities

The tangent space $T(P)$ at a point $P \in \widehat{C}_n(m)$ consists of a set of vectors $v_i \in \mathbb{E}^3$; $i = \overline{1, n}$ under the following conditions

i) $(v_i, a_i) = 0, \forall i$;

ii) $v_1 + v_2 + \cdots + v_n = 0$;

iii) *Two systems of vectors v_i and u_i define the same tangent vector iff there exists $w \in \mathbb{E}^3$ such that $u_i = v_i + [w, a_i], \forall i = \overline{1, n}$.*

Here $[u, v]$ is the vector product. The vectors v_i may be defined as velocity $v_i = \frac{da_i}{dt}$ where $a_i = a_i(t)$ is a curve in the space of polygons. The first equation follows from the constancy of length $|a_i| = m_i$ and the second – from the relation (2). The gauge transformation iii) is an infinitesimal motion of the polygon as a whole.

To eliminate the transformations iii) one has to fix an appropriate gauge. The best way is to demand that the sum

$$\frac{v_1^2}{m_1} + \frac{v_2^2}{m_2} + \cdots + \frac{v_n^2}{m_n} = \min \tag{4}$$

would be minimal among all equivalent systems $u_i = v_i + [w, a_i]$. An extremum condition leads to the equation

$$\frac{[a_1, v_1]}{m_1} + \frac{[a_2, v_2]}{m_2} + \cdots + \frac{[a_n, v_n]}{m_n} = 0. \tag{5}$$

Proposition 1.1.1 *Suppose that not all vectors a_i are collinear. Then in each class of equivalence iii) there is the only n-tuple of vectors v_i which satisfies the gauge equation (5).* □

Corollary 1.1.2 *The variety $\widehat{\mathcal{C}}_n(m)$ has only isolated singularities corresponding to the polygons, degenerated in a line segment. It is nonsingular if all sums $m_1 \pm m_2 \pm \cdots \pm m_n$ are nonzero.*

Let us look at a singular point. A relation $\sum_i \varepsilon_i m_i = 0; \varepsilon_i = \pm$ corresponds to a degenerated polygon with the vector-sides $a_i = \varepsilon_i m_i e$ for some unit vector e. The degeneration forces us to take into account the velocities $v_i = \frac{da_i}{dt}$ as well as accelerations $w_i = \frac{dv_i}{dt}$. From the equations $|a_i| = m_i$ and $\sum_i a_i = 0$ it follows:

$$(v_i, a_i) = 0; \quad (w_i, a_i) + (v_i, v_i) = 0; \tag{6}$$

$$v_1 + v_2 + \cdots + v_n = 0; \quad w_1 + w_2 + \cdots + w_n = 0. \tag{7}$$

All the velocities v_i are in the same plane Π orthogonal to the vector e. Π will be treated as a complex plane. In view of the equation (7), the gauge (5) implies

$$\sum_{\varepsilon_i = 1} m_i = \sum_{\varepsilon_j = -1} m_j = 0 \tag{8}$$

The restriction (8) reduces the gauge freedom to a rotation of the plane Π.

There is an additional quadratic relation, which follows from the acceleration equations of (6), (7):

$$0 = (w_i, a_i) + (v_i, v_i) = \varepsilon_i m_i (w_i, e) + (v_i, v_i);$$

$$\sum_i \varepsilon_i \frac{(v_i, v_i)}{m_i} = 0. \tag{9}$$

We may summarize the results as follows.

Proposition 1.1.3 *Let $\sum_i \varepsilon_i m_i = 0$; $\varepsilon_i = \pm$. Then in the previous notations a neighborhood of the corresponding singular point of the variety $\widehat{C}_n(m)$ is homeomorphic to the factor of the real quadratic cone in a complex space $\mathbb{C}^n$, defined by the equation*

$$\sum_i \varepsilon_i \frac{|z_i|^2}{m_i} = 0; \quad \sum_{\varepsilon_i=1} z_i = \sum_{\varepsilon_j=-1} z_j = 0, \tag{10}$$

by a circle action $z_i \mapsto \zeta z_i; |\zeta| = 1$.

There is a natural complex structure on the singular manifold (10).

Corollary 1.1.4 *In the previous notations let*

$$k = \#\{i \mid \varepsilon_i = 1\}, \quad l = \#\{j \mid \varepsilon_j = -1\}.$$

Then the singularity (10) is equivalent to the factor of the complex cone over the product $\mathbb{P}^{k-2} \times \mathbb{P}^{l-2} \subset \mathbb{P}^{kl-k-l}$ by a central symmetry $z \mapsto -z$. $\square$

In the subsequent section we will see that there exists a global complex structure on the variety of polygons $\widehat{C}_n(m)$.

It is worth while to mention that the varieties of spatial triangles and quadrangles are nonsingular for any lengths m_i (in fact the first is a point, and the second is 2-sphere). The varieties of higher polygons may be singular.

1.2   Complex and Kähler structure

The main advantage of the gauge (5) is that it allows to introduce an almost complex structure on the variety of polygons $\widehat{C}_n(m)$. Namely, let us define a linear operator I in the tangent space of the variety $\widehat{C}_n(m)$ by the formula:

$$I : v_i \mapsto u_i = \frac{[a_i, v_i]}{m_i}. \tag{11}$$

It is easy to see that if the vectors v_i satisfy the tangent space equations i) and ii) and the gauge condition (5) then the same is true for the vectors u_i. Hence I is correctly defined.

From the geometrical point of view the operator I rotates the vector v_i over the axis a_i by an angle $\frac{\pi}{2}$ in anti-clockwise direction. It follows that $I^2 = -1$. So there is a natural almost complex structure on the variety of polygons $\widehat{C}_n(m)$.

Theorem 1.2.1 *This almost complex structure on $\widehat{C}_n(m)$ is integrable.* □

The theorem may be proved by a direct checking of Newlander-Nirenberg integrability condition [N-N 57]. Another proof follows from an identification of the varieties $\widehat{C}_n(m)$ and $\overline{C}_n(m)$ (see below).

Now we may define a symplectic and Kähler structure on the variety of polygons $\widehat{C}_n(m)$.

Proposition 1.2.2 *Let $u = (u_1, u_2, \ldots, u_n)$ and $v = (v_1, v_2, \ldots, v_n)$ satisfy the tangent space equations i), ii) on page 2. Then*

1. The pairing

$$\omega(u, v) = \sum_i \frac{(u_i, v_i, a_i)}{(a_i, a_i)} \tag{12}$$

is invariant under the gauge transformations iii) and defines a symplectic structure in the nonsingular part of the variety $\widehat{C}_n(m)$.

2. Moreover the form $g(u, v) = \omega(Iu, v)$ is symmetric, positive definite and may be given explicitly by the formula

$$g(u, v) = \sum_i \frac{u_i v_i}{m_i} \tag{13}$$

if at least one of the arguments u, v satisfies the gauge equation (5).

3. The form

$$\Omega(u, v) = g(u, v) + i\omega(u, v) = \omega(Iu, v) + i\omega(u, v) \tag{14}$$

is a Kähler metric on the complex variety $\widehat{C}_n(m)$.

The proof of the formula (13), in view of the definition (11) of the operator I, is straightforward:

$$g(u, v) = \omega(Iu, v) = \sum_i \frac{([u_i, a_i], v_i, a_i)}{m_i^3} = \sum_i \frac{(v_i, [a_i[u_i a_i]])}{m_i^3} =$$

$$= \sum_i \frac{(v_i, u_i)(a_i, a_i)}{m_i^3} - \sum_i \frac{(v_i, a_i)(a_i, u_i)}{m_i^3} = \sum_i \frac{(v_i, u_i)}{m_i}.$$

This implies that the form $\omega(u, v)$ is nondegenerate. It may be written as follows

$$\omega := \sum_i \frac{(u_i, v_i, a_i)}{m_i^2} = \sum_i \frac{dS_i}{m_i} \tag{15}$$

where dS_i is the surface element of the sphere $S_i : |a_i| = m_i$. The forms dS_i are closed, hence the same is true for ω. □

1.3 Algebraicity

The following theorem is an immediate consequence of the formula (15).

Theorem 1.3.1 *If lengths of all sides $|a_i| = m_i$ are rational and $m_1 \pm m_2 \pm \cdots \pm m_n \neq 0$ then $\widehat{C}_n(m)$ is a nonsingular projective algebraic variety.*

Really, as the volume of the sphere S_i is equal to

$$\int_{S_i} dS_i = 4\pi m_i^2$$

then the formula (15) implies that an integral of the form $\omega/4\pi$ over any 2-cycle Σ is rational. It means that class of the form $\omega/4\pi$ is rational and $\widehat{C}_n(m)$ is a Hodge variety. Hence, by a theorem of Kodaira [Kod 54, Che 56], $\widehat{C}_n(m)$ is a projective algebraic variety. □

The rationality condition in the theorem may be removed.

The previous theorem establishes an algebraicity of the variety of polygons $\widehat{C}_n(m)$ by a roundabout way. An algebraic nature of this variety remains quiet mysterious. In the following section we clarify the situation.

1.4 Stable configurations

Let us consider a configuration of n points $p_i \in \mathbb{P}^1$ in the projective line $\mathbb{P}^1$. Suppose that for each point p_i there is given a positive *weight* (or *multiplicity*) m_i. The configuration of weighted points is called *semistable* (resp. *stable*) if sum of the weights of equal points does not exceed (resp. less then) half the weight of all points. It is easy to see that the notion of (semi)stability depends only of linear equalities and inequalities with integer coefficient connecting the weights. So, if it is needed, one may suppose the weights to be rational or integer.

From the Hilbert-Mumford stability criteria [Mum 65] it follows that there exists a nonsingular geometric factor of the space of stable configurations with respect to a natural action of the group $PSL_2(\mathbb{C})$. It will be denoted by $C_n(m)$, where $m = (m_1, m_2, \ldots, m_n)$ is the vector of weights. By definition $C_n(m)$ is nonempty iff the weights satisfy the following polygon inequalities

$$m_i < m_1 + m_2 + \cdots + \widehat{m_i} + \cdots + m_n; \quad i = \overline{1, n}. \tag{16}$$

In a similar way there exists a categorical factor of the space of the semistable configurations. It will be denoted by $\overline{C}_n(m)$.

If all the weights are equal to one we reduce the notations to $\overline{C}_n$ and C_n respectively.

Under condition (16) the variety $\overline{C}_n(m)$ is a projective compactification of the $C_n(m)$ by a finite number of points. Its ample sheaf $\mathcal{O}(1)$ and the corresponding line bundle $\mathcal{L}$ may be described as follows. Let $T(p_i)$ be a tangent space at the point $p_i \in \mathbb{P}^1$. Then $\mathcal{L}$ is a line bundle on $\overline{C}_n(m)$ with the fiber

$$\mathcal{L}(p) = T(p_1)^{\otimes m_1} \otimes T(p_2)^{\otimes m_2} \otimes \cdots \otimes T(p_n)^{\otimes m_n}, \tag{17}$$

at a point $p = (p_1, p_2, \ldots, p_n) \in \overline{C}_n(m)$.

If all semistable configurations of weight m are stable then $\overline{C}_n(m) = C_n(m)$ is a nonsingular projective variety of dimension $n - 3$. It is the case if all sums $m_1 \pm m_2 \pm \cdots \pm m_n$ are nonzero. For example $\overline{C}_n = C_n$ is nonsingular projective variety for odd n.

Theorem 1.4.1 *The algebraic variety of polygons $\widehat{C}_n(m)$ is biregular equivalent to the variety $\overline{C}_n(m)$ of semistable configurations of points in the projective line.*

Here is an explicit construction of this equivalence. Let us identify a projective line $\mathbb{P}^1 = \mathbb{C} \cup \infty$ with a unit sphere $S^2 \in \mathbb{E}^3$ by means of stereographic projection. It turns out that for every semistable configuration of points $a_i \in P^1$ there exists a configuration $b_i \in \mathbb{P}^1$ in the closure of its orbit such that

$$\sum_i m_i b_i = 0$$

(the summands considered as *vectors* in Euclidean space E^3). The system of vectors b_i is uniquely determined up to an orthogonal transformation. The map $\{a_i\} \mapsto \{m_i b_i\}$ gives rise a one-to-one correspondence between $\overline{C}_n(m)$ and $\widehat{C}_n(m)$. $\square$

This construction is closely related to the ideas of Kempf, Ness and Kirwan [K-N 78, Nes 84, Kir 85].

There are several sources of interest in the varieties $\overline{C}_n(m)$.

First of all they are closely related to the classical invariant theory of binary forms. Such a form of degree n up to proportionality is determined by its divisor of zeros in $\mathbb{P}^1$. Hence a coarse moduli space of semistable forms of degree n may be identified with the factor $\overline{C}_n/S_n$ with respect to a natural action of the symmetric group S_n by permutation of points. This means that a projective coordinate ring of the variety $\overline{C}_n/S_n$ is the classical ring $S(2, n)$ of invariant of binary forms of degree n:

$$Proj(S(2, n)) = \overline{C}_n/S_n \tag{18}$$

An explicit description of this ring is a classical open problem. Springer [Spr 77] find out the Poincaré series of these rings for all n. Nevertheless its complete structure is known only for degrees less then 7 [Pop 77, Mum 65].

The varieties $\overline{C}_n(m)$ also naturally arise in the theory of vector bundles and torsion free sheaves [Kly 91, Kly 89]. Let us consider for example the moduli space $\overline{\mathcal{M}}(D)$ of stable sheaves of rank two and discriminant $D = c_1^2 - 4c_2$ on the projective plane $\mathbb{P}^2$. The variety $\overline{\mathcal{M}}(D)$ is projective and nonsingular for D odd and $D \equiv 4 \bmod 8$. The structure of the moduli space $\overline{\mathcal{M}}(D)$ in essential is determined by a fixed point space $\overline{\mathcal{M}}(D)^T$ with respect to a natural action of a maximal torus $T \subset \mathbb{P}GL_2$. For odd discriminant D all connected components of the space $\overline{\mathcal{M}}(D)^T$ are products of the projective lines [Kly 89]. In the case $D \equiv 4 \bmod 8$ the connected components are of the form $\overline{C}_n(m)$. So the structure of the varieties $\overline{C}_n(m)$ is a key for investigation of the moduli space $\overline{\mathcal{M}}(D)$.

In conclusion let us consider several examples.

Example 1.4.2 *Spectrum of the invariant ring $S(2, d)$ of binary forms.*

As we have seen in the introduction (cf. (18)) there is an equality

$$Proj(S(2, d)) = \overline{C}_d/S_d.$$

From the theorem 1.4.1 it follows that the right hand side may be treated as a variety of *unordered* systems of unit vectors $e_i \in E^3$; $i = \overline{1, d}$ with zero sum $e_1 + e_2 + \cdots + e_n = 0$.

The system $\{e_i\}$ of rank 3 corresponds to a nonreal class of form. On the other hand, according to Minkowski theorem there is the only convex polyhedron P in E^3 with e_i as its normal vector to the face $F_i \subset P$ and $Area(F_i) = |e_i|$ (to be more precise, if $e_i \uparrow\uparrow e_j; \forall i, j \in I$ then $F_i = F_j$ and $Area(F_i) = \sum_{i \in I} |e_i|$).

There is a similar interpretation for the systems of rank 2 and 1.

So the topology of the variety $Proj(S(2, d))$ depends on the structure of the space of polyhedrons with fixed areas of faces. This space has a natural stratification by combinatorial types of polyhedrons.

Question 1.4.3 *Is it true that the space of polyhedrons of given combinatorial type and areas of faces is a topological cell?*

The affirmative answer gives a natural cell decomposition of the varieties $\overline{C}_d$ and $\overline{C}_d/S_d$. In the following section we construct another cell decomposition of the varieties of polygons.

Example 1.4.4 *Structure of the variety of polygons $\widehat{C}_n(m)$ for small n.*

From the definition it follows that this variety is empty for $n < 3$. It is reduced to a point for $n = 3$. As the only invariant of four points in $\mathbb{P}^1$ is the cross ratio then $\widehat{C}_4(m) = \overline{C}_4(m) = \mathbb{P}^1$.

So the first nontrivial case is $n = 5$. The variety $\widehat{C}_5 = \overline{C}_5$ is a nonsingular unirational (and hence rational) projective surface. We claim that it is del Pezzo surface of degree 5 (obtained from P^2 by blowing up four points in general position). It follows from two facts, which will be proved in the next section.

1. *The Picard group $Pic(\overline{C}_5)$ is free of rank 5.*

2. *The group of invariant $Pic(\overline{C}_5)^{S_5}$ is of rank one.*

The canonical class and a polarization $\mathcal{O}(1)$ of the surface $\overline{C}_5$ are invariant under symmetric group S_5. Hence by item 2 they are proportional. So $\overline{C}_5$ is del Pezzo surface with Picard number 5. $\square$

In fact all the varieties $\overline{C}_{2n+1}$ are Fano (i.e. the anticanonical class is ample).

2 Homology of the varieties of polygons

In this section there is given a recurrent formula for Betti numbers of the varieties of polygons $\widehat{C}_n(m)$. The calculations make use special Hamilton vector fields on $\widehat{C}_n(m)$. In the case of polygons with equal sides it is found the homology representation of the symmetric group.

2.1 Hamilton vector fields

Recall that $\widehat{C}_n(m)$ is the variety of polygons with vector sides $a_i; i \in \overline{1,n}$ of given lengths m_i. For a subset $I \subset \overline{1,n}$ let us denote by V_I a vector field on $\widehat{C}_n(m)$ with the components (cf. section 1.1)

$$v_i = \begin{cases} [a_i, a_I] & \text{if } i \in I \\ 0 & \text{otherwise,} \end{cases} \tag{19}$$

where $a_I = \sum_{i \in I} a_i$. The components v_i satisfy the equations i) and ii) of the tangent space on page 2. So V_I is correctly defined. The corresponding one parameter group rotates the vector sides $a_i; i \in I$ of the polygon with angular velocity a_I. This implies

$$[V_I, V_J] = 0 \text{ for } I \cap J = \emptyset.$$

Proposition 2.1.1 *The V_I is a Hamilton vector field with respect to the imaginary part of the Kähler metric (proposition 1.2.2). Its Hamilton function is given by the formula*

$$H_I = -\frac{1}{2}|a_I|^2 \tag{20}$$

where $a_I = \sum_{i \in I} a_i$.

Recall that the imaginary part of the Kähler metric is of the form

$$w(u, v) = \sum_i (u_i, w_i, a_i)/m_i^2.$$

The one parameter group of the field V_I rotates the vectors u_i, w_i, a_i by the same angle. It follows that the form $\omega(u, v)$ is invariant and the V_I is a Hamilton field.

To find out the Hamilton function let us consider a 1-form Ω_I dual to the field V_I. Its value at a tangent vector w is equal

$$\Omega_I(w) = \omega(V_I, w) = \sum_{i \in I} \frac{([a_i, a_I], w_i, a_i)}{(a_i, a_i)} = \sum_{i \in I} \frac{(w_i, [a_i[a_i, a_I]])}{(a_i, a_i)} =$$

$$= \sum_{i \in I} \frac{(w_i, a_i)(a_i, a_I) - (w_i, a_I)(a_i, a_i)}{(a_i, a_i)} = -(w_I, a_I)$$

where there are used the notation $w_I = \sum_{i \in I} w_i$ and the tangent vector equation $(w_i, a_i) = 0$ (section 1.1). This formula may be rewritten as follows

$$\Omega_I = -(da_I, a_I) = -\frac{1}{2} d(a_I, a_I). \tag{21}$$

It implies $H_I = -\frac{1}{2}(a_I, a_I)$. $\square$

One may look on the vector field V_I as a differential operator. Then its action on the vector function a_i is given by the formula

$$V_I(a_i) = \begin{cases} [a_i, a_I] & \text{if } i \in I \\ 0 & \text{otherwise.} \end{cases} \tag{22}$$

We apply this equation in order to calculate the commutator $[V_I, V_J]$.

Proposition 2.1.2 *The commutator of the vector fields (19) is given by the formula*

$$[V_I, V_J](a_k) = \begin{cases} [a_k[a_{I \setminus J}, a_{J \setminus I}]] & \text{if } k \in I \cap J \\ [a_k[a_{J \setminus I}, a_{I \cap J}]] & \text{if } k \in I \setminus J \\ [a_k[a_{I \cap J}, a_{I \setminus J}]] & \text{if } k \in J \setminus I \\ 0 & \text{if } k \notin I \cup J. \end{cases} \tag{23}$$

In view of (22) the proof is straightforward. For example if $k \in I \cap J$ then

$$
\begin{aligned}
[V_I, V_J](a_k) &= V_I V_J(a_k) - V_J V_I(a_k) = V_I([a_k, a_J]) - V_J([a_k, a_I]) = \\
&= [[a_k, a_I]a_J] + [a_k[a_{I \cap J}, a_I]] - [[a_k, a_J]a_I] - [a_k[a_{I \cap J}, a_J]] = \\
&= [a_k[a_I, a_J]] + [a_k[a_{I \cap J}, a_{I \setminus J}]] - [a_k[a_{I \cap J}, a_{J \setminus I}]] = \\
&= [a_k[a_{I \setminus J}, a_{J \setminus I}]].
\end{aligned}
$$

In a similar way for $k \in I \setminus J$ the following chain of equalities hold

$$
\begin{aligned}
[V_I, V_J](a_k) &= V_I V_J(a_k) - V_J V_I(a_k) = -V_J([a_k, a_I]) = \\
&= -[a_k[a_{I \cap J}, a_J]] = [a_k[a_{J \setminus I}, a_{I \cap J}]]. \square
\end{aligned}
$$

Corollary 2.1.3 *If* $K = (I \cap J) \backslash (I \cup J)$ *is the symmetric difference then*

$$[V_I, V_J] = [V_J, V_K] = [V_K, V_I].$$

In particular for any triplet of different indices i, j, k *the following equalities are valid*

$$[V_{ij}, V_{jk}] = [V_{jk}, V_{ki}] = [V_{ki}, V_{ij}]. \tag{24}$$

The commutator relations (24) and

$$[V_{ij}, V_{kl}] = 0; \quad V_{ij} = V_{ji}, \tag{25}$$

where $i, j, k, l \in \overline{1, n}$ are different, define a very interesting Lie algebra $\mathcal{G}_n$. Cartier [Car 90] calls it the Lie algebra of the braid group B_n (or the colored braid group P_n). For any linear representation ρ of the algebra $\mathcal{G}_n$ an operator differential equation

$$\frac{\partial F}{\partial z_j} = \sum_{i(\neq j)} \frac{\rho(V_{ij})F}{(z_i - z_j)}$$

is completely integrable. Thus we get a monodromy representation of the fundamental group

$$\pi\{z \in \mathbb{C}^n \,|\, z_i \neq z_j\} = P_n.$$

The interpretation of the V_{ij} as a vector field on the variety $\overline{C}_n(>)$ allows to construct linear representations of the algebra $\mathcal{G}_n$ by the geometric quantization method. This gives a new source of linear representations of braid groups, solutions of the classical Yang-Baxter equation and so on.

The vector fields V_I are not periodical. One may try to make it periodical by multiplication on a scalar function $\lambda = \lambda(a_1, a_2, \ldots, a_n)$. The proof of the proposition 2.1.1 shows that if λ depends of only $a_i, i \in I$ then λV_I is a Hamilton vector field. Thus the field

$$v_I = \frac{V_I}{|a_I|} \tag{26}$$

is Hamiltonean and periodical. It makes sense of course only if $a_I \neq 0$.

Proposition 2.1.4 *The Hamilon function of the field* v_I *is equal to*

$$h_I = -|a_I|. \tag{27}$$

Let Ω_I and ω_I are 1-forms dual to the vector fields V_I and v_I. Then using the formula (21) we obtain

$$\omega_I = \frac{\Omega_I}{|a_I|} = -\frac{(da_I, a_I)}{|a_I|} = -d|a_I|.$$

Therefore the Hamilton function is $h_I = -|a_I|$. $\square$

2.2   Cell decomposition and Betti numbers

A periodical vector field v on a compact Kähler variety X (i.e. an action of a unit circle $T \subset \mathbb{C}^*$ on X) gives a stratification of X [Bia 73, Gin 87]. Namely let

$$X^T = \coprod_i X_i^T$$

be the decomposition of the fix points variety X^T (=zeros of v) in the connected components. Let $X_i \subset X$ be a set of points attracted by the field Iv to the fixed component X_i^T. Then

$$X = \coprod_i X_i. \tag{28}$$

Moreover each component X_i is a complex vector bundle over X_i^T of rank

$$\nu(X_i) = \nu(X_i^T) = \begin{pmatrix} \text{the number of negative weights in the representation of} \\ T \text{ in a tangent space to the component } X_i^T \end{pmatrix} \tag{29}$$

and the stratification (28) induces the decomposition of the cohomology spaces

$$H^{pq}(X) = \bigoplus_i H^{p-\nu(X_i), q-\nu(X_i)}(X). \tag{30}$$

We may apply this theory to the variety of polygons $\widehat{C}_n(m)$ and the vector field v_I (26) if the last is regular, i.e. $a_I \neq 0$. It is the case for $I = \{i, j\}$ and $m_i \neq m_j$ so that $a_{ij} = a_i + a_j \neq 0$.

Let T be a one parameter group of the vector field v_{ij}. A typical element $t_\varphi \in T$ rotates the vector sides a_i, a_j of a polygon $P = (a_1, a_2, \ldots, a_n)$ by the angle φ over the axis $a_{ij} = a_i + a_j$. In order the polygon P to be a fix point of T one of the following conditions have to hold:

1. $a_i \uparrow\uparrow a_j$;

2. $a_i \uparrow\downarrow a_j$;

3. *the vectors* a_k, $k \neq i, j$ *are collinear.*

The polygons of the first type form a variety of $(n-1)$-gons with one side of length $m_i + m_j$ instead of two sides of lengths m_i and m_j. This variety will be denoted by $\widehat{C}_{n-1}(m_{ij}^+)$. In a similar way the polygons of the second type have one side of length $|m_i - m_j|$ in place of two sides of lengths m_i, m_j. They form a variety $\widehat{C}_{n-1}(m_{ij}^-)$.

The polygons of the third kind are in fact triangles and thus are isolated zeros of the field v_{ij}. They are in bijection with solutions of the inequalities

$$m_i + m_j > \sum_{k \neq i, j} \pm m_k > |m_i - m_j|. \tag{31}$$

Now we are in position to find out the indices (29).

Proposition 2.2.1 *The critical components of the vector field v_{ij} on the variety $\widehat{C}_n(m)$ have the following indices.*

1. $\nu(\widehat{C}_{n-1}(m_{ij}^+)) = 0$;

2. $\nu(\widehat{C}_{n-1}(m_{ij}^-)) = 1$;

3. *the index of the isolated zero corresponding to the solution (31) is equal to the number of "−" signs.*

A typical normal eigenvector to the $\widehat{C}_{n-1}(m_{ij}^+)$ has only two nonzero components v_i and $v_j = -v_i$. The operator t_φ rotates both of them in positive direction by angle φ. Thus eigenvalue of t_φ in the normal bundle is $e^{i\varphi}$ and $\nu = 0$.

In a similar way in the second case the eigenvalue is $e^{-i\varphi}$ and $\nu = 1$.

To prove the rest let us consider in the tangent space to an isolated fix point a T-invariant subspace of codimension one, defined by the following conditions

i) $v_i = v_j = 0$;

ii) *the sum of components v_k with $a_k \uparrow\uparrow a_{ij}$ is zero;*

iii) *the sum of components v_k with $a_k \uparrow\downarrow a_{ij}$ is zero.*

The gauge (5) is automatically holds, so one may look at v_k as a complex numbers. The action of the operator t_φ in the coordinates v_k is diagonal $t_\varphi : v_k \mapsto e^{\mp i\varphi} v_k$ where the sign − or + is taken according to $a_k \uparrow\uparrow a_{ij}$ or $a_k \uparrow\downarrow a_{ij}$. There is an additional negative weight in the normal space to i)-iii). This proves the item 3. □

Corollary 2.2.2 i) *The cohomology space $H^{pq}(\widehat{C}_n(m))$ is nonzero only for $p = q$ and an odd Betti numbers of the variety $\widehat{C}_n(m)$ vanish.*

ii) *The Poincaré polynomial*

$$P_q(\widehat{C}_n(m)) = \sum_i h^{ii}(\widehat{C}_n(m))q^i$$

satisfy the following recurrent equation

$$P_q(\widehat{C}_n(m)) = P_q(\widehat{C}_{n-1}(m_{ij}^+)) + qP_q(\widehat{C}_{n-1}(m_{ij}^-)) + \sum_{\substack{\text{solutions of (31)}}} q^{(\text{number of signs } -)}. \qquad (32)$$

There are several forms of an explicit formula for the Poincaré polynomial. We begin with an auxiliary one based on a notion of stable partition. A partition of weights $m = (m_1, m_2, \ldots, m_n)$ is called to be *stable* if the weight of any part less then half the weight of m ($= m_1 + m_2 + \cdots + m_n$). Let

$$p_k(m) = \begin{pmatrix} \text{the number of stable parti-} \\ \text{tions of } m \text{ in } k \text{ blocks} \end{pmatrix}. \tag{33}$$

Proposition 2.2.3 *If the variety of polygons $\widehat{C}_n(m)$ is nonsingular then*

$$P_q(\widehat{C}_n(m)) = \frac{1}{(1+q)^{[3]}} \sum_{k \geq 3} (1+q)^{[k]} p_k(m), \tag{34}$$

where $a^{[k]} = a(a-1)\ldots(a-k+1)$. $\square$

The proposition may be proved directly by counting the number of stable configurations of points in the projective line over finite fields $\mathbb{F}_q$. The formula (34) is interesting only as the first step to the following theorem.

Theorem 2.2.4 *In the previous notations and assumptions the Poincaré polynomial of the variety $\widehat{C}_n(m)$ is equals*

$$P_q(\widehat{C}_n(m)) = \frac{1}{q(q-1)} \left((1+q)^{n-1} - \sum_{m_I \leq m/2} q^{|I|} \right), \tag{35}$$

where $m = m_1 + m_2 + \cdots + m_n$; $m_I = \sum_{i \in I} m_i$.

The proof based on the properties of the *Stirling numbers* of the first and the second kind

$$x^{[n]} = \sum_{k=0}^{n} s(n,k) x^k; \quad x^n = \sum_{k=0}^{n} \sigma(n,k) x^{[k]} \tag{36}$$

and the following combinatorial interpretation [Rio 58]

$$\sigma(n,k) = \begin{pmatrix} \text{the number of partition} \\ \text{of } n \text{ in } k \text{ blocks} \end{pmatrix}.$$

This interpretation allows us to express the number of stable partitions $p_k(m)$ by the Stirling numbers:

$$p_k(m) = \sigma(n,k) - \sum_{m_I \leq n/2} \sigma(|I|, k-1). \tag{37}$$

The sum in the right hand side is equal to the number of unstable partitions. For each such partition there is the only part which is greater then $m/2$, so that the sum of the rest parts is less or equal to $n/2$. The sum in (37) counts this "small parts" of unstable partitions.

Now we can transform the formula (34) in the following way

$$P_q(\widehat{C}_n(m)) \overset{(37)}{=} \frac{1}{(1+q)^{[3]}} \sum_{k \geq 3} (1+q)^{[k]} \left(\sigma(n,k) - \sum_{m_I \leq n/2} \sigma(|I|, k-1) \right) =$$

$$\overset{(36)}{=} \frac{1}{q(q^2-1)} \left((q+1)^n - (q+1) \sum_{m_I \leq m/2} q^{|I|} \right) =$$

$$= \frac{1}{q(q-1)} \left((q+1)^{n-1} - \sum_{m_I \leq m/2} q^I \right). \square$$

Corollary 2.2.5 *Under the previous assumptions the following formulae for the Betti numbers are valid*

$$h^{2p}(\widehat{C}_n(m)) - h^{2(p-1)}(\widehat{C}_n(m)) = \binom{n-1}{p} - \sharp\{I \mid m_I > m/2; |I| = p+1\} =$$

$$= \sharp\{I \mid m_I < m/2; |I| = p+1\} - \binom{n-1}{p+1}.$$

In particular there is the following formula for Picard number of the variety $\widehat{C}_n(m)$:

$$rk \, Pic\, \widehat{C}_n(m) = n - \sharp\{i,j \mid m_i + m_j > m/2\}.$$

The proof immediately follows from (35) by multiplication on $q-1$ and comparing the coefficients of q^p. $\square$

As another corollary we get the following formulae for the Euler characteristic $\chi = \sum_p h^{2p}$ and the signatur $\tau = \sum_p (-1)^p h^{2p}$:

$$\chi(\widehat{C}_n(m)) = (n-1)2^{n-2} - \sum_{m_I \leq m/2} |I|; \tag{38}$$

$$\tau(\widehat{C}_n(m)) = \frac{1}{2} \sum_{m_I \leq m/2} (-1)^{1+|I|}. \tag{39}$$

We mention also a strange equation for the value of the Poincaré polynomial at $q = 2$, that follows from (34)

$$P_2(\widehat{C}_n(m)) = \binom{\text{the number of triangles that may be com-}}{\text{posed of } n \text{ segments of lengths } m_i}.$$

In fact for every n we have only finite number of different varieties $\widehat{C}_n(m)$. They depends only of the set of $I \subset \overline{1,n}$ for which $m_I < m/2$. This may be stated in a geometrical way. Let $C^n = \{x \in \mathbb{R}^n \mid |x_i| \leq 1\}$ be a cube in $\mathbb{R}^n$ and $C^n(m)$ be its section by a central hyperplane

$$m_1 x_1 + m_2 x_2 + \cdots + m_n x_n = 0.$$

Then the variety $\widehat{C}_n(m)$ depends only of combinatorial type of the polyhedron $C^n(m)$. All the results of this section on structure of the variety $\widehat{C}_n(m)$ may be stated in terms of the polyhedron $C^n(m)$.

In the case of equal weights all the formulae may be simplified.

Proposition 2.2.6 *For odd $n = 2k + 1 \geq 3$ the Betti numbers, the Euler characteristic and the signature of the variety $\widehat{C}_n$ are given by the formulas*

$$h^{2p}(\widehat{C}_n) = \sum_{0 \leq i \leq p} \binom{2k}{i}, \quad p < k; \tag{40}$$

$$\chi(\widehat{C}_n) = \sum_{0 \leq i < k} \binom{2k}{i} (2k - 2i - 1); \tag{41}$$

$$\tau(\widehat{C}_n) = \frac{(-1)^{k-1}}{2} \binom{2k}{k}. \tag{42}$$

Here are some examples of the Poincaré polynomials:

$$P_q(\widehat{C}_5) = q^2 + 5q + 1;$$
$$P_q(\widehat{C}_7) = q^4 + 7q^3 + 22q^2 + 7q + 1;$$
$$P_q(\widehat{C}_9) = q^6 + 9q^5 + 37q^4 + 93q^3 + 37q^2 + 9q + 1.$$

2.3   Cohomological representation

The symmetric group S_n naturally acts on the variety $\widehat{C}_n$ and on the cohomology spaces $H^{2p}(\widehat{C}_n)$. This cohomological representations are important for many applications, e.g. in calculation of Betti numbers of the factor $\widehat{C}_n/S_n$

$$\dim H^{2p}(\widehat{C}_n/S_n) = \dim H^{2p}(\widehat{C}_n)^{S_n}.$$

In the following theorem we use the notation $[n, m]$ for a representation of the symmetric group S_{n+m} induced by the trivial character of the subgroup $S_n \times S_m \subset S_{n+m}$.

Theorem 2.3.1 *For any odd $n = 2k + 1$ there is the following decomposition of the cohomological representation*

$$H^{2p}(\widehat{C}_n) \cong \bigoplus_{0 \leq 2i \leq p} [n - p + 2i, p - 2i]; \quad p < k. \tag{43}$$

This theorem may be proved in the same way as the theorem 2.2.4. One has only to use instead of the (34) the number of points in *a twisted form* $\overline{C}_n^\pi$ of the variety $\overline{C}_n$ over finite field $\mathbb{F}_q$. That is the number of points on which Frobenius acts as a given substitution $\pi \in S_n$. Then according to Weil and Deligne

$$\sum_i q^i \mathrm{Tr}(\pi | H^{2i}(\overline{C}_n)) = |\overline{C}_n^\pi(\mathbb{F}_q)|.$$

So we get a character formula for the cohomological representation of the symmetric group. Then the theorem may be deduced by purely combinatorial methods using the Stirling numbers in the same way as above.

Corollary 2.3.2 dim $H^{2p}(\widehat{C}_n/S_n) = 1 + \left[\frac{p}{2}\right]$.

The corollary hints that the cohomology ring $H^*(\widehat{C}_n/S_n)$ has two generators of degree 2 and 4 and no relations of degree less or equal $n - 3$.

Corollary 2.3.3 *All the varieties* $\widehat{C}_n$ *are Fano (i.e. their anticanonical class is ample).*

Really the previous corollary implies that up to proportionality there is the only class in Pic $\widehat{C}_n$ which is invariant under the action of the symmetric group S_n. The canonical class is certainly invariant, and besides there is an invariant polarization. Hence the canonical class is proportional to the polarization. As the canonical class is ineffective, the coefficient of proportionality is negative. $\square$

References

[Bia 73]   Bialinicki-Birula A. *Some theorems on actions of algebraic groups.* Ann. Math., **98** (1973), 480-497.

[Car 90]   Cartier P. *Développements récents sur les groupes de tresses: applications à la topologie et à l'algèbre.* Astérisque, **189-190** (1990), 17-67.

[Che 56]   Chern S.S. *Complex manifolds.* The University of Chicago, 1956.

[Gin 87]   Ginsburg V.A. *Equivariant cohomology and Kähler's geometry.* Funct. Analysis and its Appl., **21** (1987), 271-283.

[K-N 78]   Kempf G., Ness L. *Lengths of vectors in representation spaces.* Lect. Notes in Math., **732**, Springer 1978.

[Kir 85]   Kirwan F.G. *The cohomology of quotients in simplectic and algebraic geometry.* Math. Notes, **31**(1985), Prinston Univ. Press.

[Kly 91]   Klyachko A.A. *Vector bundles and torsion free sheaves on the projective plane.* Max-Planck-Institute für Math. Preprint, 1991, MPI/91-59.

[Kly 89]   Klyachko A.A. *Equivariant vector bundles on toric varieties.* Math. USSR Izv., **53** (1989), 1001–1039. (Russian)

[Kod 54]   Kodaira K. *On Kähler varieties of restricted type (an intrinsic characterization of algebraic varieties).* Ann. Math., **60** (1954), 28–48.

[Mum 65]   Mumford D. *Geometric invariant theory.* Springer 1965.

[Nes 84] Ness L. *A stratification of the null cone via the moment map.* Amer. J. Math., **106** (1984), 1281–1387.

[N-N 57] Newlander A., Nierenberg L. *Complex analytic co-ordinates in almost complex manifolds.* Ann. Math., **65** (1957), 391–404.

[Pop 77] Popp H. *Moduli theory and classification of algebraic varieties.* Springer 1977.

[Rio 58] Riordan J. *An introduction to combinatorial analysis.* Wiley, New-York 1958.

[Spr 77] Springer T.A. *Invariant theory.* Lect. Notes in Math., v. 585, Springer 1977.

Supported by a grant from Russian Ministry of Education.

Rigid Sheaves on Surfaces

S.A.Kuleshov

Sheaves on surfaces can be divided into two main classes: sheaves with deformations and sheaves without deformations. Among the latter there are sheaves that have niether global nor infinitesimal deformations, in other words, the group $\operatorname{Ext}^1(F,F)$ is trivial. These sheaves, called rigid, are the subject of the present paper.

It is known that if the anti-canonical class of a surface is ample, any rigid torsion free sheaf is locally free, and any rigid sheaf is an extension of a rigid bundle and torsion sheaf. Moreover a support of the rigid torsion sheaf is a curve with the negative cup square. A cup square of any curve on $\mathbb{P}^2$, $\mathbb{P}^1 \times \mathbb{P}^1$ is positive. Therefore any rigid sheaf on these surfaces is locally free. That is why we can limit the consideration to the rigid bundles.

The prime rigid sheaves are exceptional bundles, that is rigid bundles E satisfying the condition $\operatorname{Hom}(E,E) = \mathbb{C}$.

For the first time the exceptional bundles have appeared in preprint [2], describing the possible Chern classes which can have a stable bundle on $\mathbb{P}^2$. Later, they have been thoroughly studied at the Moscow seminar of Prof. A. Rudakov - see [3], [4], [6], [10], [11].

In the present paper it is shown that any rigid bundle F on a surface with an ample anticanonical class decomposes into direct sum of exceptional bundles. These very bundles are indecomposable in view of their simplicity. The same statement is proved for the case when F is a semi-stable rigid bundle on an general $K3$-surface.

Notations

Let us begin with some notations and recollections. In this artical we will work with a smooth complete algebraic surface X with condition $h^1(\mathcal{O}_X) = 0$.

On any surface X there is a divisor class K_X that is called the canonical class of X. Denote the divisor $-K_X$ by H, the rank of coherent sheaf F by $r(F)$, the Chern classes of F by $c_1(F)$ and $c_2(F)$ and

$$q(F) = \left(c_1^2(F) - 2c_2(F)\right)/2r(F).$$

For any divisor A and nontorsion sheaf F the rational number

$$\left(c_1(F) \cdot A\right)/r(F)$$

is called the slope of F with respect to A and denoted by $\mu_A(F)$.

For coherent sheaves E and F on X we define

$$\chi(E, F) = \sum_i (-1)^i \dim \operatorname{Ext}^i(E, F).$$

We will write down the Serre duality theorem and the theorem of Riemann-Roch.

Let $\mathcal{O}_X(A)$ be a line bundle corresponding to divisor A. We will use the notation

$$F \otimes \mathcal{O}_X(A) = F(A).$$

The Serre duality theorem.

$$\operatorname{Ext}^i(F, E) \simeq \operatorname{Ext}^{2-i}\left(E, F(K_X)\right)^*.$$

The Riemann-Roch Theorem. *For any two coherent sheaves E and F on the surface X*

$$\chi(E, F) =$$

$$r(E)r(F) \left[p_X + \frac{\mu_H(F) - \mu_H(E)}{2} + q(F) + q(E) - \frac{1}{r(F)r(E)}\left(c_1(F) \cdot c_1(E)\right) \right],$$

where $p_X = \chi(\mathcal{O}_X, \mathcal{O}_X)$.

It is easy to see that $\chi(E, F)$ is a bilinear function on a $\mathbb{Z}$-module $K_0(X)$. It can be decomposed into symmetric and antisymmetric parts

$$\chi(E, F) = \chi_+(E, F) + \chi_-(E, F),$$

where

$$\chi_-(E, F) = \frac{1}{2}r(E)r(F)\left(\mu_H(F) - \mu_H(E)\right). \tag{1.1}$$

1. Stable sheaves

In this section we will recall the basic definitions and the properties of semi-stable and stable sheaves, which will be necessary for the proof of the main results.

Definition. Let C be a smooth complete algebraic curve and E is a bundle on it. The rational number

$$\mu(E) = \frac{\deg\left(c_1(E)\right)}{r(E)}$$

is called the slope of E.

Definition. For an ample divisor A and a nontorsion sheaf E the polynomial

$$\gamma_A(E,n) = \frac{\chi(\mathcal{O}_X(-nA), E)}{r(E)}$$

is called the Gieseker slope of E with respect to A.

It is easy to prove:

$$\gamma_A(E,n) = a_1(X,A)n^2 + a_2(X,A)\mu_A(E)n + a_3(X,A,E), \tag{1.2}$$

where the coefficients $a_1(X,A)$ and $a_2(X,A)$ are independent on E.

For example, if X is $K3$-surface then

$$\gamma_A(E,n) = \frac{A^2}{2}n^2 + \mu_A(F)n + q(F) + 2. \tag{1.3}$$

Let us say that the Gieseker slope of E is grater than the Gieseker slope of F if the inequality $\gamma_A(E,n) > \gamma_A(F,n)$ holds true for any large number n. We will use the notation $\gamma_A(E,n) = \gamma_A(E)$.

Remark 1.1. The inequality $\gamma_A(E) > \gamma_A(F)$ is possible both when $\mu_A(E) > \mu_A(F)$ or $\mu_A(E) = \mu_A(F)$. If $\gamma_A(E) = \gamma_A(F)$ then $\mu_A(E) = \mu_A(F)$.

Let $\eta(F)$ be either $\mu_A(F)$ or $\gamma_A(F)$.

Definition. One calls a sheaf E η-stable if E has no torsion and for any subsheaf F such that both F and E/F have positive ranks, we have

$$\eta(F) < \eta(E).$$

One calls a sheaf $E\eta$-semi-stable if E has no torsion and for any subsheaf F such that both F and E/F have positive ranks the inequality

$$\eta(F) \le \eta(E)$$

is true.

There is the equivalent definition of stability.

Proposition 1.2. *A torsion free sheaf E is η-(semi)-stable if for any its quotient sheaf F the inequality*

$$\eta(E) < \eta(F) \qquad (\eta(E) \le \eta(F))$$

is true.

The equivalence of these definitions follows from

Lemma 1.3. (about swing). *Let*

$$0 \longrightarrow F_1 \longrightarrow F \longrightarrow F_2 \longrightarrow 0$$

be an exact sequence of sheaves on a surface X or a curve C. Then

1. $\eta(F_1) > \eta(F)$ if and only if $\eta(F) > \eta(F_2)$;

2. $\eta(F_1) < \eta(F)$ if and only if $\eta(F) < \eta(F_2)$;

3. $\eta(F_1) = \eta(F)$ if and only if $\eta(F) = \eta(F_2)$.

In case $\eta(F) = \mu(F)$ one says about Mumford-Takemoto stability. In other case $\big(\eta(F) = \gamma(F)\big)$ one says about Gieseker stability. These stabilities on surfaces are connected in the following way.

Proposition 1.4. *1. Any μ_A-stable sheaf is γ_A-stable.*

2. Any γ_A-semi-stable sheaf is μ_A-semi-stable.

It follows from remark 1.1 and definition of η-stability that this proposition holds true.

For the following discussion we will need the base properties of stable sheaves.

Proposition 1.5. *1. For a pair of η-(semi)-stable sheaves E and F with $\eta(E) > \eta(F)$ the space $\mathrm{Hom}(E, F)$ is equal to zero.*

2. If the anticanonical class of surface X is ample and slopes of η-semi-stable sheaves E and F satisfy the inequality $\eta(E) \leq \eta(F)$ then $\mathrm{Ext}^2(E, F) = 0$.

3. If $K_X = 0$ and the slope of η-(semi)-stable sheaf E is less than the slope of η-(semi)-stable sheaf F then $\mathrm{Ext}^2(E, F) = 0$.

4. Any η-stable sheaf F is simple, i.e. $\mathrm{Hom}(F, F) = \mathbb{C}$.

5. If the sheaf F is γ_A-(semi)-stable, the sheaf E is γ_A-stable and $\gamma_A(F) = \gamma_A(E)$, then any nonzero map $F \longrightarrow E$ is epimorphism.

6. Two γ_A-stable sheaves with equal Gieseker slopes γ_A either are isomorphic or have no nonzero maps one to another.

7. Two μ_A-stable bundles with equal Mumford-Takemoto slopes μ_A either are isomorphic or have no nonzero maps one to another.

8. Let $0 \longrightarrow F_1 \longrightarrow F \longrightarrow F_2 \longrightarrow 0$ be an exact sequence of η-semi-stable sheaves and $\eta(F_1) = \eta(F_2) = \eta$ then F is η-semi-stable and $\eta(F) = \eta$.

The proof of these properties can be found in [9].

2. Unstabilizing filtrations

Next we will construct the filtrations of sheaves without torsion. These filtrations and the spectral sequence associated with them are essential for the proof of the main results.

Proposition 2.1. *Any torsion free sheaf F on a surface X or a curve C has the canonical filtration*

$$0 = F_{n+1} \subset F_n \subset \ldots \subset F_2 \subset F_1 = F$$

with η-semi-stable quotients $G_i = F_i/F_{i+1}$, the slopes of F_i satisfy the inequalities $\eta(F_i) > \eta(F_j)$ and $\eta(F_i) > \eta(G_i) > \eta(G_j)$ for $i > j$. Moreover if anticanonical class of the surface X is ample or equal to zero, then $\mathrm{Hom}(F_i, G_j) = 0$ and $\mathrm{Ext}^2(G_j, F_i) = 0$ for $i > j$.

Proof. If the sheaf F is η-semi-stable then the filtration is trivial. Let there be torsion free quotient $F \longrightarrow G \longrightarrow 0$ and $r(F) > r(G)$, $\eta(G) < \eta(F)$. Let G_1 minimizes the slope $\eta(G_1)$ among all quotient sheaves of F and hence maximizes $r(G_1)$ among these quotients. We can say that the quotient G_1 is η-semi-stable. Conversely the slope of G_1 is not minimal. Denote F_2 the kernel of the epimorphism $F \longrightarrow G_1 \longrightarrow 0$.

Assume that there is a part of filtration
$$F_k \subset \ldots \subset F_2 \subset F_1 = F$$
satisfying conditions of the proposition. Then assume that each of the quotients G_i is the sheaf with the minimal slope and the maximal rank among all torsion free quotient sheaves of F_i.

If the sheaf F_k is η-semi-stable, then the filtration is constructed. In other case, there is torsion free quotient sheaf G_k of F_k satisfying the same conditions as G_i. Let F_{k+1} be the kernel of the epimorphism $F_k \longrightarrow G_k \longrightarrow 0$. It follows from lemma about swing that
$$\eta(F_{k+1}) > \eta(F_k) > \eta(G_k).$$
Besides G_k is η-semi-stable.

For the proof of inequality $\eta(G_k) > \eta(G_{k-1})$ let us consider the commutative diagram

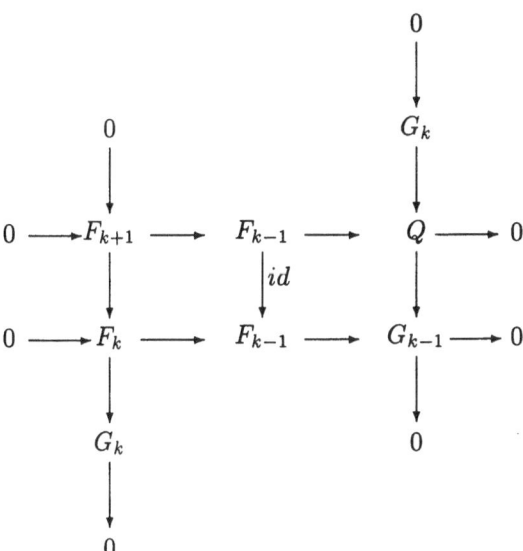

with exact rows and columns. The sheaves Q and G_{k-1} are quotient sheaves of F_{k-1}. By virtue of $r(Q) > r(G_{k-1})$, $\eta(G_{k-1}) < \eta(Q)$ by our choice of G_{k-1}. It follows from lemma about swing that

$$\eta(G_k) > \eta(Q) > \eta(G_{k-1}).$$

Now we remark that this filtration must stabilize because the ranks of the terms of the filtration decrease and sheaves of rank 1 are η-stable.

Let us proof the equalities

$$\mathrm{Hom}(F_i, G_j) = 0 \quad and \quad \mathrm{Ext}^2(G_j, F_i) = 0 \quad for \quad i > j.$$

By construction of filtration $G_n = F_n$, hence there is short exact sequence

$$0 \longrightarrow G_n \longrightarrow F_{n-1} \longrightarrow G_{n-1} \longrightarrow 0. \tag{2.1}$$

Let us apply the functor $\mathrm{Ext}(\,\cdot\,, G_j)$ to it for $j < n - 1$:

$$0 \longrightarrow \mathrm{Hom}(G_{n-1}, G_j) \longrightarrow \mathrm{Hom}(F_{n-1}, G_j) \longrightarrow \mathrm{Hom}(G_n, G_j) \longrightarrow \cdots$$

From proposition 1.5.1 by virtue of stability of the quotients of the filtration and the inequalities for the slopes of the quotients it follows that the spaces $\mathrm{Hom}(G_{n-1}, G_j)$ and $\mathrm{Hom}(G_n, G_j)$ are trivial. Therefore $\mathrm{Hom}(F_{n-1}, G_j) = 0$.

We apply the functor $\mathrm{Ext}(G_j, \cdot)$ to (2.1) to obtain

$$\cdots \longrightarrow \mathrm{Ext}^2(G_j, G_n) \longrightarrow \mathrm{Ext}^2(G_j, F_{n-1}) \longrightarrow \mathrm{Ext}^2(G_j, G_{n-1}).$$

It follows from propositions 1.5.2 and 3 that the spaces $\mathrm{Ext}^2(G_j, G_n)$ and $\mathrm{Ext}^2(G_j, G_{n-1})$ are equal to zero. Hence $\mathrm{Ext}^2(G_j, F_{n-1}) = 0$.

Applying the functors $\mathrm{Ext}(\,\cdot\,, G_j)$ and $\mathrm{Ext}(G_j, \cdot)$ in succession to the short exact sequence $0 \longrightarrow F_{i+1} \longrightarrow F_i \longrightarrow G_i \longrightarrow 0$ for $i > j$, we will finish the proof of the proposition.

Proposition 2.2. *Let F be a Gieseker γ_A-semi-stable sheaf. Then*

1) there is a filtration

$$0 = F_{n+1} \subset F_n \subset \ldots \subset F_2 \subset F_1 = F \tag{2.2}$$

such that every successive quotient $G_i = F_i/F_{i-1}$ is γ_A-semi-stable and has the same slope as F, i.e.

$$\gamma_A(G_i) = \gamma_A(F_i) = \gamma_A(F) \qquad for \qquad i = 1, 2, \ldots, n.$$

2) any sheaf G_i has a filtration

$$0 = G_i^{k_i+1} \subset \ldots \subset G_i^2 \subset G_i^1 = G_i \tag{2.3}$$

such that each G_i^j/G_i^{j-1} is isomorphic to γ_A-stable sheaf E_i and

$$\gamma_A(G_i) = \gamma_A(E_i) = \gamma_A(F).$$

3) $\mathrm{Hom}(F_{i+1}, G_i) = \mathrm{Hom}(G_{i+1}, G_i) = \mathrm{Hom}(G_i, G_{i+1}) = 0.$

Proof. 1) If the sheaf F is γ_A-stable then the filtration is trivial. Assume the converse. Then there is torsion free quotient sheaf E of F such that $r(E) < r(F)$ and $\gamma_A(E) = \gamma_A(F)$. Let E_1 minimize $r(E)$ among all these quotients of F. Obviously it is γ_A-stable. We denote the kernel of epimorphism by F_2^1.

Since any subsheaf of F_2^1 is subsheaf of F and $\gamma_A(F_2^1) = \gamma_A(F)$ (lemma about swing), the sheaf F_2^1 is γ_A-semi-stable.

Assume that there is a morphism $\phi : F_2^1 \longrightarrow E_1$. It follows from proposition 1.5.5 that it is an epimorphism. Let us denote by F_2^2 the kernel of ϕ.

Let us repeat it until we construct a subsheaf $F_2^{k_2} = F_2$, such that

$$\mathrm{Hom}(F_2, E_1) = 0.$$

If F_2 equals zero then the filtration is constructed. Conversely we find the sheaf F_3 in the same way, etc., till we find $F_{n+1} = 0$.

2) Let us prove that G_i has the filtration (2.3); for example, take $i = 1$.

Denote $Q_j = F_1/F_2^j$. Then for any j we have a commutative diagram with exact rows and columns

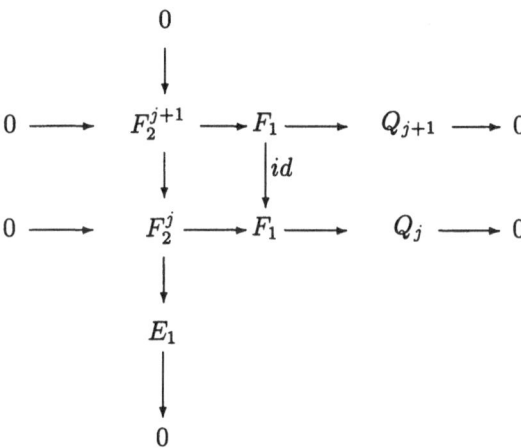

Therefore the sheaf Q_j can be included into a short exact sequence

$$0 \longrightarrow E_1 \longrightarrow Q_{j+1} \longrightarrow Q_j \longrightarrow 0.$$

Remark that $Q_1 = E_1$ and $Q_{k_2} = G_1$. Hence for any number j there is a commutative diagram with exact rows and columns

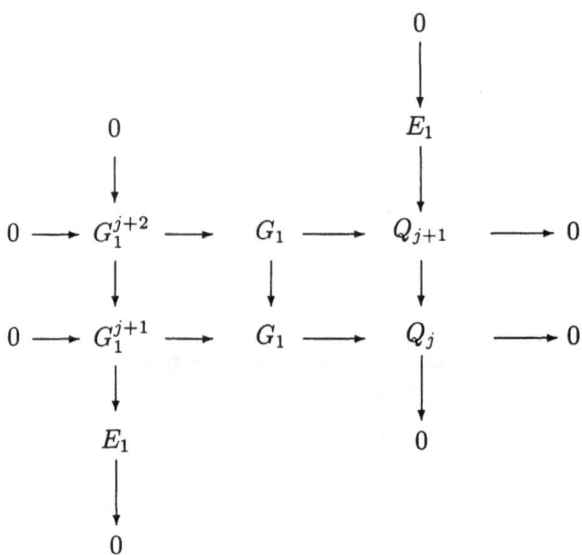

We see that the sheaves G_1^j are the members of filtration (2.3).

3) By the construction of filtration (2.3) the space $\mathrm{Hom}(F_{i+1}, E_i)$ is trivial. We apply the functor $\mathrm{Hom}(F_{i+1}, \cdot)$ to exact sequences

$$0 \longrightarrow E_i \longrightarrow G_i^{k_i-1} \longrightarrow E_i \longrightarrow 0,$$

$$0 \longrightarrow G_i^{k_i-1} \longrightarrow G_i^{k_i-2} \longrightarrow E_i \longrightarrow 0,$$

$$\cdots\cdots\cdots\cdots$$

$$0 \longrightarrow G_i^2 \longrightarrow G_i \longrightarrow E_i \longrightarrow 0$$

to obtain $\mathrm{Hom}(F_{i+1}, G_i) = 0$.

Next, each quotient G_i can be included into the short exact sequence

$$0 \longrightarrow F_{i+1} \longrightarrow F_i \longrightarrow G_i \longrightarrow 0.$$

Hence the long sequence

$$0 \longrightarrow \mathrm{Hom}(G_i, E_{i-1}) \longrightarrow \mathrm{Hom}(F_i, E_{i-1}) \longrightarrow \cdots$$

is exact. From the equality $\mathrm{Hom}(F_i, E_{i-1}) = 0$ it follows that $\mathrm{Hom}(G_i, E_{i-1}) = 0$. In the same way the equality $\mathrm{Hom}(E_i, E_{i-1}) = 0$ can be proved.

Now the proof of the proposition follows from lemma:

Lemma 2.3. *Let torsion free sheaves Q and G have filtrations*

$$Q = \mathrm{Gr}(E_1, \ldots, E_1), \qquad G = \mathrm{Gr}(E_2, \ldots, E_2),$$

i.e.

$$Q = Q_1 \supset Q_2 \ldots Q_m \supset Q_{m+1} = 0, \qquad E_1 = Q_i/Q_{i+1};$$
$$G = G_1 \supset G_2 \ldots G_n \supset G_{n+1} = 0, \qquad E_2 = G_i/G_{i+1}.$$

If the sheaves E_1 and E_2 are γ_A-stable, $\gamma_A(E_1) = \gamma_A(E_2)$ and $\mathrm{Hom}(E_1, E_2) = 0$, then the spaces $\mathrm{Hom}(Q, G)$ and $\mathrm{Hom}(G, Q)$ are trivial.

Proof of lemma. Remark, that E_1 is not isomorphic to E_2 (conversely $\mathrm{Hom}(E_1, E_2) = \mathbb{C}$). It follows from proposition 1.5.6 that $\mathrm{Hom}(E_1, E_2) = 0$.

By the symmetry of the statement it is enough to prove that $\mathrm{Hom}(Q, G) = 0$.

By the assumption there are short exact sequences

$$0 \longrightarrow E_2 \longrightarrow G_{n-1} \longrightarrow E_2 \longrightarrow 0,$$
$$0 \longrightarrow G_{m-1} \longrightarrow G_{m-2} \longrightarrow E_2 \longrightarrow 0,$$
$$\cdots\cdots\cdots$$
$$0 \longrightarrow G_2 \longrightarrow G \longrightarrow E_2 \longrightarrow 0.$$

We apply the functor $\mathrm{Hom}(E_1, \cdot)$ to them to obtain the equality $\mathrm{Hom}(E_1, G) = 0$. Next we apply the functor $\mathrm{Hom}(\cdot, G)$ to the following exact sequences:

$$0 \longrightarrow E_1 \longrightarrow Q_{m-1} \longrightarrow E_1 \longrightarrow 0,$$
$$0 \longrightarrow Q_{m-1} \longrightarrow Q_{m-2} \longrightarrow E_1 \longrightarrow 0,$$
$$\cdots\cdots\cdots$$
$$0 \longrightarrow Q_2 \longrightarrow Q \longrightarrow E_1 \longrightarrow 0.$$

Hence the lemma follows. Proposition 2.2 is proved.

3. Spectral sequence and exceptional bundles

With any filtration of a sheaf F there is associated a spectral sequence, which is convergent to spaces $\mathrm{Ext}^i(F, F)$. The main results will be proved by virtue of studying this spectral sequence. Besides, in this section we will formulate all necessary properties of exceptional bundles.

Proposition 3.1. *Let sheaf F have a filtration*

$$F = \mathrm{Gr}(G_n, G_{n-1}, \ldots, G_1).$$

Then there is a spectral sequence E^{pq}. Its E_1-part is

$$E_1^{pq} = \bigoplus_i \mathrm{Ext}^{p+q}(G_i, G_{i+p}).$$

And it is convergent to

$$\bigoplus_{p+q=i} E_\infty^{pq} = E_\infty^i = \mathrm{Ext}^i(F, F).$$

The proof of this fact can be found in [5].

Using this proposition we will prove μ-stability of simple bundle on an elliptic curve.

Proposition 3.2. *Let C be a smooth elliptic curve and E be a simple bundle on it* $(\mathrm{Hom}(E,E) = \mathbb{C})$. *Then E is μ-stable.*

Proof. Remark that if E is μ-semi-stable then it is μ-stable (the rank and the degree of the simple bundle are coprime).

Suppose the simple bundle E is not μ-semi-stable. Consider the unstabilizing filtration of it (proposition 2.1)
$$E = \mathrm{Gr}(G_n, G_{n-1}, \ldots, G_1),$$
where G_i are μ-semi-stable and $\mu(G_i) > \mu(G_j)$ for $i > j$.

Since a curve is 1-dimensional variety, then $\mathrm{Ext}^2(F, G) = 0$ for any sheaf on it. From μ-semi-stability of the quotients, inequalities $\mu(G_i) > \mu(G_j)$ and proposition 1.5.1 it follows that $\mathrm{Hom}(G_i, G_j) = 0$ for $i > j$. On the other hand, from the Serre duality on an elliptic curve we have $\mathrm{Ext}^1(G_j, G_i) = \mathrm{Hom}(G_i, G_j)^* = 0$ for $i > j$. Therefore the E_1-part of spectral sequence, associated with the filtration, has the form

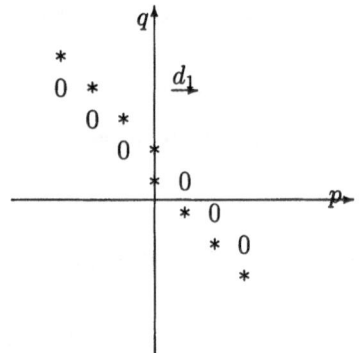

It is easy to see that $E_1^{pq} = E_\infty^{pq}$. Hence $\mathrm{Hom}(E, E) \supset \bigoplus_i \mathrm{Hom}(G_i, G_i)$. The dimension of each $\mathrm{Hom}(G_i, G_i)$ equals at least 1 and
$$\dim \mathrm{Hom}(E, E) = 1.$$
It is possible iff the filtration is trivial, i.e. E is μ-semi-stable.

Let us recall the definition and the base properties of exceptional bundles.

Definition. We shall call a sheaf E exceptional if it is simple and rigid, i.e.
$$\mathrm{Hom}(E, E) = \mathbb{C}, \qquad \mathrm{Ext}^1(E, E) = 0.$$

Proposition 3.3. *Let X be a regular smooth surface with ample or zero anticanonical class. If E is an exceptional torsion free sheaf on X then*

1. E is locally free.

2. If $H = -K_X$ is ample then E is μ_H-stable and the restriction of E to elliptic curve $C \in |-K_X|$ is a simple bundle.

3. If X is $K3$-surface and $\mathrm{Pic}(X) = \mathbb{Z}$, then E is μ_H-stable with respect to the generator H of the group $\mathrm{Pic}(X)$. The bundle E is determined by its slope in the unique way.

The proof of this fact can be found in [7], [3], [8].

4. Rigid sheaves on Del Pezzo surfaces

In this section we shall denote by X a smooth Del Pezzo surface, i.e. a regular algebraic surface with an ample anticanonical class H.

Theorem 1. *Let X be a complete regular $(q = 0)$ smooth algebraic surface, the divisor $H = -K_X$ is ample, F is γ_H-semi-stable rigid $\left(\mathrm{Ext}^1(F, F) = 0\right)$ bundle on X. Then there are exceptional bundles $E_1, \dots, E_n$ on X such that*

$$F = E_1 \oplus \dots \oplus E_1 \oplus E_2 \oplus \dots \oplus E_2 \oplus E_n \oplus \dots \oplus E_n.$$

Proof. We prove this theorem by induction on $r(F)$. Since a line bundle on a regular surface is exceptional, the base of induction holds true. Consider the filtration (2.2) of the bundle F

$$F = \mathrm{Gr}(G_n, G_{n-1}, \dots, G_1). \tag{4.1}$$

Step 1. The quotients G_i and the terms of filtration F_i are rigid and $\mathrm{Ext}^1(G_n, G_1) = 0$.

Proof of step 1. Consider the exact triple

$$0 \longrightarrow F_{i+1} \longrightarrow F_i \longrightarrow G_i \longrightarrow 0, \tag{4.2}$$

which arises from the filtration of F. By proposition 2.2

$$\mathrm{Hom}(F_{i+1}, G_i) = 0. \tag{4.3}$$

Since the sheaves F_{i+1} and G_i are γ_H-semi-stable, then from $\gamma_H(G_i) = \gamma_H(F_{i+1})$ it follows that

$$\mathrm{Ext}^2(G_i, F_{i+1}) = 0. \tag{4.4}$$

Now let us formulate the Mukai lemma.

Lemma (Mukai, [7]). *If the terms of the exact triple (4.2) satisfy the conditions (4.3) and (4.4), then the following inequality holds*

$$\dim \mathrm{Ext}^1(F_i, F_i) \geq \dim \mathrm{Ext}^1(F_{i+1}, F_{i+1}) + \dim \mathrm{Ext}^1(G_i, G_i).$$

From this lemma it follows that F_{i+1} and G_i are rigid, as well as F_i. Since $F_1 = F$ is rigid then F_i and G_i are rigid for any i.

From γ_H-stability of quotients G_i and $\gamma_H(G_i) = \gamma_H(G_j)$ for any i and j it follows, that $\mathrm{Ext}^2(G_i, G_j) = 0$ for any i and j. Besides, by construction of filtration

$$\mathrm{Hom}(G_i, G_{i+1}) = \mathrm{Hom}(G_{i+1}, G_i) = 0 \quad for \quad i = 1, 2, 3, \ldots, n-1,$$

(proposition 2.2). Therefore the E_1-part of spectral sequence, associated with the filtration, has the form

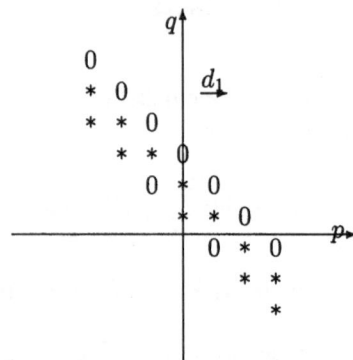

It is easy to see that

$$E_1^{1-n,n} = E_\infty^{1-n,n}.$$

But $E_\infty^{1-n,n} \subset E_\infty^1 = \mathrm{Ext}^1(F, F) = 0$. Therefore

$$\mathrm{Ext}^1(G_n, G_1) = E_1^{1-n,n} = 0.$$

The step 1 is proved.

Step 2. For any $i = 1, 2, \ldots, n$ there is an exceptional bundle E_i such that

$$G_i = E_i \oplus E_i \oplus \ldots \oplus E_i \oplus E_i.$$

Proof of the second step. By virtue of proposition 2.2 the quotients G_i have the filtration

$$G_i = \mathrm{Gr}(E_i, \ldots, E_i),$$

where E_i is γ_H-stable.

By proposition 2.2 the E_1-part of spectral sequence, associated with the filtration, has the form

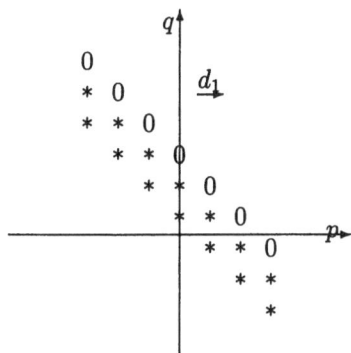

This means that

$$E_1^{1-n,n} = E_\infty^{1-n,n} \subset \operatorname{Ext}^1(G_i, G_i).$$

By step 1

$$\operatorname{Ext}^1(G_i, G_i) = 0.$$

But $E_1^{1-n,n}$ is equal to the space $\operatorname{Ext}^1(E_1, E_1)$, hence

$$\operatorname{Ext}^1(E_1, E_1) = 0.$$

Now the proof of the second step follows from

Lemma 4.1. *If a sheaf M has a filtration*

$$M = M_1 \supset M_2 \supset M_3 \supset \dots \supset M_{n-1} \supset M_n = 0,$$

the quotients of which $Q_i = M_i/M_{i+1}$ satisfy the condition

$$\operatorname{Ext}^1(Q_i, Q_j) = 0 \quad for \quad i < j,$$

then

$$M = Q_1 \oplus Q_2 \oplus \dots \oplus Q_n.$$

Proof of lemma. Consider the exact triple

$$0 \longrightarrow Q_n \longrightarrow M_{n-1} \longrightarrow Q_{n-1} \longrightarrow 0$$

which arises from the filtration. By condition $\operatorname{Ext}^1(Q_{n-1}, Q_n) = 0$. Hence this sequence splits and $M_{n-1} = Q_{n-1} \oplus Q_n$.

Assume that the equality $M = \bigoplus_{i=2}^{n} Q_i$ is proved and consider the short exact sequence

$$0 \longrightarrow M_2 \longrightarrow M \longrightarrow Q_1 \longrightarrow 0.$$

By the conditions of lemma and induction hypothesis

$$\operatorname{Ext}^1(Q_1, M_2) = \bigoplus_{i=2}^{n} \operatorname{Ext}^1(Q_1, Q_i) = 0.$$

It means that $M = M_2 \oplus Q_1 = \bigoplus_{i=1}^{n} Q_i$. The lemma is proved.

Step 3. The space $\text{Ext}^1(G_1, G_n) = 0.$

Proof. Using the first and the second steps we get
$$0 = \text{Ext}^1(G_n, G_1) = \text{Ext}^1(E_n, E_1) \oplus \ldots \oplus \text{Ext}^1(E_n, E_1) = 0.$$
Therefore $\text{Ext}^1(E_n, E_1) = 0.$

On the other hand,
$$\text{Ext}^1(G_1, G_n) = \text{Ext}^1(E_1, E_n) \oplus \ldots \oplus \text{Ext}^1(E_1, E_n).$$
Let us prove that $\text{Ext}^1(E_1, E_n) = 0.$ Recall, that the sheaves E_1 and E_2 are γ_H-stable and have the same Gieseker slopes as F. It follows from proposition 1.5.6 that either $E_1 \cong E_n$ or $\text{Hom}(E_1, E_n) = \text{Hom}(E_n, E_1) = 0.$

In the first case $\text{Ext}^1(E_n, E_1) = \text{Ext}^1(E_1, E_n) = 0.$ In the second
$$\chi(E_n, E_1) = - \dim \text{Ext}^1(E_n, E_1),$$
$$\chi(E_1, E_n) = - \dim \text{Ext}^1(E_1, E_n),$$
(*)

$\left(\text{Hom}(E_1, E_2) = \text{Hom}(E_2, E_1) = 0 \text{ by assumption, } \text{Ext}^2(E_n, E_1) = \text{Ext}^2(E_1, E_n) = 0\right.$ by proposition 1.5). From remark 1.1. and equality $\gamma_H(E_1) = \gamma_H(E_n)$ it follows that $\mu_H(E_1) = \mu_H(E_n)$. Hence $\chi_-(E_1, E_n) = 0$ (see (1.1)), i.e.
$$\chi(E_1, E_n) = \chi(E_n, E_1).$$

From the last equality and (*) it follows that
$$\dim \text{Ext}^1(E_n, E_1) = \dim \text{Ext}^1(E_1, E_n).$$

Step of induction. Suppose that all quotients G_i of filtration (2.2) of rigid bundle F decompose into the direct sum of exceptional bundles and $\text{Ext}^1(G_1, G_n) = 0.$

Consider the exact triple
$$0 \longrightarrow F_2 \longrightarrow F \longrightarrow G_1 \longrightarrow 0,$$
which arises from the filtration of F. By the first step the bundle F_2 is rigid. By induction hypothesis the sheaf F_2 may be decomposed into direct sum of exceptional bundles. It is easy to see that these bundles are the direct summands of G_i for $i = 2, 3, \ldots, n$. In particular,
$$F_2 = G_n \oplus G_{n-1} \oplus \ldots \oplus G_2.$$

Let us show that the sheaf G_n is a direct summand of F. For this purpose consider the extension
$$0 \longrightarrow G_{n-1} \oplus \ldots \oplus G_2 \longrightarrow \mathcal{F} \longrightarrow G_1 \longrightarrow 0,$$
which is determined by the same element of
$$\text{Ext}^1(G_1, G_{n-1} \oplus \ldots \oplus G_2) = \text{Ext}^1(G_1, G_n \oplus G_{n-1} \oplus \ldots \oplus G_2)$$
as the sequence (4.2). There is a commutative diagram with exact rows and columns:

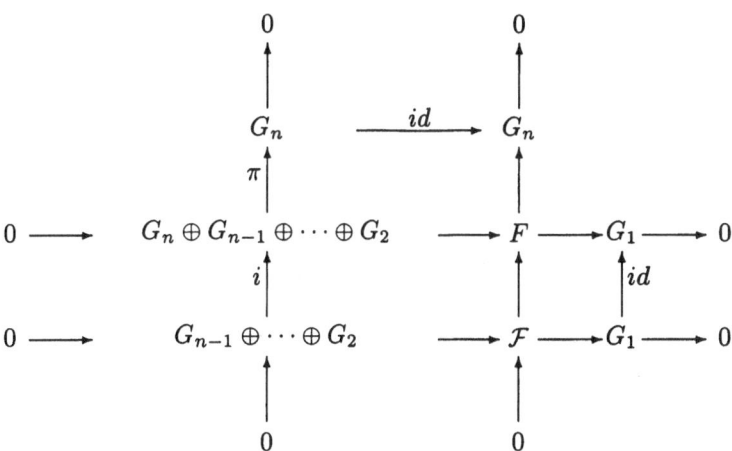

where i is the standard inclusion and π is the standard projection. Therefore $F_n = G \oplus \mathcal{F}$.

It is easy to see that $\mathcal{F}$ is also a rigid sheaf. Its γ_H-semi-stability follows from proposition 1.5.8. By the induction hypothesis and step 2 we get the statement of theorem.

In the following we need

Lemma 4.2. *Let E_1 and E_2 be exceptional bundles on a Del Pezzo surface. If $\gamma_H(E_2) > \gamma_H(E_1)$ and $\mathrm{Ext}^1(E_2, E_1) = 0$ then $\mathrm{Ext}^1(E_1, E_2) = 0$.*

Proof. By remark 1.1 the inequality from lemma 4.2 gives that

$$either$$

$$or$$

$$\mu_H(E_2) > \mu(E_1),$$

$$\mu_H(E_2) = \mu(E_1).$$

(*)

First, let us restrict the bundle $E_2^* \otimes E_1$ to an elliptic curve $C \in |-K_X|$:

$$0 \longrightarrow E_2^* \otimes E_1 \otimes K_X \longrightarrow E_2^* \otimes E_1 \longrightarrow (E_2^* \otimes E_1)\big|_C \longrightarrow 0.$$

Then the corresponding exact cohomology sequence is

$$H^0(E_2^* \otimes E_1) \longrightarrow H^0\big((E_2^* \otimes E_1)\big|_C\big) \longrightarrow H^1(E_2^* \otimes E_1 \otimes K_X) \longrightarrow H^1(E_2^* \otimes E_1) \ldots$$

(4.5)

The sheaves E_1 and E_2 are locally free. Hence

$$H^0\big((E_2^* \otimes E_1)\big|_C\big) = \mathrm{Hom}\big(E_2\big|_C, E_1\big|_C\big);$$

$$H^1(E_2^* \otimes E_1 \otimes K_X) = \mathrm{Ext}^1\big(E_2, E_1(K_X)\big);$$

$$H^1(E_2^* \otimes E_1) = \mathrm{Ext}^1(E_2, E_1).$$

It follows from proposition 3.3.2 that the exceptional bundles E_1 and E_2 are μ_H-stable and $E_i \big|_C$ is simple for $i = 1, 2$. The degree of the bundle $E_i \big|_C$ is equal to $(c_1(E_i) \cdot C)$ and $r(E_i \big|_C) = r(E_i)$. Therefore

$$\mu(E_i \big|_C) = \mu_H(E_i).$$

In particular,

$$\mu(E_2 \big|_C) > \mu(E_1 \big|_C). \tag{**}$$

It follows from μ-stability of simple bundle on elliptic curve and the inequalities (*) and (**), that the spaces

$$\operatorname{Hom}(E_2, E_1) \quad and \quad \operatorname{Hom}(E_2 \big|_C, E_1 \big|_C)$$

are trivial.

Moreover by condition of lemma, $\operatorname{Ext}^1(E_2, E_1) = 0$. Hence from the exact sequence (4.5) it follows that $\operatorname{Ext}^1(E_2, E_1(K_X)) = 0$. Applying the Serre duality theorem we get $\operatorname{Ext}^1(E_1, E_2) = 0$.

Next, assume that

$$\mu_H(E_2) = \mu_H(E_1). \tag{***}$$

From here it follows that either $E_1 \cong E_2$ (then the lemma is true) or

$$\operatorname{Hom}(E_1, E_2) = \operatorname{Hom}(E_2, E_1) = 0.$$

Suppose that the last equalities are true. On the other hand,

$$\operatorname{Ext}^2(E_1, E_2) = \operatorname{Ext}^2(E_2, E_1) = 0,$$

i.e.

$$\chi(E_2, E_1) = -\dim \operatorname{Ext}^1(E_2, E_1),$$
$$\chi(E_1, E_2) = -\dim \operatorname{Ext}^1(E_1, E_2).$$

Then from (***) we get $\chi(E_2, E_1) = \chi(E_1, E_2)$. The lemma is proved.

Theorem 2. *Any rigid bundle F on Del Pezzo surface X can be decomposed into direct sum of exceptional bundles.*

Proof. Let H be the anticanonical class of X. Consider the filtration $F = \operatorname{Gr}(G_n, \ldots, G_1)$ from proposition 2.1. By virtue of proposition 1.5 the quotients G_i satisfy the conditions

$$\operatorname{Hom}(G_i, G_j) = 0 \quad for \quad i > j$$

and

$$\operatorname{Ext}^2(G_i, G_j) = 0 \quad for \quad i \le j.$$

Hence the E_1-part of spectral sequence, associated with the filtration of F, has the form:

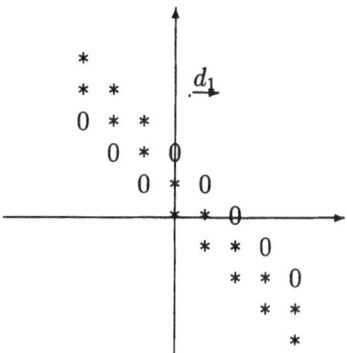

From here it follows that $E_\infty^{-1,2} = E_1^{-1,2} = \bigoplus_i \mathrm{Ext}^1(G_i, G_{i-1})$ and

$$E_\infty^{0,1} = E_1^{0,1} = \bigoplus_i \mathrm{Ext}^1(G_i, G_i).$$

But

$$E_\infty^{0,1} \oplus E_\infty^{-1,2} \subset E_\infty^1 = \mathrm{Ext}^1(F, F) = 0.$$

Therefore for any i the equalities

$$\mathrm{Ext}^1(G_i, G_{i-1}) = \mathrm{Ext}^1(G_i, G_i) = 0 \qquad (*)$$

hold true. This means that the γ_H-semi-stable quotients G_i of the filtration are rigid.

It follows from theorem 1 that $G_i = \bigoplus_s E_i^s$, where E_i^s are exceptional bundles. From the first part of (*) we obtain

$$0 = \mathrm{Ext}^1(G_i, G_{i-1}) = \bigoplus_{s,k} \mathrm{Ext}^1(E_i^s, E_{i-1}^k).$$

Therefore $\mathrm{Ext}^1(E_i^s, E_{i-1}^k) = 0$. But $\gamma_H(E_i^s) = \gamma_H(G_i) > \gamma_H(G_{i-1}) = \gamma_H(E_{i-1}^k)$ and we can apply lemma 4.2. By this lemma $\mathrm{Ext}^1(E_{i-1}^k, E_i^s) = 0$. It means that

$$\mathrm{Ext}^1(G_{i-1}, G_i) = \bigoplus_{s,k} \mathrm{Ext}^1(E_{i-1}^k, E_i^s) = 0. \qquad (4.6)$$

The bundles F and G_1 are included into the short exact sequence

$$0 \longrightarrow F_2 \longrightarrow F \longrightarrow G_1 \longrightarrow 0.$$

It follows from proposition 2.1 that

$$\mathrm{Hom}(F_2, G_1) = \mathrm{Ext}^2(G_1, F_2) = 0.$$

Applying the Mukai lemma to this sequence we get

$$\dim \operatorname{Ext}^1(F, F) \geq \dim \operatorname{Ext}^1(F_2, F_2) + \dim \operatorname{Ext}^1(G_1, G_1).$$

It means that F_2 is rigid.

By induction hypothesis F_2 is a direct sum of exceptional bundles. It is easy to prove that $F_2 = G_n \oplus G_{n-1} \oplus \ldots \oplus G_2$. Since $\operatorname{Ext}^1(G_1, G_2) = 0$ one can prove that $F = F' \oplus G_2$, where F' is rigid. Now the theorem 2 follows from induction hypothesis.

5. Rigid bundles on the general $K3$-surface

In this section we denote a $K3$-surface with $\operatorname{Pic}(S) \cong \mathbb{Z}$ by S and generator of $\operatorname{Pic}(S)$ by H. Recall that a $\mathbb{Z}$-module

$$H^0(S, \mathbb{Z}) \oplus H^2(S, \mathbb{Z}) \oplus H^4(S, \mathbb{Z})$$

with scalar product

$$(r, a, s)(r', a', s') = aa' - rs' - r's$$

is called the Mukai lattice of S. For any sheaf F on the $K3$-surface S its Mukai vector can be constructed in the following way

$$v(F) = \big(r(F), c_1(F), s(F)\big),$$

where

$$s(F) = r(F) - c_2(F) + \frac{c_1^2(F)}{2}.$$

Moreover the following formula is true:

$$\chi(F, F') = -v(F)v(F').$$

It is easy to prove the following lemma:

Lemma 5.1. *Let F, E be sheaves on S. Then*

1. If $\operatorname{Ext}^1(E, E) = 0$, then $v^2(E) < 0$.

2. If E is simple (i.e. $\operatorname{Hom}(E, E) \cong \mathbb{C}$) and $v^2(E) < 0$ then E is exceptional and $v^2(E) = -2$.

3. If $\gamma_H(E) = \gamma_H(F)$ then there is a rational number α such that $v(E) = \alpha v(F)$.

Theorem 3. *If F is γ_H-semi-stable rigid sheaf on the general $K3$-surface, then there exist exceptional bundles $E_1, \ldots, E_n$ such that*

$$F = E_1 \oplus \ldots \oplus E_1 \oplus \ldots \oplus E_n \oplus \ldots \oplus E_n.$$

Proof. Consider the filtrations (2.2) and (2.3) of F

$$F = \operatorname{Gr}(E_n, \ldots, E_n, E_{n-1}, \ldots, E_{n-1}, \ldots, E_1, \ldots, E_1).$$

The sheaves E_i are γ_H-stable and $\gamma_H(E_i) = \gamma_H(F)$ for each i. From here and lemma 5.1 we can see that there are rational numbers α_i such that $v(E_i) = \alpha_i v(F)$.

It follows from lemma 5.1 that the scalar square of the Mukai vector of rigid sheaf F is negative. Hence $v^2(E_i) < 0$ for each i.

From γ_H-stability of E_i we have $\operatorname{Hom}(E_i, E_i) \cong \mathbb{C}$. Therefore E_i are exceptional bundles. Since $\gamma_H(E_i) = \gamma_H(F)$, then $\mu_H(E_i) = \mu_H(F)$ for any i. But exceptional bundle E is determined by its slope $\mu(E)$. This means that

$$E_1 \cong E_2 \cong \ldots \cong E_n \cong E.$$

We have proved that $F = \operatorname{Gr}(E, \ldots, E)$, where E is an exceptional bundle. Now the theorem easily follows from the condition $\operatorname{Ext}^1(E, E) = 0$.

Next let us construct the rigid bundle E on the general $K3$-surface S, which does not decompose into direct sum of exceptional bundles.

We denote the line bundle corresponding to the generator of $\operatorname{Pic}(S)$ by $\mathcal{O}(1)$ and structure sheaf of S by $\mathcal{O}$. The bundles $\mathcal{O}$ and $\mathcal{O}(1)$ are exceptional.

Consider the ordered pair of exceptional bundles $(\mathcal{O}, \mathcal{O}(1))$. It is known that the canonical map $H^0(\mathcal{O}(1)) \otimes \mathcal{O} \longrightarrow \mathcal{O}(1)$ is epimorphism and

$$\operatorname{Ext}^1(\mathcal{O}, \mathcal{O}(1)) = \operatorname{Ext}^2(\mathcal{O}, \mathcal{O}(1)) = \operatorname{Ext}^1(\mathcal{O}(1), \mathcal{O}) = \operatorname{Hom}(\mathcal{O}(1), \mathcal{O}) = 0.$$

An ordered pair of exceptional bundles (A, B) satisfying the same conditions is called exceptional. The following proposition is proved in [10], p. 106:

Proposition. *If (A, B) is an exceptional pair on S then the kernel of the canonical map $\operatorname{Hom}(A, B) \otimes A \longrightarrow B$ is an exceptional bundle.*

Let

$$0 \longrightarrow L \longrightarrow H^0(\mathcal{O}(1)) \otimes \mathcal{O} \longrightarrow \mathcal{O}(1) \longrightarrow 0 \qquad (5.1)$$

be an exact triple. Then L is an exceptional bundle.

Let us show that $\operatorname{Ext}^1(\mathcal{O}(1), L) \cong \mathbb{C}$. From the long exact sequence

$$H^0(\mathcal{O}(1))^* \otimes \operatorname{Hom}(\mathcal{O}(1), \mathcal{O}) \longrightarrow \operatorname{Hom}(\mathcal{O}(1), \mathcal{O}(1)) \longrightarrow \operatorname{Ext}^1(\mathcal{O}(1), L) \longrightarrow$$
$$\longrightarrow H^0(\mathcal{O}(1))^* \otimes \operatorname{Ext}^1(\mathcal{O}(1), \mathcal{O}) \longrightarrow \ldots$$

and

$$\operatorname{Hom}(\mathcal{O}(1), \mathcal{O}) = \operatorname{Ext}^1(\mathcal{O}(1), \mathcal{O}) = 0$$

it follows that

$$\operatorname{Hom}(\mathcal{O}(1), \mathcal{O}(1)) \cong \operatorname{Ext}^1(\mathcal{O}(1), L) \cong \mathbb{C}.$$

By the Serre duality theorem the space $\operatorname{Ext}^1(L, \mathcal{O}(1))$ is 1-dimensional.

Construct the universal extension

$$0 \longrightarrow \mathcal{O}(1) \longrightarrow E \longrightarrow \mathbb{C} \otimes L \longrightarrow 0. \qquad (5.2)$$

The map δ in the long exact sequence, associated with (5.2),

$$0 \longrightarrow \operatorname{Hom}(L, \mathcal{O}(1)) \longrightarrow \operatorname{Hom}(L, E) \longrightarrow \mathbb{C} \otimes \operatorname{Hom}(L, L) \overset{\delta}{\longrightarrow} \operatorname{Ext}^1(L, \mathcal{O}(1)) \longrightarrow$$
$$\longrightarrow \operatorname{Ext}^1(L, E) \longrightarrow \operatorname{Ext}^1(L, \mathcal{O}(1)) \otimes \operatorname{Ext}^1(L, L) \longrightarrow \ldots$$

is isomorphism. From here it follows that

$$\mathrm{Ext}^1(L, E) = 0. \tag{5.3}$$

Let us apply another Mukai lemma, which states that the inequality

$$\dim \mathrm{Ext}^1(L, L) \geq \dim \mathrm{Ext}^1(E, E) + \dim \mathrm{Ext}^1\big(\mathcal{O}(1), \mathcal{O}(1)\big)$$

holds true if the sheaves of exact sequence (5.2) satisfy the condition (5.3) ([7]). By this lemma E is rigid.

But the exact sequence (5.2) does not split. Hence E does not decompose into the direct sum of exceptional bundles.

References

[1]   M.F. ATIYAH. *Vector Bundles over an Elliptic Curve*, Proc. Lond. Math. Soc., VII (1957), 414-452.

[2]   J.-M. DREZET, J. LE POTIER. *Fibrés Stables et Fibrés Exceptionelles sur* $\mathbb{P}_2$, An. Ecol. Norm. Sup., 18 (1985), 193-244.

[3]   A.L. GORODENTSEV. *Exceptional Bundles on Surfaces with a Moveable Anticanonical Class*, Math. USSR Izv, 33 (1989), 67-83 (Russian).

[4]   A.L. GORODENTSEV, A.N. RUDAKOV. *Exceptional Vector Bundles on Projective Surface*, Duke Math. J., 54 (1987), 115-130.

[5]   PH. GRIFFITHS, J. HARRIS. *Principles of algebraic geometry*, A Wiley-interscience series of texts, New York 1978.

[6]   S.A. KULESHOV. *The Existence Theorem of Exceptional Bundles on* $K3$-*surfaces*, Math. USSR Isv, 54 (1989), 363-378 (Russian).

[7]   S. Mukai. *On the Module Spaces of Bundles on* $K3$-*surfaces. I*, in Vector Bundles ed. Atiyah et al, Oxford Univ. Press, Bombey, (1986), 341-413.

[8]   D.Yu. NOGIN, S.K. ZUBE. *Computing Invariants for Exceptional Bundles on a Quadric*, Lond. Math. Soc., Lect. Note Ser., v.148 (1990) Cambridge Univ. Press, 23-32.

[9]   Ch. OKONEK, M. SCHNEIDER, H. SPINDLER. *Vector Bundles on Complex Projective Spaces*, Birkäuser (1980).

[10]   A.N. RUDAKOV et al. in *"Helices and Vector Bundles: Seminaire Rudakov"*, Lond. Math. Soc., Lect. Note Ser., v.148 (1990) Cambridge Univ. Press.

[11]   A.N. RUDAKOV. *Exceptional Bundles on a Quadric*, Math. USSR Isv., 33 (1989), 115-138 (Russian).

The Alexander Polynomials of Algebraic Curves in C^2

Vic. S.Kulikov

Let D be a plane algebraic curve in $\mathbb{C}^2$ and $f(x,y) = 0$ be an equation of D, $d = \deg D$. We shall assume for simplicity that the polynomial $f(x,y)$ does not contain multiple factors. The purpose of this paper is to describe some properties of the fundamental group $\pi_1(\mathbb{C}^2 \setminus D)$.

1. The polynomial $f(x,y)$ defines a morphism $F : \mathbb{C}^2 \longrightarrow \mathbb{C}^1$ by equation $z = f(x,y)$. The curve D is the fibre of F over $0 \in \mathbb{C}^1$, and F defines a structure of fibering on $X = \mathbb{C}^2 \setminus D$:
$$F : X = \mathbb{C}^2 \setminus D \longrightarrow \mathbb{C}^* = \mathbb{C}^1 \setminus \{0\}.$$
We shall assume that D satisfies the condition

*) a generic fibre $Y_z = F^{-1}(z)$ is connected.
If D is connected in $\mathbb{C}^2$ then D satisfies the condition *). In particular, if D is an irreducible curve then D satisfies the condition *).

The morphism F defines a rational map $F : \mathbb{P}^2 \cdots\cdots\rightarrow \mathbb{P}^1$. Let $\sigma : \bar{\mathbb{P}}^2 \longrightarrow \mathbb{P}^2$ be a sequence of monoidal transformations such that in commutative diagram

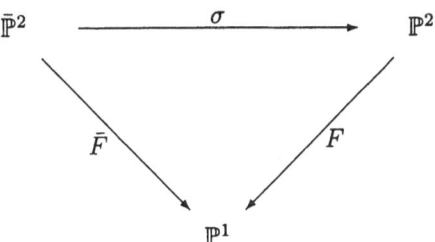

$\bar{F}$ is a morphism. The morphism $\bar{F}$ has several singular fibres. Let
$$\mathrm{Sing}\bar{F} = \{z_1, \ldots, z_n\} \subset \mathbb{C}^* \subset \mathbb{P}^1$$

be the set of points over which the morphism $\bar{F}$ is not smooth and such that

$$F : X \setminus F^{-1}(\operatorname{Sing}\bar{F}) \longrightarrow \mathbb{C}^* \setminus \operatorname{Sing}\bar{F}$$

is a locally C^∞-trivial fibering.

For each $z_i \in \operatorname{Sing}\bar{F}$ choose a small disk $B_i \subset \mathbb{C}^*$ and choose two points $z_{i,1}$ and $z_{i,2}$ belonging to the boundary ∂B_i of the disk B_i. Points $z_{i,1}$ and $z_{i,2}$ divide ∂B_i into two arcs $\gamma_{i,1}$ and $\gamma_{i,2}$. Choose nonintersecting paths γ_i connecting the points $z_{i,1}$ and $z_{i+1,2}(z_{n+1,\cdot} = z_{1,\cdot})$, and let $\gamma_{i,1}$ be such part of ∂B_i that $l_{\text{in}} = (\cup\gamma_{i,1}) \cup (\cup\gamma_i)$ is a boundary of restricted set V containing the origin $0 \in \mathbb{C}^1$, and V does not contain $\operatorname{Sing}\bar{F}$. Let l_{ex} be the boundary of the set $V \cup (\cup B_i)$.

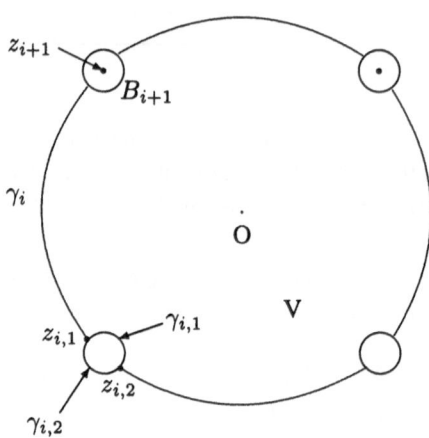

Fig. 1

Let $T = (\cup B_i) \cup (\cup\gamma_i)$ and $Z = F^{-1}(T)$. The set Z will be called a *necklace* of the curve D. Since T is a retract of $\mathbb{C}^*$ and over $\mathbb{C}^* \setminus T$ the fibering $F : X \setminus Z \longrightarrow \mathbb{C}^* \setminus T$ is a locally C^∞-trivial fibering, we have the following

Proposition 1. $X = \mathbb{C}^2 \setminus D$ and necklace Z of the curve D are homotopic.

Thus $\pi_1(\mathbb{C}^2 \setminus D) \simeq \pi_1(Z)$ *and what is more we have the following commutative diagram*

$$
\begin{array}{ccc}
\pi_1(\mathbb{C}^2 \setminus D) & \simeq & \pi_1(Z) \\
\Big\downarrow F_* & & \Big\downarrow F_* \\
\pi_1(\mathbb{C}^*) & \simeq & \pi_1(T) \quad \simeq \quad \mathbb{F}_1
\end{array}
$$

where $\mathbb{F}_1$ is a free group, $\mathrm{rg}\mathbb{F}_1 = 1$.

Note that

$$F_* : \pi_1(\mathbb{C}^2 \setminus D) \longrightarrow \pi_1(\mathbb{C}^*) \tag{1}$$

is an epimorphism if D satisfies the condition *).

Choose a point $z_0 \in \gamma_1$, let $Y = F^{-1}(z_0)$, and choose a point $y_0 \in Y$. We have a homomorphism $\psi : \pi_1(Y, y_0) \longrightarrow \pi_1(Z, y_0)$ defined by the injection $Y \subset Z$.

Let N be the kernel of $F_* : \pi_1(Z, y_0) \longrightarrow \pi_1(T, z_0)$. Then $\mathrm{Im}\psi \subset N$, because Y is a fibre of F.

Theorem 1. *If D satisfies the condition *), then*

$$\mathrm{Im}\psi = N.$$

This theorem can be proved using the following

Lemma. *If D satisfies the condition *), then*

$$\psi_i : \pi_1(Y_i) \longrightarrow \pi_1(F^{-1}(B_i))$$

are epimorphisms, where $Y_i = F^{-1}(z_{i,1})$ and ψ_i is defined by injections $Y_i \subset F^{-1}(B_i)$.

As a consequence of theorem 1 we obtain

Theorem 2. *If D is an irreducible curve, then the commutator subgroup $G' = [G, G]$ is a finitely generated group, where $G = \pi_1(\mathbb{C}^2 \setminus D)$.*

Indeed, in the case of irreducible D it is easy to show that $\mathrm{Ker}F_*$ coincides with the commutator subgroup of $\pi_1(\mathbb{C}^2 \setminus D)$.

2. Let $Z_{\mathrm{in}} = F^{-1}(l_{\mathrm{in}})$ and $Z_{\mathrm{ex}} = F^{-1}(l_{\mathrm{ex}})$. The diagram

$$
\begin{array}{ccc}
Z_{\mathrm{in}} & \xrightarrow{\ F_*\ } & l_{\mathrm{in}} \\
\cap & & \cap \\
Z & \xrightarrow{\ F_*\ } & T \\
\cup & & \cup \\
Z_{\mathrm{ex}} & \xrightarrow{\ F_*\ } & l_{\mathrm{ex}}
\end{array}
$$

gives us the following commutative diagram

$$
\begin{array}{ccccccccc}
1 & \longrightarrow & \pi_1(Y, y_0) & \longrightarrow & \pi_1(Z_{\mathrm{in}}, y_0) & \overset{F_*}{\longrightarrow} & \pi_1(l_{\mathrm{in}}, z_0) & \longrightarrow & 1 \\
 & & \downarrow \psi & & \downarrow \alpha_{\mathrm{in}} & & \downarrow \beta_{\mathrm{in}} \,\cong & & \\
1 & \longrightarrow & N & \longrightarrow & \pi_1(Z, y_0) & \longrightarrow & \pi_1(T, z_0) & \longrightarrow & 1 \\
 & & \uparrow \psi & & \uparrow \alpha_{\mathrm{ex}} & & \uparrow \beta_{\mathrm{ex}} \,\cong & & \\
1 & \longrightarrow & \pi_1(Y, y_0) & \longrightarrow & \pi_1(Z_{\mathrm{ex}}, y_0) & \longrightarrow & \pi_1(l_{\mathrm{ex}}, y_0) & \longrightarrow & 1
\end{array}
$$

$$(2)$$

The rows in this diagram are exact. We shall identify the groups $\pi_1(l_{\mathrm{in}})$, $\pi_1(T)$ and $\pi_1(l_{\mathrm{ex}})$. Let t be a generator of free group $\pi_1(T) \simeq \mathbb{F}_1$.

Choose elements $t_{\mathrm{in}} \in \pi_1(Z_{\mathrm{in}})$, $t_1 \in \pi_1(Z)$ and $t_{\mathrm{ex}} \in \pi_1(Z_{\mathrm{ex}})$ such that $\alpha_{\mathrm{in}}(t_{\mathrm{in}}) = \alpha_{\mathrm{ex}}(t_{\mathrm{ex}}) = t_1$ and such that $F_*(t_1) = t$. Then these elements define automorphisms $\tau_{\mathrm{in}} \in \mathrm{Aut}\pi_1(Y)$, $\tau_{\mathrm{ex}} \in \mathrm{Aut}\pi_1(Y)$, $\tau \in \mathrm{Aut}N$ by the rule $\tau(g) = t_1^{-1} g t$ for each $g \in N$, and $\tau_{\mathrm{in}}(g) = t_{\mathrm{in}}^{-1} g t_{\mathrm{in}}$, $\tau_{\mathrm{ex}}(g) = t_{\mathrm{ex}}^{-1} g t_{\mathrm{ex}}$ for each $g \in \pi_1(Y)$. It follows from diagram (2) that

$$
\psi \tau_{\mathrm{ex}}(g) = \tau(\psi(g)),
$$
$$(3)$$
$$
\psi \tau_{\mathrm{in}}(g) = \tau(\psi(g)).
$$

The automorphisms τ_{in}, τ_{ex}, τ induce the automorphisms h_{in}, $h_{\mathrm{ex}} \in \mathrm{Aut}H_1(Y, \mathbb{Z})$ and $\tilde{h} \in \mathrm{Aut}N/N'$, where $N' = [N, N]$. From the property (3) it follows the similar property for automorphisms h_{ex}, h_{in} and $\tilde{h}$.

It is easy to see that h_{in} is a monodromy acting on $H_1(Y)$ defined by circuit around the fibre D, and h_{ex} is a monodromy defined by circuit around the fibre $F^{-1}(\infty)$.

Let $N/N' = (N/N')_{\mathrm{Free}} \oplus (N/N')_{\mathrm{Tor}}$ be a decomposition into the free part and the torsion group. Automorphism $\tilde{h}$ defines the automorphism $h \in \mathrm{Aut}(N/N')_{\mathrm{Free}}$. Let $\Delta_D(t) = \det(h - t\mathrm{Id})$ be the characteristic polynomial of h, $\Delta_{\mathrm{in}}(t) = \det(h_{\mathrm{in}} - t\mathrm{Id})$, $\Delta_{\mathrm{ex}}(t) = \det(h_{\mathrm{ex}} - t\mathrm{Id})$ be the characteristic polynomials of h_{in} and h_{ex} respectively. The polynomial $\Delta_D(t)$ is called the Alexander polynomial of the curve D (cf. [4], [5], [2]). The polynomial $\Delta_{\mathrm{in}}(t)$ will be called the internal Alexander polynomial and $\Delta_{\mathrm{ex}}(t)$ will be called the external Alexander polynomial of the curve D.

From the property (3) it follows

Theorem 3. *The Alexander polynomial $\Delta_D(t)$ of a curve D satisfying the condition *) divides the polynomials $\Delta_{\mathrm{in}}(t)$ and $\Delta_{\mathrm{ex}}(t)$.*

It is well-known that the roots of the characteristic polynomial of monodromy are the roots of unit. Thus from theorem 3 we obtain

Theorem 4. *For a curve D satisfying the condition* *) *the roots of the Alexander polynomial* $\triangle_D(t)$ *are roots of unit.*

3. There exists another approach in defining Alexander polynomial of plane algebraic curve D. This definition coincides with the definition of Alexander polynomials of knots and links [1]. Remind this definition.

Let $I_n = \{1, 2, \ldots, n\}$ be a segment of natural numbers, $I_n^3 = \{\alpha = (\alpha_1, \alpha_2, \alpha_3) \mid \alpha_i \in I_n\}$, and let M be a subset of I_n^3. Let $m = |M|$ be the number of elements of M. A group G is called C-group of type $M \subset I_n^3$ if G admits a corepresentation

$$G = \langle x_1, \ldots, x_n \mid R_\alpha(x), \ \alpha \in M \subset I_n^3 \rangle, \tag{4}$$

where each relation $R_\alpha(x) = x_{\alpha_1} x_{\alpha_2} x_{\alpha_3}^{-1} x_{\alpha_3}^{-1}$ is a conjugation.

Examples of C-groups:

1) Free group $\mathbb{F}_n$,

2) Free abelian group Ab_n,

3) Groups of knots and links (Wirtinger corepresentation),

4) The fundamental group $\pi_1(\mathbb{C}^2 \setminus D)$ of the complement of plane algebraic curve D (a corepresentation from [3]).

To each subset $M \subset I_n^3$ we can associate an oriented graph Γ_M with the vertices $v_1, \ldots, v_n$ and with the edges e_α, $\alpha \in M$. The edge e_α connects the vertex v_{α_1} with v_{α_3}, where $\alpha = (\alpha_1, \alpha_2, \alpha_3)$.

It is easy to prove the following

Lemma. *Let G be a C-group of type M and* $G' = [G, G]$. *Then* $G/G' = \mathbb{Z}^k$, *where k is the number of connected components of the graph* Γ_M.

A C-group G of type M is called an irreducible C-group, if its graph Γ_M is connected.

The C-corepresentation (4) defines an epimorphism

$$F_* : G \longrightarrow \mathbb{F}_1, \tag{5}$$

defined by $F_*(x_i) = t$ for each generator x_i of the corepresentation (4), where t is a generator of free group $\mathbb{F}_1$. It is easy to see that if $G = \pi_1(\mathbb{C}^2 \setminus D)$ and C-corepresentation of G is the corepresentation from [3] then homomorphism (5) coincides with homomorphism (1).

The C-corepresentation of group G and homomorphism F_* defines a matrix

$$A = F_* \left(\frac{\partial R_\alpha(x)}{\partial x_i} \right)$$

with the elements in the ring of the Laurent polynomials with the integral coefficients $\mathbb{Z}[t, t^{-1}]$, where the differentiation $\frac{\partial R_\alpha(x)}{\partial x_i}$ uniquely defined by the properties

$$1) \ \frac{\partial x_j}{\partial x_i} = \delta_{ij},$$

$$2) \ \frac{\partial (fg)}{\partial x_i} = \frac{\partial f}{\partial x_i} + f \frac{\partial g}{\partial x_i}.$$

Let us consider the ideals

$$\ldots \subset E_k \subset E_{k+1} \subset \ldots$$

of the ring $\mathbb{Z}[t, t^{-1}]$ such that

$$E_k = (0), \qquad \text{if} \quad n - k > m,$$

$$E_k = \mathbb{Z}[t, t^{-1}], \qquad \text{if} \quad n - k < 0,$$

and E_k is generated by all $(n - k)$-minors of the matrix A. These ideals are called the elementary ideals. The ideals E_k do not depend on a corepresentation of group G and depend only on the morphism F_* [1].

The ring $\mathbb{Z}[t, t^{-1}]$ is the unique factorial domain. Thus for each E_k there exists uniquely determined minimal principal ideal $(\triangle_{k,G}(t))$ containing E_k. We can choose a generator $\triangle_{k,G}(t)$ such that

$$\triangle_{k,G}(t) = \sum_{i=0}^{s} a_i t^i \in \mathbb{Z}[t]$$

and $a_0 \neq 0$ if $\triangle_{k,G}(t) \neq 0$. The polynomial $\triangle_{k,G}(t)$ is called the k-th Alexander polynomial of C-group G.

It is possible to prove the following

Proposition 2. *If* $G = \pi_1(\mathbb{C}^2 \setminus D)$, *then* $\triangle_D(t)$, *defined earlier, coincides with* $\triangle_{1,G}(t)$.

Similar to theory of knots it is easy to prove the following

Theorem 5. *If* G *is an irreducible* C-group, then

$$\triangle_{1,G}(1) = \pm 1.$$

From this theorem we have:

Theorem 6. *Let* D *be an irreducible curve and let* t_0 *be a root of* $\triangle_D(t)$. *Then for each prime number* p *and integer* n

$$t_0^{p^n} \neq 1.$$

Corollary. *Let* D *be an irreducible curve. Then*

$$rk \left(G'/G'' \right)_{\text{Free}} = \deg \triangle_D(t) = 2q$$

is an even number, where $G = \pi_1(\mathbb{C}^2 \setminus D)$, $G' = [G, G]$ *and* $G'' = [G', G']$.

Let $\bar{D}$ be the closure of D in $\mathbb{P}^2$. It is well-known that if $\bar{D}$ intersects transversally with the line at infinity $L_\infty = \mathbb{P}^2 \setminus \mathbb{C}^2$, then

$$\triangle_{\text{ex}}(t) = (t^d - 1)^{d-2}(t - 1),$$

where $d = \deg D$.

From this and from theorems 3 and 6 it follows

Theorem 7. *Let $\bar{D}$ be an irreducible curve intersecting transversally with L_∞, $\deg D = p^n$, where p is a prime number. Then G'/G'' is a finite group.*

Note that if D is a curve of degree 6 with six cuspidal singular points lying on a conic, then G'/G'' is not a finite group [6].

4. Let $\mathcal{C}$ be the set of irreducible C-groups; $\mathcal{A}$ be the set of the fundamental groups $\pi_1(\mathbb{C}^2 \setminus D)$ for irreducible curves $D \subset \mathbb{C}^2$; $\mathcal{K}$ be the set of knot groups; $\mathcal{A}_{\text{loc}}$ be the set of groups of knots which can be obtained as the intersection of an algebraic curve in $\mathbb{C}^2$ with a small sphere $S^3 \subset \mathbb{C}^2$; $\mathcal{U}$ be the set of irreducible C-groups G for which the roots of Alexander polynomial $\triangle_{1,G}(t)$ are the roots of unit; $\mathcal{C}_{\text{rec}}$ be the set of irreducible C-groups G, for which the Alexander polynomial $\triangle_{1,G}(t)$ is reciprocal, i.e. the polynomial $\triangle_{1,G}(t)$ satisfies the property

$$\triangle_{1,G}(t) = (t)^{\deg \triangle_{1,G}} \triangle_{1,G}(t^{-1}).$$

It is well-known that we have two posibilities for $G' = [G, G]$ of knot group G : either G' is a free group or G' is not finitely generated. Thus from theorem 2 it follows that $\mathcal{K} \not\subset \mathcal{A}$.

It is easy to see that $\mathcal{A} \not\subset \mathcal{K}$. For example, the group $G = \pi_1(\mathbb{C}^2 \setminus D)$ does not belong to $\mathcal{K}$, where D is an irreducible curve with 3 cuspidal singular points, $\deg D = 4$. For this curve $G'/G'' \simeq \mathbb{Z}/3\mathbb{Z}$.

In the end of this note I would like to formulate the following questions:

Q_1. Is it true that $\mathcal{A} \cap \mathcal{K} = \mathcal{A}_{\text{loc}}$?

Q_2. Is it true that $\mathcal{A} = \mathcal{U}$?

Q_3. Is it true that $\mathcal{A} \cup \mathcal{K} = \mathcal{C}_{\text{rec}}$?

References

[1]   R.H. CROWELL, R.H. FOX. *Introduction to knot theory*, Boston, 1963.

[2]   T. KOHNO. *An algebraic computation of the Alexander polynomial of a plane algebraic curve*, Proc. Japan Acad. Ser. A59, 94-97 (1983).

[3]   VIC.S. KULIKOV. *On the fundamental group of the complement of a hypersurface in $\mathbb{C}^n$*, Lect. Notes in Math., 1479, 122-130 (1991), Springer-Verlag.

[4]   A. LIBGOBER. *Alexander polynomial of plane algebraic curves and cyclic multiple planes*, Duke Math. J. 49, 833-851 (1982).

[5]   R. RANDELL. *Milnor fibres and Alexander polynomials of plane curves*, Proc. Symp. Pure Math. 40, 415-419 (1983).

[6]   O. ZARISKI. *Algebraic surfaces*, 1971, Springer-Verlag.

On the Brauer Group of Real Algebraic Surfaces

Viacheslav V. Nikulin

Dedicated to Professor Igor R. Shafarevich on the occasion of his seventieth birthday

0 Formulation of basic results

In the paper of R. Sujatha and the author [N-S], the Brauer group of a real Enriques surface was studied. Here we continue the study of Brauer group with the remark that most of the results of these paper generally valid for an arbitrary smooth projective real algebraic surface.

Let X be a projective algebraic variety over the field $\mathbb{R}$ of real numbers. Let

$$Br'(X) = H^2_{et}(X; G_m)$$

denote the cohomological Brauer group of X. See a definition in the book of J. Milne [Mi], for example. We mention that the cohomological Brauer group is very closely related with the more interesting classical Brauer group $Br(X)$ classifying Azumaya algebras over X (see the papers of A. Grothendieck [Gr2] and the book [Mi]). For example, it is known that $Br(X) \subset Br'(X)$. For curves and smooth surfaces it gives an isomorphism. But we will only consider here the cohomological Brauer group $Br'(X)$.

Let $X(\mathbb{R})$ denote the space of $\mathbb{R}$-rational points of X with the Euclidean topology and s denote the number of real connected components of this space. Let $_2Br'(X)$ denote the group of elements of order two in $Br'(X)$. If $P \in X(\mathbb{R})$ is a real point of X, we get a natural map $_2Br'(X) \to {_2}Br'(P) \cong \mathbb{Z}/2$. It is shown in [CT-P] that this map does not depend from a choice of the point P in a connected component of $X(\mathbb{R})$. Thus, the canonical map

$$_2Br'(X) \to (\mathbb{Z}/2)^s \qquad (0-1)$$

is defined.

We mention that a studying of the map (0–1) and a description of the Brauer group of X is very important for calculation of a such interesting group connected with X as the Witt group $W(X)$. See R. Sujatha [Su].

It is shown in the paper of J.-L. Colliot-Thélène and R. Parimala [CT-P] that the map (0–1) is epimorphic if X is a smooth projective algebraic surface and $H^3(X(\mathbb{C}); \mathbb{F}_2) = 0$ (here $\mathbb{F}_2 = \mathbb{Z}/2$). This is a generalization of the old result of E. Witt [W] about curves (compare with Remark 1.8 below). There doesn't seem to be any known example of a surface where the map (0–1) fails to be epimorphic.

This paper is devoted to studying of this map (0–1) and also calculating of dim $_2Br'(X)$. Our idea is to interpret $_2Br'(X)$ and the map (0–1) purely topologically, and apply to them topological considerations.

We prove here the following basic result where G is the group of order two generated by the antiholomorphic involution g of $X(\mathbb{C})$ defined by the structure of real algebraic variety on X.

Theorem 0.1. *Let $X/\mathbb{R}$ be an algebraic projective manifold (smooth) over the field $\mathbb{R}$ of real numbers.*

Then, the homomorphism (0–1) is epimorphic if $H^3(X(\mathbb{C})/G; \mathbb{F}_2) = 0$. More generally, the homomorphism (0–1) is epimorphic if the kernel of the homomorphism

$$i^* : H^3(X(\mathbb{C})/G; \mathbb{F}_2) \to H^3(X(\mathbb{R}); \mathbb{F}_2)$$

is equal to zero. Here $i : X(\mathbb{R}) \subset X(\mathbb{C})/G$ denote the embedding.

Proof. See Theorem 1.6 below. In fact, in Theorem 1.6, we give a precise topological obstruction to epimorphicity of the map (0–1). This obstruction is zero if the kernel of the homomorphism i^* above is zero.

We mention that for smooth curves X the group $H^3(X(\mathbb{C})/G; \mathbb{F}_2) = 0$, thus the map (0–1) is epimorphic (it is well-known, compare with [W]). For surfaces X the group $H^3(X(\mathbb{R}); \mathbb{F}_2) = 0$, and $\mathrm{Ker}\,i^* = H^3(X(\mathbb{C})/G; \mathbb{F}_2)$.

Now, let us show that, from Theorem 0.1, the result of J.-L. Colliot-Thélène and R. Parimala mentioned above follows.

Let X be a smooth projective algebraic surface and $H^3(X(\mathbb{C}); \mathbb{F}_2) = 0$. Then, by Poincaré duality, we have $H_1(X(\mathbb{C}); \mathbb{F}_2) = 0$. If $X(\mathbb{R}) \neq \emptyset$, then any loop in $X(\mathbb{C})/G$ with the beginning on $X(\mathbb{R})$ has a lifting to a loop on $X(\mathbb{C})$. It follows that the canonical homomorphism $H_1(X(\mathbb{C}); \mathbb{F}_2) \to H_1(X(\mathbb{C})/G; \mathbb{F}_2)$ is epimorphic. Thus, $H_1(X(\mathbb{C})/G; \mathbb{F}_2) = 0$. For the dimension 2 the quotient space $X(\mathbb{C})/G$ is homeomorphic to a smooth compact 4-dimensional manifold. By Poincaré duality, we then get that $H^3(X(\mathbb{C})/G; \mathbb{F}_2) = 0$. This proves the statement.

We apply the Theorem 0.1 to real Enriques surfaces.

By a complex Enriques surface Y over $\mathbb{C}$, we mean a non-singular minimal projective algebraic surface $Y/\mathbb{C}$ such that the invariants $\kappa(Y) = p_g(Y) = q(Y) = 0$. These are equivalent to irregularity $q(Y) = 0$ and $2K_Y = 0$ but $K_Y \neq 0$ where K_Y is the canonical class of Y. One may find all information about Enriques surfaces we need in the books [A] and [C-D].

By a real Enriques surface $Y/\mathbb{R}$, we mean a projective algebraic surface $Y/\mathbb{R}$ such that

$Y \underset{\mathbb{R}}{\otimes} \mathbb{C}$ is a complex Enriques surface. Universal covering complex surface of an Enriques surface $Y(\mathbb{C})$ is a $K3$-surface $X(\mathbb{C})$ (see [A] and [C-D]) which twice covers the Enriques surface $Y(\mathbb{C})$. We denote by τ the holomorphic involution on $X(\mathbb{C})$ of this covering. There are precisely two liftings σ and $\tau\sigma$ on $X(\mathbb{C})$ of the antiholomorphic involution θ of $Y(\mathbb{C})$ corresponding to the real structure on Y. Besides, one can see very easily that if $Y(\mathbb{R}) \neq \emptyset$, then both σ and $\tau\sigma$ are antiholomorphic involutions of $X(\mathbb{C})$. Thus, σ and $\tau\sigma$ define two real structures X_σ and $X_{\tau\sigma}$ on the $K3$-surface X. We denote by

$$X_\sigma(\mathbb{R}) = X(\mathbb{C})^\sigma, \quad X_{\tau\sigma}(\mathbb{R}) = X(\mathbb{C})^{\tau\sigma}$$

the real parts of the real $K3$-surfaces X_σ and $X_{\tau\sigma}$ corresponding to these real structures respectively. Since τ has no fixed points on $X(\mathbb{C})$, it follows that the sets $X_\sigma(\mathbb{R})$ and $X_{\tau\sigma}(\mathbb{R})$ have an empty intersection. From the Theorem 0.1, we get

Corollary 0.2. *Let Y be a real Enriques surface with the antiholomorphic involution θ, and the real part $Y(\mathbb{R}) \neq \emptyset$. Suppose that the real parts $X_\sigma(\mathbb{R})$ and $X_{\tau\sigma}(\mathbb{R})$ of both liftings σ and $\tau\sigma$ of θ to the universal covering $K3$-surface $X(\mathbb{C})$ are non-empty.*

Then the canonical map (0–1) corresponding to the real Enriques surface Y is epimorphic.

Proof.
$$Y(\mathbb{C})/\{\mathrm{id}_{Y(\mathbb{C})}, \theta\} = X(\mathbb{C})/\{\mathrm{id}_{X(\mathbb{C})}, \tau, \sigma, \tau\sigma\} =$$
$$(X(\mathbb{C})/\{\mathrm{id}_{X(\mathbb{C})}, \sigma\})/\{\mathrm{id}, \tau\sigma\,\mathrm{mod}\{\mathrm{id}_{X(\mathbb{C})}, \sigma\}\}.$$

Here involutions σ and $\tau\sigma \bmod\{\mathrm{id}_{X(\mathbb{C})}, \sigma\}$ have non-empty sets of fixed points because real parts of both involutions σ and $\tau\sigma$ of $X(\mathbb{C})$ are non-empty and are not coincided. Since for a $K3$-surface X, the group $H_1(X(\mathbb{C}); \mathbb{F}_2) = 0$, it follows like above that $H_1(X(\mathbb{C})/\{\mathrm{id}_{X(\mathbb{C})}, \sigma\}; \mathbb{F}_2) = 0$ and $H_1(Y(\mathbb{C})/\{\mathrm{id}_{Y(\mathbb{C})}; \theta\}; \mathbb{F}_2) = 0$. The topological space $Y(\mathbb{C})/\{\mathrm{id}_{Y(\mathbb{C})}; \theta\}$ is isomorphic to a smooth compact 4-dimensional manifold. By Poincaré duality, then $H^3(Y(\mathbb{C})/\{\mathrm{id}_{Y(\mathbb{C})}, \theta\}; \mathbb{F}_2) = 0$. By Theorem 0.1, the map (0–1) is epimorphic for the real Enriques surface Y.

The same considerations show that if one of involutions σ or $\tau\sigma$ has an empty set of real points, then

$$H_1(Y(\mathbb{C})/\{\mathrm{id}_{Y(\mathbb{C})}, \theta\}; \mathbb{F}_2) = \mathbb{F}_2 \text{ and } H^3(Y(\mathbb{C})/\{\mathrm{id}_{Y(\mathbb{C})}, \theta\}; \mathbb{F}_2) = \mathbb{F}_2.$$

– If $X_\sigma(\mathbb{R}) \neq \emptyset$ but $X_{\tau\sigma}(\mathbb{R}) = \emptyset$ then the surface $X(\mathbb{C})/\{\mathrm{id}_{X(\mathbb{C})}, \sigma\}$ is a 2-sheeted universal covering of the surface $Y(\mathbb{C})/\{\mathrm{id}_{Y(\mathbb{C})}, \theta\}$. Thus, the Corollary 0.2 gives precisely the case when the Theorem 0.1 may be applied to real Enriques surfaces.

We discuss in Remark 1.7 below a chance of constructing a counterexample to epimorphisity of the map (0–1) using real Enriques surfaces Y above with $H^3(Y(\mathbb{C})/G; \mathbb{F}_2) = \mathbb{F}_2$ (equivalently, with $X_\sigma(\mathbb{R}) \neq \emptyset$ but $X_{\tau\sigma}(\mathbb{R}) = \emptyset$). It is not difficult to show that real Enriques surfaces with the condition $H^3(Y(\mathbb{C})/G; \mathbb{F}_2) = \mathbb{F}_2$ do exist. For the most part of real Enriques surfaces (from the point of view of the number of connected components of the moduli space) both involutions σ and $\tau\sigma$ have a non-empty set of real points. But for some real Enriques surfaces one of these involutions may have an empty set of real points.

The following results are devoted to a calculation of the dimension of étale cohomology groups with coefficients $\mathbb{F}_2$, and $_2Br'(X)$.

We begin with the following general remark about a real algebraic variety X.

We recall that the Kummer sequence

$$0 \to \mu_2 \to G_m \to G_m \to 0 \qquad (0-2)$$

yields the exact sequence

$$0 \to \text{Pic } X/2\text{Pic } X \to H^2_{et}(X; \mu_2) \to_2 Br'(X) \to 0. \qquad (0-3)$$

If $X(\mathbb{R})$ is non-empty, $\text{Pic } X = (\text{Pic } (X \otimes \mathbb{C}))^G$ (it is well-known [Ma] and not difficult to see). Thus, from (0–3), we have

$$\dim {}_2Br'(X) = \dim H^2_{et}(X; \mu_2) - \dim (\text{Pic } (X \otimes \mathbb{C}))^G / 2(\text{Pic } (X \otimes \mathbb{C}))^G. \quad (0-4)$$

The dimension of the étale cohomology group $H^2_{et}(X; \mu_2) = H^2_{et}(X; \mathbb{F}_2)$ is estimated using the Serre-Hochschild spectral sequence where $G = Gal(\mathbb{C}/\mathbb{R})$,

$$E_2^{p,q} = H^p(G; H^q_{et}(X \otimes \mathbb{C}; \mathbb{F}_2)) \Longrightarrow H^{p+q}_{et}(X; \mathbb{F}_2), \qquad (0-5)$$

where for a complex manifold $X \otimes \mathbb{C}$ we have $H^q_{et}(X \otimes \mathbb{C}; \mathbb{F}_2) = H^q(X(\mathbb{C}); \mathbb{F}_2)$ (see [Mi], for example).

In the §2, we prove the following results which permit to calculate the dimension of the étale cohomology groups with coefficients $\mathbb{F}_2$ and the 2-torsion of the Brauer group for surfaces satisfying to the condition of Theorem 0.1. These results show that the class of real smooth projective surfaces X satisfying to the condition of Theorem 0.1 is very nice (easy to work with).

Theorem 0.3. *Let $X/\mathbb{R}$ be a real smooth projective algebraic surface such that $X(\mathbb{R}) \neq \emptyset$ and $H^3(X(\mathbb{C})/G; \mathbb{F}_2) = 0$. Then the Serre-Hochschild spectral sequence (0–5) degenerates and*

$$\dim H^0_{et}(X; \mathbb{F}_2) = 1;$$

$$\dim H^1_{et}(X; \mathbb{F}_2) = \dim H^1(X(\mathbb{C}); \mathbb{F}_2) + 1;$$

$$\dim H^2_{et}(X; \mathbb{F}_2) = \dim H^2(X(\mathbb{C}); \mathbb{F}_2)^G + \dim H^1(X(\mathbb{C}); \mathbb{F}_2) + 1;$$

$$\dim H^3_{et}(X; \mathbb{F}_2) \;=\; 2 \dim H^2(X(\mathbb{C}); \mathbb{F}_2)^G - \dim H^2(X(\mathbb{C}); \mathbb{F}_2)$$

$$+ \;\; 2 \dim H^1(X(\mathbb{C}); \mathbb{F}_2) + 1$$

$$\dim H^k_{et}(X; \mathbb{F}_2) \;=\; 2 \dim H^2(X(\mathbb{C}); \mathbb{F}_2)^G - \dim H^2(X(\mathbb{C}); \mathbb{F}_2)$$

$$+ \;\; 2 \dim H^1(X(\mathbb{C}); \mathbb{F}_2) + 2$$

for $k \geq 4$.

Using Theorem 0.3 and (0–4), (0–5), we get

Theorem 0.4. *Let $X/\mathbb{R}$ be a real smooth projective algebraic surface such that $X(\mathbb{R}) \neq \emptyset$ and $H^3(X(\mathbb{C})/G; \mathbb{F}_2) = 0$.*

Then

$$\dim {}_2Br'(X) = 2s - 1 + h^{2,0}(X(\mathbb{C})) + h^{1,1}_-(X(\mathbb{C})) - \rho_+(X \otimes \mathbb{C}).$$

Here $h^{1,1}_-(X(\mathbb{C})) = \dim H^{1,1}_-(X(\mathbb{C}))$ *where*

$$H^{1,1}_-(X(\mathbb{C})) = \{x \in H^{1,1}(X(\mathbb{C})) \mid g(x) = -x\}$$

is the set of potentially real algebraic cycles. And $\rho_+(X \otimes \mathbb{C}) = \dim (\mathrm{Pic}\ (X \otimes \mathbb{C}) \otimes \mathbb{C})^G$. *The characteristic class map gives an injection of* $(\mathrm{Pic}\ (X \otimes \mathbb{C}) \otimes \mathbb{C})^G$ *to* $H^{1,1}_-(X(\mathbb{C}))$.

For an Enriques surface, $h^{2,0}(Y(\mathbb{C})) = 0$ and all cycles are algebraic. For a real Enriques surface Y, we have seen above that the condition $H^3(Y(\mathbb{C})/G; \mathbb{F}_2) = 0$ is equivalent to the condition of Corollary 0.2. Thus, we get

Corollary 0.5. *Let* Y *be a real Enriques surface with the antiholomorphic involution* θ *and the real part* $Y(\mathbb{R}) \neq \emptyset$. *Suppose that the real parts* $X_\sigma(\mathbb{R})$ *and* $X_{\tau\sigma}(\mathbb{R})$ *of both liftings* σ *and* $\tau\sigma$ *of* θ *to the universal covering* $K3$-*surface* $X(\mathbb{C})$ *are not empty.*

Then the Serre-Hochschild spectral sequence (0–5) degenerates and

$$\dim {}_2Br'(Y) = 2s - 1$$

where s *is the number of real connected components of* $Y(\mathbb{R})$.

We mention that for a real rational surface Z with a non-empty set of real points $Z(\mathbb{R})$ the same results were known: The map (0–1) is epimorphic and $\dim {}_2Br'(Z) = 2s - 1$. It is also known that the Witt group $W(Z) \cong (\mathbb{Z})^s \oplus (\mathbb{Z}/2)^{s-1}$. See [Su]. Perhaps, the last result about the Witt group also valid for real Enriques surfaces with non-empty sets $X_\sigma(\mathbb{R})$ and $X_{\tau\sigma}(\mathbb{R})$.

Using results of [N-S], we may prove some additional results about Brauer groups of real Enriques surfaces which also valid if one of the sets $X_\sigma(\mathbb{R})$, $X_{\tau\sigma}(\mathbb{R})$ is empty.

In [N-S], the important invariants $b(Y)$ and $\epsilon(Y)$ of a real Enriques surface Y with an antiholomorphic involution θ were introduced. Here

$$b(Y) = \dim H^2(Y(\mathbb{C}); \mathbb{F}_2)^\theta - \dim (\mathrm{Pic}\ Y \otimes \mathbb{C})^\theta / 2(\mathrm{Pic}\ Y \otimes \mathbb{C})^\theta + 1.$$

The invariant $\epsilon(Y) = 1$ if the differential $d_2^{0,2}$ of the Hochschild-Serre spectral sequence (0–5) is zero, and $\epsilon(Y) = 0$ otherwise. We have the following results from [N-S] about these invariants:

$$\dim {}_2Br'(Y) = b(Y) + \epsilon(Y) \text{ if } Y(\mathbb{R}) \neq \emptyset, \tag{0 – 6}$$

$$b(Y) \geq 2s - 2 \text{ for any} Y. \tag{0 – 7}$$

Thus, by (0–6) and (0–7), for any real Enriques surface Y,

$$\dim {}_2Br'(Y) \geq 2s - 2 + \epsilon(Y). \tag{0 – 8}$$

Additionally to Corollary 0.5 and (0–6)–(0–8), we prove

Theorem 0.6. *Let* Y *be a real Enriques surface. Then:*

(i) The inequality (0–7) is an equality, i.e. $b(Y) = 2s - 2$, *iff the Hochschild-Serre spectral sequence (0–5) degenerates. In particular, by Corollary 0.5,* $b(Y) = 2s - 2$ *if* $X_\sigma(\mathbb{R}) \neq \emptyset$ *and* $X_{\tau\sigma}(\mathbb{R}) \neq \emptyset$.

(ii) dim $_2Br'(Y) = 2s - 1$ *if the Hochschild-Serre spectral sequence (0–5) degenerates and* $Y(\mathbb{R}) \neq \emptyset$. *In particular, it is true if* $X_\sigma(\mathbb{R}) \neq \emptyset$ *and* $X_{\tau\sigma}(\mathbb{R}) \neq \emptyset$.

(iii) dim $_2Br'(Y) \geq 2s - 1$.

In §3, we give an application of results of [N-S] and Corollary 0.5 and Theorem 0.6 to a topological studying of real Enriques surfaces Y. Let Y be a real Enriques surface with the antiholomorphic involution θ and $Y(\mathbb{R}) \neq \emptyset$. Let s_{nor} be the number of non-orientable connected components of $Y(\mathbb{R})$. We denote by

$$\Gamma = \{\mathrm{id}, \tau, \sigma, \tau\sigma\} \cong (\mathbb{Z}/2)^2$$

the group acting on the $K3$-surface $X(\mathbb{C})$ (we use notation above). Let us suppose that the both real parts $X_\sigma(\mathbb{R})$, $X_{\tau\sigma}(\mathbb{R})$ are non-empty. Then we give a formula connecting the number s_{nor} with some invariants of the action of the group Γ on the lattice $H^2(X(\mathbb{C}); \mathbb{Z})$ with the intersection pairing. We mention that it is not clear that one can express the number s_{nor} using the action of Γ on the lattice $H^2(X(\mathbb{C}); \mathbb{Z})$. This formula is very important for the topological classification of real Enriques surfaces (see [N4]). See §3 for details.

I am grateful to O. Gabber for assuring me that the statement of Proposition 1.1 below should be true. I am grateful to R. Sujatha for very useful discussions, in particular, for pointing out me on the results of E. Witt from [W].

A preliminary variant of this paper [N3] was written during my stay in the University of Notre Dame (USA) at 1991–1992. I am grateful to the University of Notre Dame for hospitality.

1 The proof of the Theorem 0.1

We recall that if X is a topological space with an action of a group G and $\mathcal{A}$ is a G-sheaf of groups on X, then the group $H^k(X; G, \mathcal{A})$ of equivariant cohomology (or Galois cohomology) is defined. See A. Grothendieck [Gr1, Ch. 5]. It is the right derived functor $R^k\Gamma^G$ to the functor $\mathcal{A} \mapsto \Gamma(X; \mathcal{A})^G$ of G-invariant sections. This composition of functors $\mathcal{A} \mapsto \Gamma(X; \mathcal{A})$ and $M \mapsto M^G$ for a G-module M defines two spectral functors which tend to this cohomology:

$$I_2^{p,q} = H^p(X/G; \mathcal{H}^q(G; \mathcal{A})) \Longrightarrow H^{p+q}(X; G, \mathcal{A})$$

where $\mathcal{H}^q(G; \mathcal{A}) = R^q f_*^G \mathcal{A}$ is the sheaf corresponding to the presheaf on X/G:

$$U \longmapsto H^q(\pi^{-1}(U); G, \mathcal{A}). \qquad (1-1)$$

Here $\pi : X \to X/G$ is the quotient map. And

$$II_2^{p,q} = H^p(G; H^q(X; \mathcal{A})) \Longrightarrow H^{p+q}(X; G, \mathcal{A}). \qquad (1-2)$$

The following statement is fundamental for us. This is analogous to the well-known connection between étale and ordinary cohomology for a complex algebraic manifold Z and a finite abelian group B (see [Mi, Ch. III, §3]): The morphism of the ordinary site (Euclidean topology) to the étale site gives rise an isomorphism

$$H^k_{et}(Z; B) \cong H^k(Z(\mathbb{C}); B). \tag{1-3}$$

Proposition 1.1. *Let $X/\mathbb{R}$ be a real algebraic manifold and $G = Gal(\mathbb{C}/\mathbb{R})$. Let $\mathcal{A}$ be a sheaf in étale topology of X such that this sheaf is a constant G-sheaf A on $X \otimes \mathbb{C}$, where A is a finite abelian group with an action of the group G.*

Then there exists the canonical isomorphism $H^k_{et}(X; A) \cong H^k(X(\mathbb{C}); G, A)$ together with the canonical isomorphism of the Hochschild-Serre spectral sequence and the Grothendieck spectral sequence II, which is defined by the canonical isomorphism

$$E^{p,q}_2 = H^p(G; H^q_{et}(X \otimes \mathbb{C}; A)) \cong II^{p,q}_2 = H^p(G; H^q(X(\mathbb{C}); A))$$

induced by the canonical isomorphism (1–3).

Proof. We follow to the proof of the isomorphism (1–3) for complex algebraic manifolds. See [Mi,Ch. III, §3], for example.

Let $X(\mathbb{C})_{cx}$ be a small site $X(\mathbb{C})^{an}_E$ of morphisms of complex analytic spaces over $X(\mathbb{C})^{an}$, which are local isomorphisms. (We use notation of [Mi].) An open subset $U \subset X(\mathbb{C})$ is a local isomorphism. It follows that we have the morphism of sites $X(\mathbb{C})_{cx} \to X(\mathbb{C})$. Every covering of $X(\mathbb{C})$ in the site $X(\mathbb{C})_{cx}$ has a refinement covering of $X(\mathbb{C})$ in the site $X(\mathbb{C})$. It follows that the morphism $X(\mathbb{C})_{cx} \to X(\mathbb{C})$ gives an isomorphism of cohomology

$$H^i(X(\mathbb{C}); A) \cong H^i(X(\mathbb{C})_{cx}; A). \tag{1-4}$$

Like above, we can define G-equivariant cohomology $H^k(X(\mathbb{C})_{cx}; G, \mathcal{F})$ for the site $X(\mathbb{C})_{cx}$ as the right derived functor to the composition of functors $\mathcal{F} \to \Gamma(X(\mathbb{C})_{cx}; \mathcal{F})^G$ for a G-sheaf $\mathcal{F}$ on the site $X(\mathbb{C})_{cx}$. This equivariant cohomology also has a spectral sequence $II(X(\mathbb{C})_{cx})$ with the beginning

$$II^{p,q}_2(X(\mathbb{C})_{cx}) = H^p(G; H^q(X(\mathbb{C})_{cx}; \mathcal{F})) \Longrightarrow H^{p+q}(X(\mathbb{C})_{cx}; G, \mathcal{F}).$$

One can calculate equivariant cohomology using invariant coverings (see [Gr1, Ch.5]). Any G-invariant covering of $X(\mathbb{C})$ in site $X(\mathbb{C})_{cx}$ also contains a refinement G-invariant covering of $X(\mathbb{C})$ in the site $X(\mathbb{C})$. It follows that we also have the canonical isomorphism of equivariant cohomology:

$$H^i(X(\mathbb{C})_{cx}; G, A) \cong H^i(X(\mathbb{C}); G, A), \tag{1-5}$$

together with the isomorphism of the corresponding spectral sequences II of these cohomology

$$II^{p,q}_2(X(\mathbb{C})_{cx}) = H^p(G; H^q(X(\mathbb{C})_{cx}; A)) \cong II^{p,q}_2(X(\mathbb{C})) = H^p(G; H^q(X(\mathbb{C}); A)). \tag{1-6}$$

defined by the isomorphism (1–4).

Further, we use standard results about étale cohomology (see [Mi]). Since $X \otimes \mathbb{C} \to X$ is an étale covering with Galois group G, it follows that a sheaf F on the site X_{et} corresponds to a G-sheaf F on the site $(X \otimes \mathbb{C})_{et}$. Étale cohomology is a right derived functor to the composition of functors

$$F \to (G - mod \ \Gamma(X \otimes \mathbb{C}; F)) \text{ and } M \to M^G.$$

Here M is a G-module. The Hochschild-Serre spectral sequence

$$E_2^{p,q} = H^p(G; H_{et}^q(X \otimes \mathbb{C}; F)) \implies H_{et}^{p+q}(X; F)$$

corresponds to the composition of these functors.

Every étale morphism $Y \to X \otimes \mathbb{C}$ gives a morphism $Y(\mathbb{C}) \to X(\mathbb{C})$ in the site $X(\mathbb{C})_{cx}$. This defines the morphism of sites $X(\mathbb{C})_{cx} \to X_{et}$. From the remarks above, this morphism defines the homomorphism of cohomology

$$H_{et}^k(X; A) \to H^k(X(\mathbb{C})_{cx}; G, A) = H^k(X(\mathbb{C}); G, A) \qquad (1-7)$$

together with the homomorphism of spectral sequences

$$E_r^{p,q} \to II_r^{p,q} \qquad (1-8)$$

defined by the homomorphism

$$
\begin{aligned}
E_2^{p,q} = H^p(G; H_{et}^q(X \otimes \mathbb{C}; A)) \to II_2^{p,q} &= H^p(G; H^q(X(\mathbb{C})_{cx}; A)) \\
&= H^p(G; H^q(X(\mathbb{C}); A)).
\end{aligned}
\qquad (1-9)
$$

The last homomorphism is defined by the homomorphism

$$H_{et}^q(X \otimes \mathbb{C}; A) \to H^q(X(\mathbb{C})_{cx}; A) = H^q(X(\mathbb{C}); A), \qquad (1-10)$$

which is the isomorphism (1–3). It follows that (1–9), (1–8) and (1–7) are isomorphisms too. This finishes the proof.

We mention that D. A. Cox [C] had shown that étale homotopy type of a real algebraic manifold is defined by Euclidean topology.

We recall that for a compact manifold M and a constant sheaf of modules, one can calculate sheaf cohomology using simplicial triangulation of the manifold M. One can prove this using the canonical isomorphisms between sheaf, Čech, Alexander-Spanier, singular and simplicial triangulation cohomology for compact manifolds. See [Gr1] and [Sp], for example. Similarly, for a compact manifold M with an action of the group G and an Abelian G-group A, one can prove that equivariant sheaf, Čech, Alexander-Spanier, singular and simplicial triangulation cohomology are isomorphic. Thus, for this case, we can calculate equivariant cohomology using the following elementary procedure (this definition is used in the book [Bro], for example):

We consider some G-equivariant simplicial triangulation K of M. Thus, K is a G-equivariant simplicial complex. We may suppose that K/G is a simplicial triangulation of M/G and the fixed part K^G is a simplicial triangulation of M^G. We consider the corresponding chain complex $C_n = C_n(K; \mathbb{Z})$ and the corresponding cochain complex $C^n(K; A)$, where $C^n(K; A) = \mathrm{Hom}(C_n, A)$. We organize a cochain complex for calculation of group cohomology of $C^n = C^n(K, A)$. For the group $G = \{1, g\}$ of order two this is a complex

$$0 \longrightarrow C^n \xrightarrow{1-g} C^n \xrightarrow{1+g} C^n \xrightarrow{1-g} C^n \xrightarrow{1+g} C^n \xrightarrow{1-g} \cdots \qquad (1-11)$$

Thus, we get a double complex

$$
\begin{array}{ccccccccccc}
\vdots & & \vdots & & \vdots & & \vdots & & \vdots & & \cdots\\
d^4 \uparrow & & d^4 \uparrow & & d^4 \uparrow & & d^4 \uparrow & & d^4 \uparrow & &\\
0 \longrightarrow & C^4 & \overset{1-g}{\longrightarrow} & C^4 & \overset{1+g}{\longrightarrow} & C^4 & \overset{1-g}{\longrightarrow} & C^4 & \overset{1+g}{\longrightarrow} & C^4 & \overset{1-g}{\longrightarrow} \cdots\\
d^3 \uparrow & & d^3 \uparrow & & d^3 \uparrow & & d^3 \uparrow & & d^3 \uparrow & &\\
0 \longrightarrow & C^3 & \overset{1-g}{\longrightarrow} & C^3 & \overset{1+g}{\longrightarrow} & C^3 & \overset{1-g}{\longrightarrow} & C^3 & \overset{1+g}{\longrightarrow} & C^3 & \overset{1-g}{\longrightarrow} \cdots\\
d^2 \uparrow & & d^2 \uparrow & & d^2 \uparrow & & d^2 \uparrow & & d^2 \uparrow & &\\
0 \longrightarrow & C^2 & \overset{1-g}{\longrightarrow} & C^2 & \overset{1+g}{\longrightarrow} & C^2 & \overset{1-g}{\longrightarrow} & C^2 & \overset{1+g}{\longrightarrow} & C^2 & \overset{1-g}{\longrightarrow} \cdots\\
d^1 \uparrow & & d^1 \uparrow & & d^1 \uparrow & & d^1 \uparrow & & d^1 \uparrow & &\\
0 \longrightarrow & C^1 & \overset{1-g}{\longrightarrow} & C^1 & \overset{1+g}{\longrightarrow} & C^1 & \overset{1-g}{\longrightarrow} & C^1 & \overset{1+g}{\longrightarrow} & C^1 & \overset{1-g}{\longrightarrow} \cdots\\
d^0 \uparrow & & d^0 \uparrow & & d^0 \uparrow & & d^0 \uparrow & & d^0 \uparrow & &\\
0 \longrightarrow & C^0 & \overset{1-g}{\longrightarrow} & C^0 & \overset{1+g}{\longrightarrow} & C^0 & \overset{1-g}{\longrightarrow} & C^0 & \overset{1+g}{\longrightarrow} & C^0 & \overset{1-g}{\longrightarrow} \cdots\\
\uparrow & & \uparrow & & \uparrow & & \uparrow & & \uparrow & &\\
0 & & 0 & & 0 & & 0 & & 0 & &\\
\end{array}
$$

$$(1-12)$$

Equivariant cohomology $H^n(M; G, A)$ are the homology of this double complex. The spectral sequence I corresponds to the filtration from below on cohomology of this double complex, and the spectral sequence II corresponds to the filtration from left of this double complex. From this complex, it is clear (see Godement [Go, Chapter 1], for example) that

$$II_1^{p,q}(M; G, A) = H^q(M; A) \Longrightarrow H^{p+q}(M; G, A), \qquad (1-13)$$

and

$$II_2^{p,q}(M; G, A) = H^p(G; H^q(M; A)) \Longrightarrow H^{p+q}(M; G, A). \qquad (1-14)$$

Here, for a G-module R, we have

$$H^0(G; R) = R^G, \qquad (1-15)$$

$$H^p(G; R) = R^g/(1+g)R \text{for } p \text{ even and } p > 0; \qquad (1-16)$$

and

$$H^p(G; R) = R^{(-g)}/(1-g)R \text{for } p \text{ odd.} \qquad (1-17)$$

If R is a vector space over the field $\mathbb{F}_2$, the formulae (1–16) and (1–17) give the same.

Further, we calculate the equivariant cohomology for $A = \mathbb{F}_2$. For this case, $1 - g = 1 + g$, and the double complex (1–12) is periodic.

The spectral sequence I corresponds to the filtration of this complex from below. Then

$$I_1^{p,0}(M; G, \mathbb{F}_2) = (C^p)^g = C^p(M/G; \mathbb{F}_2), \qquad (1-18)$$

and

$$I_2^{p,0}(M; G, \mathbb{F}_2) = (C^p)^g/d^{p-1}((C^{p-1})^g) = H^p(M/G; \mathbb{F}_2). \qquad (1-19)$$

For $q > 0$,

$$I_1^{p,q}(M; G, \mathbb{F}_2) = (C^p)^g/(1+g)(C^p) \;\; = \;\; C^p(M/G; \mathbb{F}_2)/C^p(M/G, M^G; \mathbb{F}_2)$$
$$\text{(1--20)}$$
$$= \;\; C^p(M^G; \mathbb{F}_2).$$

And we get

$$I_2^{p,q}(M; G, \mathbb{F}_2) = H^p(M^G; \mathbb{F}_2) \text{ if } q > 0. \qquad (1-21)$$

In particular, for a real projective algebraic manifold X, the set $X(\mathbb{C})$ of complex points with the group $G = \{1, g\}$ generated by antiholomorphic involution g acting on $X(\mathbb{C})$, and the set $X(\mathbb{R}) = X(\mathbb{C})^G$ of real points, we have:

$$II_2^{p,q}(X(\mathbb{C}); G, A) = H^p(G; H^q(X(\mathbb{C}); A)) \Longrightarrow H^{p+q}(X(\mathbb{C}); G, A); \qquad (1-22)$$

$$I_2^{p,0}(X(\mathbb{C}); G, \mathbb{F}_2) = H^p(X(\mathbb{C})/G; \mathbb{F}_2); \qquad (1-23)$$

and

$$I_2^{p,q}(X(\mathbb{C}); G, \mathbb{F}_2) = H^p(X(\mathbb{R}); \mathbb{F}_2) \text{ if } q > 0. \qquad (1-24)$$

Further, we will consider this case of real projective algebraic manifold X, but, actually, all results valid for a compact manifold M with an involution.

For $X(\mathbb{R})$, the double complex (1–12) has $1 - g = 1 + g = 0$. Thus, we evidently get (see [Gr1, Corollaire 5.4.1])

Proposition 1.2. *We have a canonical isomorphism*

$$H^k(X(\mathbb{R}); G, \mathbb{F}_2) = \bigoplus_{i=0}^{k} H^i(X(\mathbb{R}); \mathbb{F}_2).$$

Besides, the spectral sequence $I(X(\mathbb{R}); G, \mathbb{F}_2)$ degenerates from I_2, and for all p, q we have $I_2^{p,q}(X(\mathbb{R}); G, \mathbb{F}_2) = H^p(X(\mathbb{R}); \mathbb{F}_2)$. Thus, all differentials

$$d_r^{p,q}(X(\mathbb{R}); G, \mathbb{F}_2)$$

of the spectral sequence I vanish for $r \geq 2$.

The same is true for a topological space with the trivial action of the group G.

Now we have (see [Kr])

Proposition 1.3. *For $k > 2 \dim X$,*

$$H^k(X(\mathbb{C}); G, \mathbb{F}_2) = H^k(X(\mathbb{R}); G, \mathbb{F}_2) = \bigoplus_{i=0}^{k} H^i(X(\mathbb{R}); \mathbb{F}_2).$$

Proof. Like above, using the double complex, we can define equivariant cohomology of a pair. We evidently have an exact sequence;

$$\ldots \to H^i(X(\mathbb{C}), X(\mathbb{R}); G, \mathbb{F}_2) \to H^i(X(\mathbb{C}); G, \mathbb{F}_2) \to H^i(X(\mathbb{R}); G, \mathbb{F}_2) \to$$
$$H^{i+1}(X(\mathbb{C}), X(\mathbb{R}); G, \mathbb{F}_2) \to \ldots.$$

The group G acts without fixed points on the pair $(X(\mathbb{C}), X(\mathbb{R}))$. – For the corresponding double complex all differentials $1 + g, 1 - g$ give an exact sequence. It follows that for $q > 0$ the spectral sequence $I_1^{p,q} = 0$ for this double complex and

$$H^i(X(\mathbb{C}), X(\mathbb{R}); G, \mathbb{F}_2) = H^i(X(\mathbb{C})/G, X(\mathbb{R}); G, \mathbb{F}_2).$$

Thus, $H^i(X(\mathbb{C}), X(\mathbb{R}); G, \mathbb{F}_2) = 0$ for $i > 2 \dim X$. It follows the statement.

From the calculation above of the beginning of the spectral sequence I and Proposition 1.2, we get

Proposition 1.4. *The embedding $\rho : X(\mathbb{R}) \subset X(\mathbb{C})$ gives the homomorphism of the spectral sequences $\rho^* : I(X(\mathbb{C}); G, \mathbb{F}_2) \to I(X(\mathbb{R}); G, \mathbb{F}_2)$. For $r = 2$ it is defined by*

$$I_2^{p,0}(X(\mathbb{C}); G, \mathbb{F}_2) = H^p(X(\mathbb{C})/G; \mathbb{F}_2) \xrightarrow{i^*} H^p(X(\mathbb{R}); \mathbb{F}_2) = I_2^{p,0}(X(\mathbb{R}); G, \mathbb{F}_2),$$

where $i : X(\mathbb{R}) \subset X(\mathbb{C})/G$ is the embedding, and by the identical isomorphism

$$I_2^{p,q}(X(\mathbb{C}); G, \mathbb{F}_2) = H^p(X(\mathbb{R}); \mathbb{F}_2) = I_2^{p,q}(X(\mathbb{R}); G, \mathbb{F}_2) \quad for \quad q > 0.$$

In particular, since the spectral sequence $I(X(\mathbb{R}); G, \mathbb{F}_2)$ degenerates, the differential $d_r^{p,q}$ of $I(X(\mathbb{C}); G, \mathbb{F}_2)$ vanish if $r \geq 2$ and $q \neq r - 1$, and $\rho^* d_r^{p,r-1} = 0$ for $r \geq 2$.

Now we can reformulate the map (0–1)

$$_2Br'(X) \to (\mathbb{Z}/2)^s.$$

Let us choose a point P_i, $i = 1, 2, ..., s$, for every connected component of $X(\mathbb{R})$. Here $i = 1, 2, \cdots, s$ numerates connected components of $X(\mathbb{R})$. From the Kummer exact sequence (0–3), from the isomorphism $Br'(\{P_i\}) = H_{et}^2(\{P_i\}; \mathbb{F}_2) = \mathbb{F}_2$ and Proposition 1.1 (also see Proposition 1.2), the image of the homomorphism (0–1) is the same as the image of the composition of the homomorphisms

$$H^2(X(\mathbb{C}); G, \mathbb{F}_2) \to H^2(X(\mathbb{R}); G, \mathbb{F}_2) \to \bigoplus_{i=1}^{s} H^2(\{P_i\}; G, \mathbb{F}_2). \qquad (1-25)$$

defined by the inclusions $\{P_i\} \subset X(\mathbb{R}) \subset X(\mathbb{C})$. By Proposition 1.2, the second homomorphism is epimorphic and has the kernel

$$H^1(X(\mathbb{R}); \mathbb{F}_2) \oplus H^2(X(\mathbb{R}); \mathbb{F}_2).$$

– It is clear from the definition of equivariant cohomology using the double complex (1–12). We remark that we then have a canonical identification

$$H^0(X(\mathbb{R}); \mathbb{F}_2) \quad = \quad H^2(X(\mathbb{R}); G, \mathbb{F}_2)/(H^1(X(\mathbb{R}); \mathbb{F}_2) \oplus H^2(X(\mathbb{R}); \mathbb{F}_2)$$

$$= \quad I_\infty^{0,2}(X(\mathbb{R}); G, \mathbb{F}_2)$$

for the first spectral sequence I of the equivariant cohomology of $X(\mathbb{R})$. It follows, that the image of the map (1–25) is the image of the canonical homomorphism

$$H^2(X(\mathbb{C}); G, \mathbb{F}_2) \to H^2(X(\mathbb{R}); G, \mathbb{F}_2) \to I_\infty^{0,2}(X(\mathbb{R}); G; \mathbb{F}_2) = H^0(X(\mathbb{R}); \mathbb{F}_2).$$

This homomorphism preserves the filtration I on $H^2(X(\mathbb{C}); G, \mathbb{F}_2)$. Thus, this image is the same as the image of the canonical homomorphism

$$\rho^* : I_\infty^{0,2}(X(\mathbb{C}); G, \mathbb{F}_2) \to I_\infty^{0,2}(X(\mathbb{R}); G, \mathbb{F}_2) = H^0(X(\mathbb{R}); \mathbb{F}_2),$$

where $\rho : X(\mathbb{R}) \subset X(\mathbb{C})$ is the embedding.

Let us look on the part of the spectral sequence $I_2^{p,q}(X(\mathbb{C}); G, \mathbb{F}_2)$ which takes part in calculation of the $I_\infty^{0,2}(X(\mathbb{C}); G, \mathbb{F}_2)$. It is the left-below corner

$H^3(X(\mathbb{C})/G; \mathbb{F}_2)$

$H^2(X(\mathbb{C})/G; \mathbb{F}_2)$ $H^2(X(\mathbb{R}); \mathbb{F}_2)$

$H^1(X(\mathbb{C})/G; \mathbb{F}_2)$ $H^1(X(\mathbb{R}); \mathbb{F}_2)$ $H^1(X(\mathbb{R}); \mathbb{F}_2)$

$H^0(X(\mathbb{C})/G; \mathbb{F}_2)$ $H^0(X(\mathbb{R}); \mathbb{F}_2)$ $H^0(X(\mathbb{R}); \mathbb{F}_2)$

where the differential hits from below to up and from right to left. In particular, we have $I_\infty^{0,2}(X(\mathbb{C}); G, \mathbb{F}_2) \subset I_2^{0,2}(X(\mathbb{C}); G, \mathbb{F}_2) = H^0(X(\mathbb{R}); \mathbb{F}_2)$. From the double complex (1–12), the canonical map

$$\rho^* : I_2^{0,2}(X(\mathbb{C}); G, \mathbb{F}_2) = H^0(X(\mathbb{R}); \mathbb{F}_2)$$
$$\to I_2^{0,2}(X(\mathbb{R}); G, \mathbb{F}_2) = I_\infty^{0,2}(X(\mathbb{R}); G, \mathbb{F}_2) = H^0(X(\mathbb{R}); \mathbb{F}_2)$$

is the identical isomorphism of $H^0(X(\mathbb{R}); \mathbb{F}_2)$. Thus, we get

Proposition 1.5. *The image of the map ρ^* (equivalently, of the map (0–1)) is equal to the image of the embedding*

$$I_\infty^{0,2}(X(\mathbb{C}); G, \mathbb{F}_2) \subset I_2^{0,2}(X(\mathbb{C}); G, \mathbb{F}_2) = H^0(X(\mathbb{R}); \mathbb{F}_2) = I_\infty^{0,2}(X(\mathbb{R}); G, \mathbb{F}_2).$$

From Proposition 1.4, the differential

$$d_2^{0,2} : H^0(X(\mathbb{R}); \mathbb{F}_2) \to H^2(X(\mathbb{R}); \mathbb{F}_2)$$

is equal to zero. By Proposition 1.4, we also have $i^* d_2^{1,1} = 0$ and $\rho^* d_3^{0,2} = 0$. We then get the basic result

Theorem 1.6. *Let $i : X(\mathbb{R}) \subset X(\mathbb{C})/G$ be the embedding and $i^* : H^3(X(\mathbb{C})/G; \mathbb{F}_2) \to H^3(X(\mathbb{R}); \mathbb{F}_2)$ the corresponding canonical homomorphism.*

Then we have: The differential $d_2^{1,1}$ is the composition

$$d_2^{1,1} : H^1(X(\mathbb{R}); \mathbb{F}_2) \to \mathrm{Ker}\, i^* \subset H^3(X(\mathbb{C})/G; \mathbb{F}_2),$$

and the image of the homomorphism ρ^* from Proposition 1.5 (equivalently, of the homomorphism (0–1) from Introduction) is equal to the kernel of the differential

$$d_3^{0,2} : H^0(X(\mathbb{R}); \mathbb{F}_2) \to \mathrm{Ker}\, i^* / d_2^{1,1}(H^1(X(\mathbb{R}); \mathbb{F}_2))$$

$$\subset H^3(X(\mathbb{C})/G; \mathbb{F}_2) / d_2^{1,1}(H^1(X(\mathbb{R}); \mathbb{F}_2)).$$

In particular, ρ^* is epimorphic if $\mathrm{Ker}i^* = 0$ or $H^3(X(\mathbb{C})/G; \mathbb{F}_2) = 0$. It is epimorphic too if the differential

$$d_2^{1,1} : H^1(X(\mathbb{R}); \mathbb{F}_2) \to \mathrm{Ker}i^*$$

is epimorphic.

As a corollary, we get the Theorem 0.1 from Introduction.

We remark that for surfaces X the dimension $\dim X(\mathbb{R}) = 2$, and $H^3(X(\mathbb{R}); \mathbb{F}_2) = 0$. Thus, $\mathrm{Ker}i^* = H^3(X(\mathbb{C})/G; \mathbb{F}_2)$.

Remark 1.7. In Introduction (after Corollary 0.2), we showed real Enriques surfaces Y such that

$$H^3(Y(\mathbb{C})/G; \mathbb{F}_2) = \mathbb{F}_2.$$

Theorem 1.6 shows that such a Y gives an example when the map (0–1) is not epimorphic, exactly if simultaneously the homomorphism

$$d_2^{1,1} : H^1(Y(\mathbb{R}); \mathbb{F}_2) \to H^3(Y(\mathbb{C})/G; \mathbb{F}_2) = \mathbb{F}_2$$

is zero, and the homomorphism

$$d_3^{0,2} : H^0(Y(\mathbb{R}); \mathbb{F}_2) \to H^3(Y(\mathbb{C})/G; \mathbb{F}_2) = \mathbb{F}_2$$

is not zero.

Of course, these homomorphisms are "the same" for Enriques surfaces which belong to one connected component of the moduli space of real Enriques surfaces. Using Global Torelli Theorem [PŠ − Š], epimorphisity of Torelli map [Ku] for $K3$-surfaces and methods developed in [N1,N2], in principle, it is possible to enumerate all this connected components using some invariants. Thus, the problem is to rewrite, using these invariants, the invariants of differentials $d_2^{1,1}, d_3^{0,2}$ above. We hope to do it later. Some important results in this direction were obtained in [N-S].

Remark 1.8. We can consider the homomorphism

$$H^1_{et}(X; \mathbb{F}_2) \to (\mathbb{Z}/2)^s \qquad (1-26)$$

which is defined like the map (0–1) using the isomorphism $H^1_{et}(\{real\ point\}; \mathbb{F}_2) = \mathbb{F}_2$.

Like above, we can interpret the right side of (1–26) as the group $H^0(X(\mathbb{R}); \mathbb{F}_2) = I^{0,1}_\infty(X(\mathbb{R}); G, \mathbb{F}_2)$, and the image of the homomorphism (1–26) as the

$$\mathrm{Ker}\{d_2^{0,1} : H^0(X(\mathbb{R}), \mathbb{F}_2) \to H^2(X(\mathbb{C})/G; \mathbb{F}_2)\}$$

for the differential $d_2^{0,1}$ of the spectral sequence $I(X(\mathbb{C}); G, \mathbb{F}_2)$.

Suppose that X is a smooth projective real curve and $X(\mathbb{R}) \neq \emptyset$. Then the quotient space $X(\mathbb{C})/G$ is a connected 2-dimensional manifold with a not-trivial boundary $X(\mathbb{R})$. By Poincaré duality, $H^2(X(\mathbb{C})/G; \mathbb{F}_2) = H_0(X(\mathbb{C})/G, X(\mathbb{R}); \mathbb{F}_2) = 0$, and the map (1–26) is epimorphic. This gives a geometrical interpretation of the result of E.Witt from [W].

We mention, that similarly to the homomorphisms (0–1) and (1–26), one may define and study the general homomorphism (1–27) below

$$H^n_{et}(X; \mathbb{F}_2) \to (\mathbb{Z}/2)^s \qquad (1-27)$$

using the isomorphism $H^n_{et}(\{real\ point\}; \mathbb{F}_2) = \mathbb{F}_2$ for $n \geq 0$.

2 The Proof of the Theorems 0.3 – 0.6

We prove the following

Theorem 2.1. *Let $X/\mathbb{R}$ be a smooth real projective algebraic surface such that $X(\mathbb{R}) \neq \emptyset$ and $H^3(X(\mathbb{C})/G; \mathbb{F}_2) = 0$.*

Then the Serre-Hochschild spectral sequence for the $H^*_{et}(X; \mathbb{F}_2)$ and the spectral sequence $II(X(\mathbb{C}); G, \mathbb{F}_2)$ for the equivariant cohomology $H^*(X(\mathbb{C}); G, \mathbb{F}_2)$ degenerate. Besides (by Proposition 1.1), $H^k_{et}(X; \mathbb{F}_2) \cong H^k(X(\mathbb{C}); G, \mathbb{F}_2)$, and we have the formulae for their dimensions:

$$\dim\ H^0(X(\mathbb{C}); G, \mathbb{F}_2) = 1;$$
$$\dim\ H^1(X(\mathbb{C}); G, \mathbb{F}_2) = \dim\ H^1(X(\mathbb{C}); \mathbb{F}_2) + 1;$$
$$\dim\ H^2(X(\mathbb{C}); G, \mathbb{F}_2) = \dim\ H^2(X(\mathbb{C}); \mathbb{F}_2)^G + \dim\ H^1(X(\mathbb{C}); \mathbb{F}_2) + 1;$$
$$\dim\ H^3(X(\mathbb{C}); G, \mathbb{F}_2) = 2 \dim\ H^2(X(\mathbb{C}); \mathbb{F}_2)^G \quad - \quad \dim\ H^2(X(\mathbb{C}); \mathbb{F}_2)$$
$$+ \quad 2 \dim\ H^1(X(\mathbb{C}); \mathbb{F}_2) + 1;$$

$$\dim\ H^k(X(\mathbb{C}); G, \mathbb{F}_2) = 2 \dim\ H^2(X(\mathbb{C}); \mathbb{F}_2)^G \quad - \quad \dim\ H^2(X(\mathbb{C}); \mathbb{F}_2)$$
$$+ \quad 2 \dim\ H^1(X(\mathbb{C}); \mathbb{F}_2) + 2$$

for $k \geq 4$.

Proof. By Proposition 1.1, we should prove that the spectral sequence $II(X(\mathbb{C}); G, \mathbb{F}_2)$ degenerates. To prove this, we use the following important result of V. A. Krasnov [Kr].

Proposition 2.2. *Let $X/\mathbb{R}$ be a real projective algebraic manifold.*

Then the spectral sequence $II(X(\mathbb{C}); G, \mathbb{F}_2)$ with the beginning

$$II_2^{p,q} = H^p(G; H^q(X(\mathbb{C}), \mathbb{F}_2)) \Longrightarrow H^{p+q}(X(\mathbb{C}); G, \mathbb{F}_2)$$

degenerates iff

$$\dim\ H^*(X(\mathbb{R}); \mathbb{F}_2) = \dim\ H^1(G; H^*(X(\mathbb{C}); \mathbb{F}_2)).$$

Proof. Let $k > 2 \dim\ X$. By Propositions 1.2 and 1.3,

$$\dim\ H^*(X(\mathbb{R}); \mathbb{F}_2) = \dim\ H^k(X(\mathbb{C}); G, \mathbb{F}_2).$$

On the other hand,

$$\bigoplus_{p+q=k} II_2^{p,q} = \bigoplus_{p+q=k} H^p(G; H^q(X(\mathbb{C}); \mathbb{F}_2)) = H^1(G; H^*(X(\mathbb{C}); \mathbb{F}_2)).$$

Thus, we have the inequality

$$\dim H^*(X(\mathbb{R}); \mathbb{F}_2) = \dim H^k(X(\mathbb{C}); G, \mathbb{F}_2) \leq \dim H^1(G; H^*(X(\mathbb{C}); \mathbb{F}_2)). \quad (2-1)$$

Besides, (2–1) gives an equality iff $II_2^{p,q} = II_\infty^{p,q}$ for $p+q = k > 2\dim X$. Thus, if the spectral sequence $II(X(\mathbb{C}); G, \mathbb{F}_2)$ degenerates, we have the equality of the Proposition.

Now we assume that the equality of Proposition holds. We have then proven that all differentials $d_r^{p,q}$ vanish for $r \geq 2$ and $p+q > 2\dim X$. From periodicity of the double complex (1–12), it follows that all differentials $d_r^{p,q}$ vanish for $r \geq 2$.

We recall the Smith exact sequence for an action of a group $G = \{1, g\}$ of order two (see [Bre, Ch. III, §3], for example).

We have the exact sequence for the chain complex $C_n(K; \mathbb{F}_2)$ above with an action of G:

$$0 \to C_n(K; \mathbb{F}_2)^G \to C_n(K; \mathbb{F}_2) \to (1+g)(C_n(K; \mathbb{F}_2)) \to 0.$$

Here we have the canonical identifications

$$(1+g)(C_n(K; \mathbb{F}_2)) = C_n(K/G; \mathbb{F}_2)/C_n(K^G; \mathbb{F}_2) = C_n(K/G, K^G; \mathbb{F}_2)$$

and

$$C_n(K; \mathbb{F}_2)^G = (1+g)C_n(K; \mathbb{F}_2) \oplus C_n(K^G; \mathbb{F}_2) = C_n(K/G, K^G; \mathbb{F}_2) \oplus C_n(K^G; \mathbb{F}_2).$$

Thus, we get the exact sequence

$$0 \to C_n(K/G, K^G; \mathbb{F}_2) \oplus C_n(K^G; \mathbb{F}_2) \to C_n(K; \mathbb{F}_2) \to C_n(K/G, K^G; \mathbb{F}_2) \to 0.$$

This gives the corresponding homological Smith exact sequence:

$$\cdots \xrightarrow{\delta_{n+1}} H_n(K/G, K^G; \mathbb{F}_2) \oplus H_n(K^G; \mathbb{F}_2) \xrightarrow{i_n} H_n(K; \mathbb{F}_2) \xrightarrow{\rho_n} H_n(K/G, K^G; \mathbb{F}_2)$$

$$\xrightarrow{\delta_n} H_{n-1}(K/G, K^G; \mathbb{F}_2) \oplus H_{n-1}(K^G; \mathbb{F}_2) \xrightarrow{i_{n-1}} \cdots .$$

Thus, for a real algebraic variety, we have the homological Smith exact sequence

$$\cdots \xrightarrow{\delta_{n+1}} H_n(X(\mathbb{C})/G, X(\mathbb{R}); \mathbb{F}_2) \oplus H_n(X(\mathbb{R}); \mathbb{F}_2) \xrightarrow{i_n} H_n(X(\mathbb{C}); \mathbb{F}_2) \xrightarrow{\rho_n}$$

$$H_n(X(\mathbb{C})/G, X(\mathbb{R}); \mathbb{F}_2) \xrightarrow{\delta_n} H_{n-1}(X(\mathbb{C})/G, X(\mathbb{R}); \mathbb{F}_2) \oplus H_{n-1}(X(\mathbb{R}); \mathbb{F}_2) \to \cdots$$

We repeat some well-known standard facts connected with the Smith exact sequence (see V.A.Rokhlin [R], for example). From this exact sequence, we get

$$\dim H_{n-1}(X(\mathbb{C})/G, X(\mathbb{R}); \mathbb{F}_2) \quad + \quad \dim H_{n-1}(X(\mathbb{R}); \mathbb{F}_2)$$

$$= \quad \dim \operatorname{Im} \delta_n + \dim \operatorname{Im} i_{n-1}.$$

Moreover,

$$\dim \operatorname{Im} \delta_n \quad = \quad \dim H_n(X(\mathbb{C})/G, X(\mathbb{R}); \mathbb{F}_2) - \dim \operatorname{Im} \rho_n$$

$$= \quad \dim H_n(X(\mathbb{C})/G, X(\mathbb{R}); \mathbb{F}_2) - \dim H_n(X(\mathbb{C}); \mathbb{F}_2) + \dim \operatorname{Im} i_n.$$

Thus, we get

$$\dim \ H_{n-1}(X(\mathbb{C})/G, X(\mathbb{R}); \mathbb{F}_2) + \dim \ H_{n-1}(X(\mathbb{R}); \mathbb{F}_2)$$

$$= \dim \ H_n(X(\mathbb{C})/G, X(\mathbb{R}); \mathbb{F}_2) - \dim \ H_n(X(\mathbb{C}); \mathbb{F}_2)$$

$$+ \dim \ \text{Im} \ i_n + \dim \ \text{Im} \ i_{n-1}.$$

Considering the sum by n, we then get

$$\dim \ H_*(X(\mathbb{R}); \mathbb{F}_2) = 2 \sum_n \dim \ \text{Im} \ i_n - \dim \ H_*(X(\mathbb{C}); \mathbb{F}_2). \qquad (2-2)$$

From the exact sequence,

$$0 \to H_*(X(\mathbb{C}); \mathbb{F}_2)^G \to H_*(X(\mathbb{C}); \mathbb{F}_2) \to (1+g)H_*(X(\mathbb{C}); \mathbb{F}_2) \to 0,$$

we have

$$\dim \ H^1(G; H_*(X(\mathbb{C}); \mathbb{F}_2)) = \dim \ H_*(X(\mathbb{C}); \mathbb{F}_2)^G - \dim \ (1+g)H_*(X(\mathbb{C}); \mathbb{F}_2)$$
$$(2-3)$$
$$= 2 \dim \ H_*(X(\mathbb{C}); \mathbb{F}_2)^G - \dim \ H_*(X(\mathbb{C}); \mathbb{F}_2).$$

It is clear that
$$\text{Im} \ i_n \subset H_n(X(\mathbb{C}); \mathbb{F}_2)^G. \qquad (2-4)$$
Thus, from (2–2), (2–3) and (2–4), we have an inequality

$$\dim \ H_*(X(\mathbb{R}); \mathbb{F}_2) \le \dim \ H^1(G; H_*(X(\mathbb{C}); \mathbb{F}_2)). \qquad (2-5)$$

Of course, the inequalities (2–1) and (2–5) are equivalent. Moreover, we see that the inequality (2–5) is an equality iff for any n we have for the Smith exact sequence

$$\text{Im} \ i_n = H_n(X(\mathbb{C}); \mathbb{F}_2)^G. \qquad (2-6)$$

Thus, to prove that the spectral sequence II degenerates, we have to prove the equalities (2–6) for $0 \le n \le 4$ for a surface X with the condition $H^3(X(\mathbb{C})/G; \mathbb{F}_2) = 0$.

It will be convenient for us using the following general statement which follows from Smith exact sequence (compare with the proof of [H, Lemma 3.7]).

Proposition 2.3. *For the Smith exact sequence,*

$$\rho_n(H_n(X(\mathbb{C}); \mathbb{F}_2)^G) = \text{Im} \ \{H_{n+1}(X(\mathbb{C})/G; \mathbb{F}_2) \to H_{n+1}(X(\mathbb{C})/G, X(\mathbb{R}); \mathbb{F}_2)$$

$$\xrightarrow{\delta_{n+1}} \text{Ker} \ \delta_n \subset H_n(X(\mathbb{C})/G, X(\mathbb{R}); \mathbb{F}_2)\}.$$

In particular, $\text{Im} \ i_n = H_n(X(\mathbb{C}); \mathbb{F}_2)^G$ *iff the image on the right is zero.*

Proof. We use the following properties of the Smith exact sequence which follow from the definition above of this sequence:

$$i_n(\rho_n \oplus 0) = id + g; \qquad (2-7)$$

and the homomorphism

$$H_n(X(\mathbb{C})/G; X(\mathbb{R}); \mathbb{F}_2) \xrightarrow{\delta_n} H_{n-1}(X(\mathbb{C})/G, X(\mathbb{R}); \mathbb{F}_2) \oplus H_{n-1}(X(\mathbb{R}); \mathbb{F}_2)$$
$$(2-8)$$
$$\xrightarrow{\pi_{X(\mathbb{R})}} H_{n-1}(X(\mathbb{R}); \mathbb{F}_2)$$

is equal to the homomorphism $\partial_n : H_n(X(\mathbb{C})/G, X(\mathbb{R}); \mathbb{F}_2) \to H_{n-1}(X(\mathbb{R}); \mathbb{F}_2)$ in the homological exact sequence of the pair $(X(\mathbb{C})/G, X(\mathbb{R}))$.

Now, let $x_n \in H_n(X(\mathbb{C}); \mathbb{F}_2)^G$. Then, from (2–7), it is equivalent to $i_n(\rho_n(x_n) \oplus 0) = 0$. By Smith exact sequence, it is equivalent to existence of an element $y_{n+1} \in H_{n+1}(X(\mathbb{C})/G, X(\mathbb{R}); \mathbb{F}_2)$ such that $\delta_{n+1}(y_{n+1}) = \rho_n(x_n) \oplus 0$. By (2–8), it is equivalent to

$$\rho_n(x_n) \in \text{Im} \{H_{n+1}(X(\mathbb{C})/G; \mathbb{F}_2) \quad \to H_{n+1}(X(\mathbb{C})/G, X(\mathbb{R}); \mathbb{F}_2)$$
$$\xrightarrow{\delta_{n+1}} H_n(X(\mathbb{C})/G, X(\mathbb{R}); \mathbb{F}_2)\}.$$

Now, we should only remark that, by Smith exact sequence, $\rho_n(x_n) \in \text{Ker } \delta_n$.

From the Proposition 2.3, we have:

Im $i_4 = H_4(X(\mathbb{C}); \mathbb{F}_2)^G$ for any surface since $H_5(X(\mathbb{C})/G; \mathbb{F}_2) = 0$.

Im $i_3 = H_3(X(\mathbb{C}); \mathbb{F}_2)^G$ since for our case $\text{Ker}\delta_3 = 0$, because $\text{Ker}\delta_3 \subset \text{Ker}\partial_3 = 0$. Here $\text{Ker}\partial_3 = 0$, because $H_3(X(\mathbb{C})/G; \mathbb{F}_2) = 0$ for our case.

Im $i_2 = H_2(X(\mathbb{C}); \mathbb{F}_2)^G$ since $H_3(X(\mathbb{C})/G; \mathbb{F}_2) = 0$ in our case.

Im $i_1 = H_1(X(\mathbb{C}); \mathbb{F}_2)^G$ since $\text{Ker}\delta_1 \subset \text{Ker}\partial_1 = 0$ in our case. Here $\text{Ker}\partial_1 = 0$ since $H_1(X(\mathbb{C})/G; \mathbb{F}_2) = 0$ in our case.

Im $i_0 = H_0(X(\mathbb{C}); \mathbb{F}_2)^G$ since $H_0(X(\mathbb{C}), X(\mathbb{R}); \mathbb{F}_2) = 0$ because $X(\mathbb{R}) \neq \emptyset$.

Thus, we proved that the spectral sequence II degenerates.

Now let us prove the formulae of Theorems 1.2 and 0.3. Since the spectral sequence II degenerates, we have

$$\dim H^k(X(\mathbb{C}); G, \mathbb{F}_2) = \bigoplus_{p+q=k} H^p(G; H^q(X(\mathbb{C}); \mathbb{F}_2)).$$

To get formulae, by Poincaré duality, we should only prove that G is trivial on $H^0(X(\mathbb{C}); \mathbb{F}_2)$ and $H_1(X(\mathbb{C}); \mathbb{F}_2)$. It is true for $H^0(X(\mathbb{C}); \mathbb{F}_2)$, since $H^0(X(\mathbb{C}); \mathbb{F}_2) \cong \mathbb{F}_2$. Since $H_1(X(\mathbb{C})/G; \mathbb{F}_2) = 0$, the homomorphism ∂_1 is injective. By (2–8), then the homomorphism δ_1 for the Smith exact sequence is injective too. Thus, the homomorphism ρ_1 is zero and $H_1(X(\mathbb{C}); \mathbb{F}_2) = \text{Im } i_1 \subset H_1(X(\mathbb{C}); \mathbb{F}_2)^G$. It follows the statement. It finishes the proof of Theorems 2.1 and 0.3.

Proof of Theorem 0.4. For $n \in N$, the exact sequence of sheafs

$$0 \to \mathbb{Z} \xrightarrow{\times n} \mathbb{Z} \to \mathbb{Z}/n \to 0$$

gives the exact sequence of cohomology (universal coefficient sequence)

$$\cdots \quad \to H^{k-1}(M;\mathbb{Z}/n) \to H^k(M;\mathbb{Z}) \xrightarrow{\times n} H^k(M;\mathbb{Z})$$

$$\to H^k(M;\mathbb{Z}/n) \to H^{k+1}(M;\mathbb{Z}) \xrightarrow{\times n} H^{k+1}(M;\mathbb{Z}) \to \tag{2-9}$$

For a compact manifold M, the beginning of this sequence gives the exact sequences

$$0 \to H^0(M;\mathbb{Z}) \xrightarrow{\times n} H^0(M;\mathbb{Z}) \to H^0(M;\mathbb{Z}/n) \to 0, \tag{2-10}$$

and

$$0 \to H^1(M;\mathbb{Z}) \xrightarrow{\times n} H^1(M;\mathbb{Z}) \to H^1(M;\mathbb{Z}/n) \to \cdots. \tag{2-11}$$

In particular, $H^1(M;\mathbb{Z})$ has no torsion. As we had mentioned in Introduction, for a smooth surface X, the quotient $X(\mathbb{C})/G$ is a smooth manifold. The group G preserves the canonical orientation of $X(\mathbb{C})$. It follows that $X(\mathbb{C})/G$ is a smooth oriented manifold. Since $H^3(X(\mathbb{C})/G;\mathbb{F}_2) = 0$, by Poincaré duality $H^1(X(\mathbb{C})/G;\mathbb{F}_2) = 0$. Thus, by (2–11), we then get that $H^1(X(\mathbb{C})/G;\mathbb{Z}) = 0$. It follows $H^1(X(\mathbb{C})/G;\mathbb{C}) = 0$. Since 2 is invertible in $\mathbb{C}$, we then have $H^1(X(\mathbb{C});\mathbb{C})^G = \pi^* H^1(X(\mathbb{C})/G;\mathbb{C}) = 0$ where $\pi : X(\mathbb{C}) \to X(\mathbb{C})/G$ is the quotient map (see [Bre, Ch. III, Theorem 2.4]).

For the Hodge decomposition $H^1(X(\mathbb{C});\mathbb{C}) = H^{1,0}(X(\mathbb{C})) + H^{0,1}(X(\mathbb{C}))$, the antiholomorphic involution g on $X(\mathbb{C})$ evidently maps $H^{1,0}(X(\mathbb{C}))$ to $H^{0,1}(X(\mathbb{C}))$. It follows that $0 = \dim H^1(X(\mathbb{C});\mathbb{C})^G = (1/2) \dim H^1(X(\mathbb{C});\mathbb{C})$. Thus, applying (2–11) to $X(\mathbb{C})$, we get

$$H^1(X(\mathbb{C});\mathbb{Z}) = 0. \tag{2-12}$$

Thus, X is a regular surface: the irregularity $q(X) = \dim H^{1,0}(X(\mathbb{C})) = 0$. It follows that the characteristic class map gives an embedding

$$\text{Pic } (X \otimes \mathbb{C}) \subset H^2(X(\mathbb{C});\mathbb{Z}), \tag{2-13}$$

and the image of this map is defined by the condition

$$\{x \in \text{Pic } (X \otimes \mathbb{C}) \mid x \cdot H^{2,0}(X(\mathbb{C})) = 0\} = \gamma^{-1}(H^{1,1}(X(\mathbb{C}))), \tag{2-14}$$

where

$$\gamma : H^2(X(\mathbb{C});\mathbb{Z}) \to H^2(X(\mathbb{C});\mathbb{C})$$

is a coefficient map. See [G-H], for example. From the definition of the characteristic class map, we have

$$g(\gamma(x)) = -\gamma(g(x)) \text{ for } x \in \text{Pic } (X \otimes \mathbb{C}).$$

We remark that $\text{Pic } (X \otimes \mathbb{C})$ contains the all torsion of $H^2(X(\mathbb{C});\mathbb{Z})$ by (2–14).

Since $H^1(X(\mathbb{C});\mathbb{Z}) = 0$ and the sequence

$$0 \to H^4(X(\mathbb{C});\mathbb{Z}) \xrightarrow{\times 2} H^4(X(\mathbb{C});\mathbb{Z}) \to H^4(X(\mathbb{C});\mathbb{F}_2) \to 0$$

is exact, from (2–9), we get the exact sequence

$$0 \quad \to H^1(X(\mathbb{C});\mathbb{F}_2) \to H^2(X(\mathbb{C});\mathbb{Z}) \xrightarrow{\times 2} H^2(X(\mathbb{C});\mathbb{Z})$$

$$\to H^2(X(\mathbb{C});\mathbb{F}_2) \to H^3(X(\mathbb{C});\mathbb{Z}) \xrightarrow{\times 2} H^3(X(\mathbb{C});\mathbb{Z}) \to H^3(X(\mathbb{C});\mathbb{F}_2) \to 0. \tag{2-15}$$

Besides, by Poincaré duality, dim $H^1(X(\mathbb{C}); \mathbb{F}_2) = \dim H^3(X(\mathbb{C}); \mathbb{F}_2)$. It follows that

$$\dim H^2(X(\mathbb{C}); \mathbb{Z})/2H^2(X(\mathbb{C}); \mathbb{Z}) = b_2 + \dim H^1(X(\mathbb{C}); \mathbb{F}_2), \qquad (2-16)$$

and

$$\dim H^2(X(\mathbb{C}); \mathbb{F}_2) = b_2 + 2\dim H^1(X(\mathbb{C}); \mathbb{F}_2) \qquad (2-17)$$

where the Betti number $b_2 = \dim H^2(X(\mathbb{C}); \mathbb{C}) = \dim H^2(X(\mathbb{C}); \mathbb{Z}) \otimes \mathbb{C}$.

Let Pic $(X \otimes \mathbb{C}) = T \oplus \mathbb{Z}^{\rho(X \otimes \mathbb{C})}$ where T is the torsion of Pic $(X \otimes \mathbb{C})$ and $\mathbb{Z}^{\rho(X \otimes \mathbb{C})} =$ Pic $(X \otimes \mathbb{C})/T$. Since $X(\mathbb{R}) \neq \emptyset$, we have

$$\text{Pic } X = \text{Pic } (X \otimes \mathbb{C})^G$$

(this is well-known, see [Ma]). Let

$$\text{Pic } X = \text{Pic } (X \otimes \mathbb{C})^G = T' \oplus \mathbb{Z}^{\rho(X)}$$

where T' is the torsion of Pic X and $\mathbb{Z}^{\rho(X)} = $ Pic X/T'. If for $a \in$ Pic $(X \otimes \mathbb{C})$ we have $g(a) = a \bmod T$, then $g(ma) = ma$ for some $m \in N$ such that $mT = 0$. It follows that

$$(\text{Pic } (X \otimes \mathbb{C})/T)^G \cong (\text{Pic } X)/T' = \mathbb{Z}^{\rho(X)}.$$

Thus,

$$\rho(X) = \rho_+(X \otimes \mathbb{C}), \qquad (2-18)$$

where $(\text{Pic } (X \otimes \mathbb{C})/T)^G \cong \mathbb{Z}^{\rho_+(X \otimes \mathbb{C})}$. We had proven above that G is trivial on $H_1(X(\mathbb{C}); \mathbb{F}_2)$. Then, it is trivial on $H^1(X(\mathbb{C}); \mathbb{F}_2) = H_1(X(\mathbb{C}); \mathbb{F}_2)^*$. From (2–15) and the remarks above, we then get that the group G is trivial on

$$\text{Ker}\{\text{Pic } (X \otimes \mathbb{C}) \xrightarrow{\times 2} \text{Pic } (X \otimes \mathbb{C})\} = H^1(X(\mathbb{C}); \mathbb{F}_2).$$

Thus,

$$\text{Ker}\{\text{Pic } X \xrightarrow{\times 2} \text{Pic } X\} = H^1(X(\mathbb{C}); \mathbb{F}_2).$$

As a result, we get that

$$\dim \text{ Pic } X/2\text{Pic } X = \rho_+(X \otimes \mathbb{C}) + \dim H^1(X(\mathbb{C}); \mathbb{F}_2). \qquad (2-19)$$

Here, by the remarks above about the map (2–13),

$$\rho_+(X \otimes \mathbb{C}) \le h_-^{1,1}(X(\mathbb{C})).$$

From Proposition 1.3 and formulae of Theorem 2.1, we have

$$\dim H^*(X(\mathbb{R}); \mathbb{F}_2) \quad = \quad \dim H^5(X(\mathbb{C}); G, \mathbb{F}_2)$$

$$= \quad 2\dim H^2(X(\mathbb{C}); G, \mathbb{F}_2) - \dim H^2(X(\mathbb{C}); \mathbb{F}_2).$$

Thus,

$$\dim H^2(X(\mathbb{C}); G, \mathbb{F}_2) = (1/2)\dim H^*(X(\mathbb{R}); \mathbb{F}_2) + (1/2)\dim H^2(X(\mathbb{C}); \mathbb{F}_2). \qquad (2-20)$$

By Proposition 1.1, $H^2(X(\mathbb{C}); G, \mathbb{F}_2) = H^2_{et}(X; \mathbb{F}_2)$. By the exact sequence (0–3), we get $\dim {}_2Br'(X) = \dim H^2_{et}(X; \mathbb{F}_2) - \dim$ Pic $X/2$Pic X. Thus, from (2–17), (2–19) and (2–20), we get

$$\dim {}_2Br'(X) = (1/2)\dim H^*(X(\mathbb{R}); \mathbb{F}_2) + b_2/2 - \rho_+(X \otimes \mathbb{C}). \qquad (2-21)$$

Let $(b_2)_+ = \dim H^2(X(\mathbb{C}); \mathbb{C})^G$ and $(b_2)_- = b_2 - (b_2)_+$. From the Lefschetz fixed point formula for the involution g (see [Sp], for example), we get

$$\chi(X(\mathbb{R})) = 2 + 2(b_2)_+ - b_2 = 2 + b_2 - 2(b_2)_-. \qquad (2-22)$$

Thus, from (2–21) and (2–22), we get

$$\dim {}_2Br'(X) = (1/2)\dim H^*(X(\mathbb{R}); \mathbb{F}_2) + (1/2)\chi(X(\mathbb{R})) - 1 + (b_2)_- - \rho_+(X \otimes \mathbb{C}). \qquad (2-23)$$

For the Hodge decomposition

$$H^2(X(\mathbb{C}); \mathbb{C}) = H^{2,0}(X(\mathbb{C})) + H^{0,2}(X(\mathbb{C})) + H^{1,1}(X(\mathbb{C})),$$

the antiholomorphic involution g sends $H^{2,0}(X(\mathbb{C})) \to H^{0,2}(X(\mathbb{C}))$ and $H^{1,1}(X(\mathbb{C})) \to H^{1,1}(X(\mathbb{C}))$. It follows that

$$(b_2)_- = \dim H^{2,0}(X(\mathbb{C})) + \dim H^{1,1}_-(X(\mathbb{C})).$$

For a connected compact surface F we have $\dim H^*(F; \mathbb{F}_2) + \chi(F) = 4$. Thus, from (2–23), we get the formula of Theorem 0.4.

Proof of Theorem 0.6. Let Y be a real Enriques surface. In [N-S], the inequality (0–7), i.e. $b(Y) \geq 2s - 2$, was proved. The proof was similar to the proof above of the Theorem 0–4 and used Lefschetz fixed-point formula and the inequality (2–1). From the proof, it follows that the equality $b(Y) = 2s - 2$ holds iff the inequality (2–1) is an equality. By Proposition 2.2 (of V.A.Krasnov), it then follows that the spectral sequence II degenerates iff $b(Y) = 2s - 2$. By Proposition 1.1, the Hochschild-Serre spectral sequence degenerates iff $b(Y) = 2s - 2$. It follows the statement (i) of Theorem 0.6. From the statement (i), the definition of the invariant $\epsilon(Y)$, and from (0–6), (0–7), the statements (ii) and (iii) of Theorem 0.6 follow.

3 Applications to topology of real Enriques surfaces

We use notation on real Enriques surfaces of Introduction. Thus, for a real Enriques surface Y, we denote by θ the antiholomorphic involution of Y, by X the universal covering $K3$-surface, by τ the holomorphic involution of the 2-sheeted universal covering $\pi : X(\mathbb{C}) \to Y(\mathbb{C})$, and by $\sigma, \tau\sigma$ two liftings of θ on $X(\mathbb{C})$. We suppose that the automorphism group

$$\Gamma = \{\mathrm{id}, \tau, \sigma, \tau\sigma\}$$

on $X(\mathbb{C})$ is isomorphic to $(\mathbb{Z}/2)^2$. In particular, it is true if $Y(\mathbb{R}) \neq \emptyset$ (see [N-S]).

First, we discuss the following problem. We have $\dim H^2(Y(\mathbb{C}); \mathbb{F}_2) = 12$. A subgroup $H^2(Y(\mathbb{C}); \mathbb{Z}) \otimes \mathbb{F}_2 \subset H^2(Y(\mathbb{C}); \mathbb{F}_2)$ has $\dim H^2(Y(\mathbb{C}); \mathbb{Z}) \otimes \mathbb{F}_2 = 11$. Thus, we can introduce the invariant

$$\beta(Y) = \dim H^2(Y(\mathbb{C}); \mathbb{F}_2)^\theta - \dim(H^2(Y(\mathbb{C}); \mathbb{Z}) \otimes \mathbb{F}_2)^\theta.$$

This invariant is very important for real Enriques surfaces, and first, we want to calculate $\beta(Y)$ in some cases. Evidently, $\beta(Y) = 0$ or 1. For the invariant

$$b(Y) = \dim\ H^2(Y(\mathbb{C}); \mathbb{F}_2)^\theta - \dim\ (\text{Pic}\ Y \otimes \mathbb{C})^\theta /2(\text{Pic}\ Y \otimes \mathbb{C})^\theta + 1,$$

(see Introduction), we have

$$b(Y) = b'(Y) + \beta(Y), \tag{3 – 1}$$

where we denote

$$b'(Y) = \dim\ (H^2(Y(\mathbb{C}); \mathbb{Z}) \otimes \mathbb{F}_2)^\theta - \dim\ (\text{Pic}\ Y \otimes \mathbb{C})^\theta /2(\text{Pic}\ Y \otimes \mathbb{C})^\theta + 1.$$

In [N-S, Theorem 3.4.7], there was obtained a formula for $b'(Y)$:

$$b'(Y) = r(\theta) - a(\theta) + \max\{1 - \alpha(\sigma),\ (\delta_{\sigma L^{\tau,\sigma}} + \delta_{\sigma L_\sigma^\tau})/2\}. \tag{3 – 2}$$

Here $r(\theta), a(\theta), \alpha(\sigma), \delta_{\sigma L^{\tau,\sigma}}, \delta_{\sigma L_\sigma^\tau}$ are some invariants of the action of Γ on the lattice L which is the lattice $H^2(X(\mathbb{C}); \mathbb{Z})$ with the intersection pairing. The invariants $r(\sigma), a(\sigma)$ are some non-negative integers, and

$$r(\sigma) \equiv a(\sigma)\bmod 2. \tag{3 – 3}$$

The invariants $\alpha(\sigma), \delta_{\sigma L^{\tau,\sigma}}, \delta_{\sigma L_\sigma^\tau}$ are equal to 0 or 1, and

$$\delta_{\sigma L^{\tau,\sigma}} = \delta_{\sigma L_\sigma^\tau}. \tag{3 – 4}$$

The precise definition of these invariants is very long, and we refer to [N-S] for their definition. Actually, these invariants are some specialization to Enriques surfaces of general invariants from [N1, N2a].

Besides, in [N-S], it was proved that the invariant $\beta(Y) = 0$ if

$$\max\{1 - \alpha(\sigma),\ (\delta_{\sigma L^{\tau,\sigma}} + \delta_{\sigma L_\sigma^\tau})/2\} = 0$$

or, equivalently, $\alpha(\sigma) = 1$ and $\delta_{\sigma L^{\tau,\sigma}} = \delta_{\sigma L_\sigma^\tau} = 0$.

We want to prove here the following result which also gives another prove of the statement about $\beta(Y)$ above, but only in the case if both real parts $X_\sigma(\mathbb{R})$ and $X_{\tau\sigma}(\mathbb{R})$ are non-empty.

Theorem 3.1. *Let Y be a real Enriques surface and both $X_\sigma(\mathbb{R})$ and $X_{\tau\sigma}(\mathbb{R})$ are non-empty.*

Then:

$$\beta(Y) = \max\{1 - \alpha(\sigma),\ (\delta_{\sigma L^{\tau,\sigma}} + \delta_{\sigma L_\sigma^\tau})/2\},$$

and

$$b(Y) = r(\theta) - a(\theta) + 2\max\{1 - \alpha(\sigma),\ (\delta_{\sigma L^{\tau,\sigma}} + \delta_{\sigma L_\sigma^\tau})/2\},$$

where $r(\theta)$, $a(\theta)$, $\alpha(\sigma)$, $\delta_{\sigma L^{\tau,\sigma}}$, $\delta_{\sigma L_\sigma^\tau}$ are some invariants of the action of the group Γ on the lattice $H^2(X(\mathbb{C}); \mathbb{Z})$ with the intersection pairing.

Proof. By Theorem 0.6, $b(Y) = 2s - 2 \equiv 0\bmod 2$. By (3–1) – (3–3), we then get

$$\beta(Y) \equiv \max\{1 - \alpha(\sigma),\ (\delta_{\sigma L^{\tau,\sigma}} + \delta_{\sigma L_\sigma^\tau})/2\}\bmod 2.$$

The right side of this congruence is equal to 0 or 1 since $\alpha(\sigma) = 0$ or 1, and $\delta_{\sigma L^{\tau,\sigma}} = \delta_{\sigma L_\sigma^\tau} = 0$ or 1. It follows the first formula since $\beta(Y)$ is equal to 0 or 1 too. From the first formula and (3–1), (3–2), the second one follows.

We don't know if this statement valid when $\max\{1 - \alpha(\sigma), (\delta_{\sigma L^{\tau,\sigma}} + \delta_{\sigma L^\tau_\sigma})/2\} = 1$ and one of $X_\sigma(\mathbb{R})$ or $X_{\tau\sigma}(\mathbb{R})$ is empty.

In [N-S, Theorems 3.5.1–3.5.3, formula (3–5–1)], there was obtained a formula for $b(Y)$ using the numbers s_{or} and s_{nor} of orientable and non-orientable connected components of $Y(\mathbb{R})$ respectively. The numbers s_{or} and s_{nor} are connected with the numbers $s(\sigma)$ and $s(\tau\sigma)$ of connected components of $X_\sigma(\mathbb{R})$ and $X_{\tau\sigma}(\mathbb{R})$ respectively by the formula $s(\sigma) + s(\tau\sigma) = 2s_{or} + s_{nor}$ (see [N-S, Lemma 3.2.1]). For the case $s(\sigma) > 0$ and $s(\tau\sigma) > 0$ (or when both sets $X_\sigma(\mathbb{R})$ and $X_{\tau\sigma}(\mathbb{R})$ are non-empty) which is necessary for us , this formula for $b(Y)$ claims that

$$b(Y) \;=\; 2s_{or} + s_{nor} - 2 + \min\left\{\alpha(\sigma), (\delta_{\sigma L^{\tau,\sigma}} + \delta_{\sigma L^\tau_\sigma})/2\right\}$$

$$+ \;\; \dim\ H(\sigma)_- - \dim\ H(\sigma)^\perp_+ \cap H(\sigma)_- + \beta(Y).$$

Here $\dim\ H(\sigma)_-$ and $\dim\ H(\sigma)^\perp_+ \cap H(\sigma)_-$ are some other invariants ot the action of the group Γ on the lattice L (see [N-S]). From the formula for $\beta(Y)$ of Theorem 3.1, we get the formula

$$b(Y) \;=\; 2s_{or} + s_{nor} - 2 + \min\{\alpha(\sigma), (\delta_{\sigma L^{\tau,\sigma}} + \delta_{\sigma L^\tau_\sigma})/2\}$$

$$+ \;\; \max\{1 - \alpha(\sigma), (\delta_{\sigma L^{\tau,\sigma}} + \delta_{\sigma L^\tau_\sigma})/2\} + \dim\ H(\sigma)_- - \dim\ H(\sigma)^\perp_+ \cap H(\sigma)_-$$
$$\text{(3 – 5)}$$

$$= \;\; 2s_{or} + s_{nor} - 1 + \alpha(\sigma)(\delta_{\sigma L^{\tau,\sigma}} + \delta_{\sigma L^\tau_\sigma} - 1)$$

$$+ \;\; \dim\ H(\sigma)_- - \dim\ H(\sigma)^\perp_+ \cap H(\sigma)_-,$$

if both $X_\sigma(\mathbb{R})$ and $X_{\tau\sigma}(\mathbb{R})$ are non-empty.

By the formula $b(Y) = 2s - 2 = 2s_{or} + 2s_{nor} - 2$ of Theorem 0.6, we get

Theorem 3.2. *Let Y be a real Enriques surface and both $X_\sigma(\mathbb{R})$ and $X_{\tau\sigma}(\mathbb{R})$ are non-empty.*

Then for the number s_{nor} of non-orientable connected components of $Y(\mathbb{R})$ we have the formula:

$$s_{nor} = 1 + \alpha(\sigma)(\delta_{\sigma L^{\tau,\sigma}} + \delta_{\sigma L^\tau_\sigma} - 1) + \dim\ H(\sigma)_- - \dim\ H(\sigma)^\perp_+ \cap H(\sigma)_-, \;\; \text{(3 – 6)}$$

where $\alpha(\sigma)$, $\delta_{\sigma L^{\tau,\sigma}}$, $\delta_{\sigma L^\tau_\sigma}$, $\dim\ H(\sigma)_-$, $\dim\ H(\sigma)^\perp_+ \cap H(\sigma)_-$ are some invariants of the action of the group Γ on the lattice $H^2(X(\mathbb{C}); \mathbb{Z})$ with the intersection pairing.

We mention that by Theorem 0.6, $b(Y) = 2s - 2$ if both $X_\sigma(\mathbb{R})$ and $X_{\tau\sigma}(\mathbb{R})$ are non-empty. Thus, by the formula for $b(Y)$ of Theorem 3.1, we also have the formula for the number $s = s_{or} + s_{nor}$ of all connected components of $Y(\mathbb{R})$ if both $X_\sigma(\mathbb{R})$ and $X_{\tau\sigma}(\mathbb{R})$ are non-empty.

Theorem 3.3. *Let Y be a real Enriques surface and both $X_\sigma(\mathbb{R})$ and $X_{\tau\sigma}(\mathbb{R})$ are non-empty.*

Then for the number s of all connected components of $Y(\mathbb{R})$ we have the formula

$$s = 1 + (r(\theta) - a(\theta))/2 + \max\{1 - \alpha(\sigma), (\delta_{\sigma L^{\tau,\sigma}} + \delta_{\sigma L^\tau_\sigma})/2\},$$

where $r(\theta)$, $a(\theta)$, $\alpha(\sigma)$, $\delta_{\sigma L^{\tau,\sigma}}$, $\delta_{\sigma L_\sigma^\tau}$ *are some invariants of the action of the group* Γ *on the lattice* $H^2(X(\mathbb{C}); \mathbb{Z})$ *with the intersection pairing.*

Of course, from the formulae for s and s_{nor} or Theorems 3.2, 3.3, we get a formula for $s_{or} = s - s_{nor}$. These Theorems 3.1–3.3 are very important for the topological classification of real Enriques surfaces–describing of all possible topological types of $Y(\mathbb{R})$ for real Enriques surfaces Y. See [N4] for these applications.

References

[A]   I. R. Shafarevich (ed.), *Algebraic surfaces*, Proc. Steklov Math. Inst. vol. 75, 1965; English transl. by A.M.S. 1969.

[Bre]   G. E. Bredon, *Introduction to compact transformation groups*, Academic Press, New York and London, 1972.

[Bro]   K. S. Brown, *Cohomology of groups*, Springer, 1982.

[C]   D. A. Cox, *The étale homotopy type of varieties over* $\mathbb{R}$, Proc. Amer. Math. Soc. vol. 76, no. 1 (1979).

[C-D]   F. R. Cossec and I. Dolgachev, *Enriques surfaces.I*, Progress in Mathematics, vol. 76, Birkhäuser, 1989.

[CT-P]   J.-L. Colliot-Thélène and R. Parimala, *Real components of algebraic varieties and étale cohomology*, Invent. math. 101 (1990), 81-99.

[Go]   R. Godement, *Topologie algébrique et théorie des faisceaux*, Hermann, Paris, 1958.

[Gr1]   A. Grothendieck, *Sur quelques points d'algèbre homologique*, Tohoku Mathem. J. 9(1957), 119-221.

[Gr2]   A. Grothendieck, *Le groupe de Brauer*, Dix Exposes sur la Cohomologie des Schemas, North-Holland, Amsterdam, 1968, 46-188.

[G-H]   P. Griffiths and J. Harris, *Principles of algebraic geometry*, John Wiley and sons, New York, 1978.

[H]   V. M. Harlamov, *Topological types of nonsingular surfaces of degree 4 in* RP^3, Funktcional. Anal. i Prilozhen. 10(1976), 55-68; English transl. in Functional Anal. Appl.

[Kr]   V. A. Krasnov, *Harnack-Thom inequalities for mappings of real algebraic varieties*, Izv. Akad. Nauk SSSR Ser. Mat. 47(1983), 268-297; English transl. in Math. USSR Izv. 22(1984), 247-275.

[Ku]   Vik. S. Kulikov, *Degenerations of* $K3$-*surfaces and Enriques surfaces*, Izv. Akad. Nauk SSSR Ser. Mat. 41(1977), 1008-1042; English transl. in Math. USSR Izv. 11(1978), 957-989.

[Ma]   Y. I. Manin, *Le groupe de Brauer-Grothendieck en Géometrie diophantienne*, Actes du Congrès Intern. Math. Nice (1970), vol. 1, Gauthier-Villars, Paris, 1971, 401-411.

[Mi]   J. Milne, *Étale cohomology*, Princeton Univ. Press, 1980.

[N1]   V. V. Nikulin, *Integral symmetric bilinear forms and some of their geometric applications*, Izv. Akad. Nauk SSSR Ser. Mat. 43(1979), 111-177; English transl. in Math. USSR Izv. 14(1980), 103-167.

[N2]   V. V. Nikulin, *Involutions of integral quadratic forms and their application to real algebraic geometry*, Izv. Akad. Nauk SSSR Ser. Mat. 47(1983), 109-188; English transl. in Math. USSR Izv. 22(1984).

[N3]   V. V. Nikulin, *Lectures on the Brauer group of real algebraic surfaces*, Preprint of University of Notre Dame, College of science, Dept. of Math. # 179 (January, 1992).

[N4]   V. V. Nikulin, *On the topological classification of real Enriques surfaces.I*, Universität Bielefeld, Sonderforschungsbereich 343, Diskrete Strukturen in der Mathematik (1992) (to appear).

[N-S]   V. V. Nikulin and R. Sujatha, *On Brauer groups of real Enriques surfaces*, Preprint IHES /M/91/33 (1991); To appear in J. reine angew. Math.

[PŠ-Š]   I. I. Pjateckiĭ-Šapiro and I. R. Šafarevič, *A Torelli theorem for algebraic surfaces of type K3*, Izv. Akad. Nauk SSSR Ser. Mat. 35(1971), 530-572; English transl. in Math. USSR Izv. 5(1971).

[R]   V. A. Rokhlin, *Congruences modulo 16 in Hilbert's sixteen problem*, Funktcional. Anal. i Prilozhen 4(1972), no. 4, 58-64; English transl. in Funct. Anal. Appl. 6(1972), 301-306.

[Sp]   E. H. Spanier, *Algebraic Topology*, McGraw-Hill Book Company, 1966.

[Su]   R. Sujatha, *Witt groups of real projective surfaces*, Math. Ann. 28(1990), 89-101.

[W]   E. Witt, *Zerlegung reeler algebraischer Funktionen in Quadrate, Schiefkörper über reellem Funktionenkörper*, J. für die reine und angew Math. 171(1934), 4-11.

Symplectic Twistors and Geometric Quantization of Strings

A.D.Popov, A.G.Sergeev

Abstract. We present the geometric quantization scheme for the bosonic string theory in twistor terms. Starting from the loop space of a Lie group we define a symplectic twistor bundle over the loop space and reformulate the geometric quantization problem in terms of this bundle. For the standard bosonic string we recover in this way the well known critical dimension condition.

In this paper motivated by Bowick–Rajeev [1] we formulate a geometric quantization scheme for the string theory based on the twistor approach. The role of the phase space for a bosonic string is played by the loop space ΩG of a Lie group G (for the standard bosonic string G is the group of translations of the d-dimensional Minkowski space $\mathbb{R}^{d-1,1}$). The loop space ΩG is an infinite-dimensional analogue of flag manifolds and in fact can be considered (for a compact G) as a universal flag manifold of G. We define a twistor bundle $\mathcal{Z} \to \Omega G$ as a bundle of invariant (w.r. to the natural action of the mapping group $LG = Map(S^1, G)$ on ΩG) complex structures on ΩG compatible with the symplectic structure of ΩG. Structures of that type in a given point of ΩG are parametrized by points of an infinite-dimensional manifold $\mathcal{S} = Diff(S^1)/S^1$. This manifold may be considered as an infinite-dimensional analogue of the Hermitian symmetric domain $Sp(2n, \mathbb{R})/U(n)$ and is closely related (as in the finite-dimensional case) to the universal Teichmüller space S consisting of univalent holomorphic functions in the unit disc in $\mathbb{C}$. The space S was considered in a series of papers [2]. We introduce an almost complex structure on the twistor space $\mathcal{Z}$ so that the natural projection $p : \mathcal{Z} \to \mathcal{S}$ is a holomorphic mapping. Thus we obtain the following twistor diagram

$$
\begin{array}{ccc}
\mathcal{Z} & \overset{p}{\longrightarrow} & \mathcal{S} \\
\pi \downarrow & & \\
\Omega G & &
\end{array}
\qquad (1)
$$

which is crucial for our considerations.

Double fibrations similar to (1) arise naturally in different problems of the twistor theory. We refer first to the non-linear graviton of Penrose [3] where instead of ΩG a Lorentz

4-manifold M is considered and the role of S is played by the complex projective line $\mathbb{CP}^1$. Then points of M are identified according to (1) with real holomorphic sections of $p : \mathcal{Z} \to \mathbb{CP}^1$. Here "real" means invariant w.r. to a real structure on $\mathcal{Z}$ having no fixed points. Another example is given by hyperKähler manifolds [4]. In particular, Hitchin has defined a double fibration of the type (1) where ΩG is substituted by a $4n$-dimensional hyperKähler manifold M and S — by $\mathbb{CP}^1$. In this case $\mathcal{Z}$ appears to be a $(2n + 1)$-dimensional complex manifold and points of M are again identified with real holomorphic sections of $\mathcal{Z} \to \mathbb{CP}^1$. The main difference between these constructions and our diagram (1) is that twistor spaces considered in [3], [4] consist of almost complex structures compatible with a Riemannian metric while ours — of complex structures compatible with a symplectic form. So it's natural to call the twistor space we are considering by a symplectic twistor space.

In terms of the diagram (1) the geometric quantization proceeds as follows. Using a construction analogous to that of Ward from the twistor theory we pull back the prequantum bundle L over ΩG to a holomorphic line bundle $\tilde{L} \to \mathcal{Z}$. Projective representations of the Lie algebra $Vect(S^1)$ of of the group $\mathrm{Diff}(S^1)$ in polarized Fock spaces on ΩG generate a connection on a Fock bundle $\tilde{H} \to S$ with the fiber $\tilde{H}_J$ in a point $J \in S$ consisting of holomorphic sections of $\tilde{L}$ over $p^{-1}(J) \subset \mathcal{Z}$. The quantization of ΩG is equivalent in twistor terms to the construction of a flat unitary connection on a quantization bundle $\tilde{\mathcal{H}} \to S$ over S. The Fock bundle $\tilde{H}$ cannot be used for $\tilde{\mathcal{H}}$ because the connection mentioned above is never flat. To circumvent this difficulty it's necessary to introduce "ghosts". In twistor terms it means that we should "twist" (tensorially multiply) $\tilde{L}$ by the square root $K^{-1/2} \to S$ of the anticanonical bundle of S. The bundle $K^{-1/2}$ can be provided with a natural spinorial connection so that the bundle $\tilde{\mathcal{H}} := \tilde{H} \otimes K^{-1/2} \to S$ has the connection being the tensor product of connections on $\tilde{H}$ and $K^{-1/2}$. This connection on $\tilde{\mathcal{H}}$ is flat in the critical dimension equal to $d = 26$ for $G = \mathbb{R}^{d-1,1}$ (the group of translations of the Minkowski space $\mathbb{R}^{d-1,1}$) and to $c(\mathbf{g}) = \frac{k \dim \mathbf{g}}{k+\kappa(\mathbf{g})}$ for a simple compact Lie group G (where $\kappa(\mathbf{g})$ is the dual Coxeter number of the Lie algebra $\mathbf{g}$, k is an integer parametrizing Kähler metrics on ΩG).

We are grateful to A.N.Tyurin for the discussion of the results of this paper.

1 Twistors on Riemannian and symplectic manifolds

1 Riemannian twistors

We recall here some basic facts about twistor spaces over Riemannian and in particular hyperKähler manifolds (cf. [5],[4]).

1.1 Hermitian structures on $\mathbb{R}^{2n}$

Let $\mathbb{R}^{2n}$ be the Euclidean real vector space of dimension $2n$. Denote by $\mathcal{J}(\mathbb{R}^{2n})$ the space of Hermitian structures on $\mathbb{R}^{2n}$, i.e. complex structures J on $\mathbb{R}^{2n}$ compatible with the Euclidean metric. It means that $J \in \mathrm{End}(\mathbb{R}^{2n})$ with $J^2 = -\mathrm{id}$ and J is an isometry of $\mathbb{R}^{2n}$. The space $\mathcal{J}(\mathbb{R}^{2n}) \approx O(2n, \mathbb{R})/U(n)$ is a compact Hermitian symmetric space having two connected components isomorphic to $SO(2n, \mathbb{R})/U(n)$. The space $SO(2n, \mathbb{R})/U(n)$ may be also considered as the space of complex structures on $\mathbb{R}^{2n}$ compatible with the metric and orientation of $\mathbb{R}^{2n}$. In particular, for $n = 2$ we have $SO(4, \mathbb{R})/U(2) \approx \mathbb{CP}^1$.

1.2 The bundle of almost Hermitian structures

Let M be a Riemannian manifold of dimension $2n$. We define the twistor bundle of M as the bundle $\pi : Z = \mathcal{J}(M) \to M$ of almost Hermitian structures on M associated with the principal $O(2n, \mathbb{R})$-bundle of orthogonal frames of M. It means that the fiber $\pi^{-1}(p)$ of $Z \to M$ over a point $p \in M$ coincides with the space $\mathcal{J}(T_pM)$ of Hermitian structures on T_pM defined above. Sections of π are identified with almost complex structures on M.

1.3 Almost complex structure of Z

While a manifold M has in general no preferred complex or almost complex structure its twistor space Z can be always provided in a natural way with an almost complex structure. In fact, let's denote by ∇ the Levi-Civita connection on M. It generates the splitting of the tangent bundle TZ into the direct sum

$$TZ = V \oplus H \tag{2}$$

of vertical and horizonthal subbundles of TZ. The fiber V_z in $z \in Z$ is tangent to the fiber $\pi^{-1}(\pi(z))$ of $Z \to M$ through the point z. Recall that the fiber of $Z \to M$ over $\pi(z)$ is identified with $\mathcal{J}(T_pM) \approx O(2n, \mathbb{R})/U(n)$ so it has a natural complex structure $\mathcal{J}^v$. Hence we can define an almost complex structure $\mathcal{J}$ on Z using the decomposition (2) by setting

$$\mathcal{J} = \mathcal{J}^v \oplus \mathcal{J}^h$$

where $\mathcal{J}^h$ is an almost complex structure equal in a point $z \in Z$ to the complex structure $\mathcal{J}_z^h$ on $T_zH \approx T_{\pi(z)}M$ given by the point z.

Thus the twistor space Z has a natural almost complex structure; using it we can analyze the real geometry of M through the complex geometry of Z. Unfortunately, this almost complex structure is almost never integrable (it is integrable $\iff$ M is conformally flat, cf. [5]) so Z is only an almost complex manifold. However, it is possible to construct complex twistor manifolds over M by the restriction of Z or, more precisely, by considering almost complex structures on M compatible with the holonomy of M and then resticting the class of admissible manifolds M. One of the most important examples of this kind is the class of hyperKähler manifolds.

1.4 HyperKähler manifolds

A Riemannian manifold M of dimension $4n$ is called hyperKähler if it has almost complex structures I, J, K such that

$$I J = K, \quad J K = I, \quad K I = J$$

which are parallel with respect to the Riemannian metric, i.e.

$$\nabla I = \nabla J = \nabla K = 0$$

for the Levi-Civita connection ∇ on M. Otherwise speaking, they are manifolds with the holonomy group Sp_n.

Now we take for the twistor bundle $Z \to M$ the bundle of almost Hermitian structures on M compatible with the holonomy of M. The fiber of $\pi : Z \to M$ can be identified with the Riemann sphere $S^2 \approx \mathbb{CP}^1$ of complex structures having the form

$$a\,I + b\,J + c\,K$$

where $(a, b, c) \in S^2 \subset \mathbb{R}^3$. The twistor manifold Z is now complex and the natural projection $p : Z \to \mathbb{CP}^1$ is holomorphic so we obtain a twistor diagram of the type (1)

$$\begin{array}{ccc} Z & \xrightarrow{\ p\ } & \mathbb{CP}^1 \\ \pi \downarrow & & \\ M & & \end{array}$$

Moreover, Z has a natural real structure given by the antipodal mapping on fibers of π for which fibers of π are real (i.e. invariant under the real structure) holomorphic sections of $p : Z \to \mathbb{CP}^1$ having the normal bundle $\mathcal{O}(1) \otimes \mathbb{C}^{2n}$.

Conversely, Hitchin (cf. [4]) proved that if a holomorphic bundle $p : Z^{2n+1} \to \mathbb{CP}^1$ has a real $(4n)$-parameter collection of real holomorphic sections with the normal bundle $\mathcal{O}(1) \otimes \mathbb{C}^{2n}$ then the parameter space M is a hyperKähler manifold.

2 Symplectic twistors

In the geometric quantization method the phase space M of a system is a symplectic manifold so it's natural to introduce for a twistor space of M a bundle of almost complex structures on M compatible with the symplectic structure of M rather than its metric.

2.1 PseudoKähler structures on $\mathbb{R}^{2n}$

Denote now by $\mathbb{R}^{2n}$ the real vector space of dimension $2n$ with coordinates $(p, q) = (p_1, \ldots, p_n, q_1, \ldots, q_n)$ and the standard symplectic structure

$$\omega_0 = \sum_{i=1}^{n} dp_i \wedge dq_i \ .$$

Let $\mathcal{S}(\mathbb{R}^{2n})$ be the space of pseudoKähler structures on $\mathbb{R}^{2n}$, i.e. complex structures J on $\mathbb{R}^{2n}$ compatible with the symplectic structure ω_0. It means that $J \in \mathrm{End}(\mathbb{R}^{2n})$ belongs to the group $\mathrm{Sp}(2n, \mathbb{R})$ of linear symplectic transformations of $\mathbb{R}^{2n}$.

The space $\mathcal{S}(\mathbb{R}^{2n}) \approx \mathrm{Sp}(2n, \mathbb{R})/\mathrm{U}(n)$ of pseudoKähler structures is an Hermitian symmetric domain of dimension $\frac{n(n+1)}{2}$ which can be identified with the Siegel unit disc S_n. This disc consists of complex $n \times n$-matrices Z subject to

$$Z^t = Z \quad , \quad I - Z^* Z \gg 0 \quad \text{(positive definite)}$$

where Z^t is the transpose matrix of Z, Z^* – its Hermitian conjugate. For the proof of the identification $\mathrm{Sp}(2n, \mathbb{R})/\mathrm{U}(n) \cong S_n$ note that the action of $\mathrm{Sp}(2n, \mathbb{R})$ on $\mathbb{R}^{2n} \approx \mathbb{C}^n$ can be given (in coordinates $z_j, \bar{z}_j$ on $\mathbb{C}^n$) by block matrices

$$g = \begin{pmatrix} A & B \\ \bar{B} & \bar{A} \end{pmatrix}$$

preserving the matrix $\begin{pmatrix} 0 & I \\ -I & 0 \end{pmatrix}$ of the symplectic form ω_0: $g\omega_0 g^t = \omega_0$, i.e. subject to the relations

$$AB^t = BA^t \quad , \quad AA^* - BB^* = I .$$

The group $\mathrm{Sp}(2n, \mathbb{R})$ acts transitively on the Siegel disc S_n by the fractional linear transformations

$$Z \longmapsto (AZ + B)(\bar{B}Z + \bar{A})^{-1}$$

and the isotropy group of the point $Z = 0$ is $\mathrm{U}(n)$.

There exists a well known relation between the Teichmüller space T_g of Riemann surfaces with genus g and the Siegel upper halfplane. The Siegel upper halfplane is biholomorphic to the Siegel unit disc by the Cayley transform thus we obtain a natural holomorphic mapping from T_g into the Siegel disc S_g, hence into $\mathcal{S}(\mathbb{R}^{2g})$.

In the simplest case $n = 1$ we have $\mathcal{S}(\mathbb{R}^2) = \mathrm{Sp}(2, \mathbb{R})/\mathrm{U}(1) = $ the unit disc in $\mathbb{C}^1$.

2.2. The bundle of almost pseudoKähler structures.

Let (M, ω) be a symplectic manifold of dimension $2n$. We define the symplectic twistor bundle of M as the bundle $\pi : Z = \mathcal{S}(M) \to M$ of almost pseudoKähler structures on M associated with the Lagrangian $\mathrm{Sp}(2n, \mathbb{R})$-bundle of M. The fiber $\pi^{-1}(p)$ in a point $p \in M$ is the space $\mathcal{S}(T_p M)$ of pseudoKähler structures on $T_p M$ defined above.

As in the Riemannian case, taking a symplectic connection D on M we can provide Z with a natural almost complex structure. This almost complex structure, similar to the Riemannian case, is integrable if and only if M has a flat symplectic connection, i.e. is symplectically flat [6].

We define next a holomorphic line bundle over Z which is crucial for applications in the geometric quantization method.

2.3 Prequantum bundle

Suppose that (M, ω) satisfies to the following quantization condition

$$\text{cohomology class of } \omega/2\pi \text{ is integral .}$$

Otherwise speaking, the integral of ω along any oriented compact 2-cycle in M is an integral multiple of 2π.

Under this condition there exists (cf. [7]) a complex line bundle $L \to M$ with a connection ∇ having the curvature

$$F_\nabla = \omega .$$

Such a bundle is called the prequantum bundle of M.

2.4 Ward's construction

Let $L \to M$ be the prequantum line bundle of M with the connection ∇. If we define the bundle $\tilde{L} \to Z$ as the pull-back of L to Z then $\tilde{L}$ will be a holomorphic line bundle over Z. It is essentially the Ward's construction from the twistor theory (cf. [9]).

To prove that assertion denote by $\tilde{\nabla}$ the pull-back of the connection ∇ to $\tilde{L}$ and define a $\bar{\partial}$-operator on sections $\tilde{s}$ of $\tilde{L} \to Z$ by setting

$$\bar{\partial}\tilde{s} := \tilde{\nabla}^{(0,1)}\tilde{s}$$

where $\tilde{\nabla}^{(0,1)}$ is the $(0, 1)$-component of $\tilde{\nabla}$ with respect to the almost complex structure on Z introduced above. The symplectic structure ω on M being compatible with all pseudoKähler structures on M has the type $(1, 1)$ w.r. to any such structure, hence the curvature F_∇ also has the type $(1, 1)$ w.r. to any pseudoKähler structure. According to the definition of almost complex structure on Z it means that the curvature $F_{\tilde{\nabla}}$ of the pulled-back connection $\tilde{\nabla}$ on $\tilde{L}$ has the type $(1, 1)$ w.r. to the almost complex structure of Z. It follows that

$$\bar{\partial}^2\tilde{s} = F_{\tilde{\nabla}}^{(0,2)}\tilde{s} = 0$$

i.e. $\tilde{L}$ is holomorphic.

Our goal is to extend these symplectic twistor constructions to infinite-dimensional loop spaces. In Sec.3 we shall define a homogeneous symplectic twistor bundle over a universal flag manifold.

2 Loop spaces

1 Loop space of a Lie group

We recall now some general facts about loop spaces of Lie groups ([8]). Let G be a Lie group with the Lie algebra $\mathbf{g}$, $G^{\mathbb{C}}$ – its complexification. We shall consider mostly compact

Lie groups having in mind also the physically interesting case when $G = \mathbb{R}^{d-1,1}$ is the group of translations of the d-dimensional Minkowski space.

1.1 Loop space of a Lie group

Define the loop space ΩG of a Lie group G as

$$\Omega G = LG/G \tag{3}$$

where $LG = \text{Map}(S^1, G)$ is the space of smooth mappings $S^1 \to G$ of the unit circle $S^1 \subset \mathbb{C}$ into the group G, the denominator being identified with the group G of constant mappings $S^1 \to g_0 \in G$. For a compact Lie group G there is a Birkhoff's theorem (cf. [8]) asserting that ΩG has also a complex representation in the form

$$\Omega G = LG^{\mathbb{C}}/L_+G^{\mathbb{C}} \tag{4}$$

where $LG^{\mathbb{C}} = \text{Map}(S^1, G^{\mathbb{C}})$ and $L_+G^{\mathbb{C}} = \text{Hol}(\Delta, G^{\mathbb{C}})$ is the subspace of $LG^{\mathbb{C}}$ consisting of maps which can be extended to holomorphic maps of the unit disc $\Delta \subset \mathbb{C}$ into $G^{\mathbb{C}}$. For $G = \mathbb{R}^{d-1,1}$ the loop space ΩG may be considered as the phase space of a bosonic string.

We shall define next a natural symplectic structure on ΩG.

1.2 Symplectic structure on ΩG

Considering LG as a group w.r. to the pointwise multiplication we see that ΩG is a homogeneous space of LG with the origin o identified with the class of the constant mapping $S^1 \to 1 \in G$. All structures on ΩG we are going to define will be invariant w.r. to the natural left action of LG on ΩG. So it's sufficient to define them only in the origin $o \in \Omega G$ and then transport to other points of ΩG using left translations of LG.

The tangent space of ΩG in the origin o is identified with the space $\Omega \mathfrak{g} = L\mathfrak{g}/\mathfrak{g}$ where $L\mathfrak{g} = \text{Map}(S^1, \mathfrak{g})$. Denote by $< \cdot, \cdot >$ an invariant inner product on $\mathfrak{g}$ and define a 2-form ω on $L\mathfrak{g}$ by the expression

$$\omega(\xi, \eta) = \frac{1}{2\pi} \int_0^{2\pi} < \xi(\theta), \eta'(\theta) > d\theta \tag{5}$$

for $\xi, \eta \in L\mathfrak{g}$. This form is closed and $\omega(\xi, \eta) = 0$ iff one of the mappings ξ, η is constant. Hence the formula (5) defines a closed non-degenerate 2-form on $\Omega \mathfrak{g}$ which can be extended to a symplectic structure on ΩG.

1.3 Complex structure of ΩG

A complex structure J^0 on ΩG for a compact Lie group G is induced from the complex representation (4). It's easy to describe explicitly its restriction to the tangent space in the origin. Any vector ξ in the complexified tangent space $T_o^{\mathbb{C}}(\Omega G) = T_o(\Omega G) \otimes \mathbb{C}$ has a Fourier decomposition

$$\xi = \sum_{k \neq 0} \xi_k z^k$$

where $\xi_k \in \mathbf{g}^{\mathbb{C}}$; it belongs to $T_o(\Omega G)$ iff $\xi_{-k} = \bar{\xi}_k$. The restriction of J^0 to $T_o(\Omega G)$ is given by the formula

$$J_o^0 \xi = -i \sum_{k \neq 0} \operatorname{sgn} k \cdot \xi_k z^k = -i \sum_{k > 0} \xi_k z^k + i \sum_{k < 0} \xi_k z^k . \tag{6}$$

Note that the definition of J^0 by (6) is valid not only for compact Lie groups but also for the string case $G = \mathbb{R}^{d-1,1}$.

1.4 PseudoKähler structure of ΩG

The introduced symplectic and complex structures on ΩG are compatible with each other in the sense that

$$\omega(J^0 \xi, J^0 \eta) = \omega(\xi, \eta) \quad , \xi, \eta \in \Omega\mathbf{g} .$$

The complex structure J^0 defines an invariant metric g^0 on ΩG given in the origin by the formula

$$g^0(\xi, \eta) := \omega(\xi, J^0 \eta) \quad , \xi, \eta \in \Omega\mathbf{g} . \tag{7}$$

This metric is Riemannian for a compact G, so it defines a Kähler structure on ΩG. In the string case $G = \mathbb{R}^{d-1,1}$ it generates a pseudoKähler structure on ΩG.

1.5 ΩG as a universal flag manifold

We have defined on ΩG (for a compact G) a structure of an infinite-dimensional Kähler manifold which has two different representations ("real" and "complex") as a homogeneous space

$$\Omega G = LG/G = LG^{\mathbb{C}}/L_+ G^{\mathbb{C}} .$$

Such representations in the finite-dimensional case are characteristic to flag manifolds F of the group G which are compact Kähler manifolds having "real" and "complex" representations of the form

$$F = G/H = G^{\mathbb{C}}/P$$

where H is a closed subgroup in G (the centralizer of a torus in G) and P is a parabolic subgroup in the complexified group $G^{\mathbb{C}}$. We can consider ΩG as a universal flag manifold because according to K. Uhlenbeck any flag manifold F of G could be holomorphically (and totally geodesically) immersed into ΩG (cf. e.g. Burstall–Rawnsley [5]).

2 Space of complex structures

We need to understand next how reparametrizations of a string influence the introduced structures (cf. [1]).

2.1 The group Diff(S^1) and its Lie algebra Vect(S^1)

Denote by Diff(S^1) the group of diffeomorphisms of S^1 preserving the orientation of S^1 (i.e. the identity component of the group of all diffeomorphisms of S^1).

This group consists of symplectomorphisms of ΩG, i.e. symplectic structure ω on ΩG is invariant under Diff(S^1) in the sense that

$$\omega(f_*\xi, f_*\eta) = \omega(\xi, \eta)$$

for $f \in \text{Diff}(S^1)$, $\xi, \eta \in L\mathfrak{g}$, where the action of f on $\xi \in L\mathfrak{g}$ is given by the usual formula $f_*\xi(\theta) := \xi(f(\theta))$.

Denote by Vect(S^1) the Lie algebra of Diff(S^1) identified with the Lie algebra of tangent vector fields on S^1. It follows from above that Vect(S^1) consists of Hamiltonian vector fields on ΩG. A base of the complexified Lie algebra $\text{Vect}^{\mathbb{C}}(S^1)$ can be given by tangent vector fields

$$L_n = ie^{in\theta}\frac{d}{d\theta} \quad , \quad n = 0, \pm 1, \pm 2, \ldots \tag{8}$$

on S^1 satisfying to the commutation relations

$$[\, L_n, L_m \,] = (n - m)L_{n+m} \, . \tag{9}$$

Using the duality between Hamiltonian vector fields and functions on a symplectic manifold we can consider Vect(S^1) also as a Poisson algebra, i.e. a Lie algebra of functions with the Poisson bracket on ΩG. As a Poisson algebra Vect(S^1) is generated by the functions λ_n, $n = 0, \pm 1, \pm 2, \ldots$, on ΩG given by the formula

$$\lambda_n(\gamma) = \frac{1}{2}\int_0^{2\pi} < \gamma'(\theta), \gamma'(\theta) > e^{-in\theta} d\theta$$

for $\gamma \in \Omega G$.

2.2 The Virasoro algebra and Virasoro–Bott group

The algebra Vect(S^1) has a well known non-trivial central extension called the Virasoro algebra vir. The corresponding central extension of the complexified Lie algebra $\text{Vect}^{\mathbb{C}}(S^1)$ is denoted by $\text{vir}^{\mathbb{C}}$ and generated by the operators L_n, $n \in \mathbb{Z}$, and a central element c. The commutation relations for $\text{vir}^{\mathbb{C}}$ have the form

$$\begin{aligned} [\, L_n, L_m \,] &= (n-m)L_{n+m} + \tfrac{n^3-n}{12}\,\delta_{n,-m}c \\ [\, L_n, c \,] &= 0 \, . \end{aligned}$$

The Lie algebra vir is assigned to a Lie group Vir called the Virasoro–Bott group which is a central extension of $\mathrm{Diff}(S^1)$. Note that no Lie groups correspond to the complexified Lie algebras $\mathrm{Vect}^{\mathbb{C}}(S^1)$ and $\mathrm{vir}^{\mathbb{C}}$.

2.3 The space $\mathcal{S} = \mathrm{Diff}(S^1)/S^1$

Consider the action of $\mathrm{Diff}(S^1)$ on the complex structure J^0 on ΩG. A diffeomorphism $f \in \mathrm{Diff}(S^1)$ transforms J^0 into a new complex structure J given by

$$J := f_*^{-1} \circ J^0 \circ f_* . \tag{10}$$

The complex structure J given by this formula coincides with J^0 if and only if f is a rotation, i.e. f belongs to the subgroup $S^1 \subset \mathrm{Diff}(S^1)$ consisting of rotations of S^1. It is clear that J defined by (10) is also an invariant complex structure on ΩG compatible with ω so it defines a pseudoKähler structure on ΩG.

We introduce now (following Bowick–Rajeev [1]) the space

$$\mathcal{S} = \mathrm{Diff}(S^1)/S^1$$

parametrizing complex structures (pseudoKähler structures) on ΩG which are invariant under the left action of LG and compatible with the symplectic structure ω. We shall call these complex structures for brevity compatible complex structures on ΩG.

2.4 Kähler structure of $\mathcal{S}$

The group $\mathrm{Diff}(S^1)$ acts on $\mathcal{S}$ by the left translations. Denote again by L_n the tangent vector fields on $\mathcal{S}$ corresponding (w.r. to the action of $\mathrm{Diff}(S^1)$ on ΩG) to the tangent vector fields L_n on ΩG. As in the case of ΩG all structures we shall define on $\mathcal{S}$ will be invariant w.r. to the left action of $\mathrm{Diff}(S^1)$ on $\mathcal{S}$ so it's sufficient to define them only in the origin $o \in \mathcal{S}$.

The tangent space $T_o\mathcal{S}$ in the origin consists of vectors

$$v = \sum_{n \neq 0} v_n L_n \quad , \quad v_n \in \mathbb{C} ,$$

where $\bar{v}_n = v_{-n}$. We define an invariant complex structure $\mathcal{I}$ on $\mathcal{S}$ by its restriction $\mathcal{I}_o$ to $T_o\mathcal{S}$ given by the formula

$$\mathcal{I}_o v = -i \sum_{n \neq 0} \mathrm{sgn}\, n \cdot v_n L_n . \tag{11}$$

It's easy to check that $\mathcal{I}$ so defined is an (integrable) complex structure on $\mathcal{S}$.

We shall define also a family of Kähler structures on $\mathcal{S}$. They are generated by invariant 2-forms on $\mathcal{S}$ whose restrictions to $T_o\mathcal{S}$ are given by the formulas (cf. [10])

$$w(L_n, L_m) = (an^3 + bn)\delta_{n,-m} .$$

These forms are non-degenerate (i.e. generate symplectic structures on S) in the following two cases: i) $a = 0$, $b \neq 0$; ii) $a \neq 0$, $-b/a$ is not a square of an integer. We restrict further on to the case ii) (it is necessary for making the sense of the Ricci curvature of S below, cf. [1]).

2.5 S as an infinite-dimensional Siegel disc

We have defined a structure of an infinite-dimensional Kähler manifold on S. This manifold is an infinite-dimensional analogue of the Hermitian symmetric domain

$$S(\mathbb{R}^{2n}) \approx \mathrm{Sp}(2n, \mathbb{R})/\mathrm{U}(n)$$

introduced in the Sec.1. In fact, we can embed S into a homogeneous infinite-dimensional manifold $\mathrm{Sp}_{\mathrm{res}}(H)/\mathrm{U}(H)$ for a suitable complex Hilbert space H (cf. [11]). For such a space we can take the space H of real functions f on S^1 with the zero mean value $\oint f(\theta)\frac{d\theta}{2\pi i\theta} = 0$ and the norm $\|f\|^2 = \oint f(\theta)^2 \frac{d\theta}{2\pi i\theta}$; a complex structure J on H being given by the Hilbert transform. The group $\mathrm{Sp}_{\mathrm{res}}(H)$ for H is defined as follows. We represent H as the direct sum of subspaces $H = W \oplus \bar{W}$ so that the complex structure J is equal to

$$J = \begin{pmatrix} i & 0 \\ 0 & -i \end{pmatrix}.$$

Now take the symplectic structure $\omega = \begin{pmatrix} 0 & 1 \\ -1 & 0 \end{pmatrix}$ and consider the group $\mathrm{Sp}(H)$ of linear transformations of H preserving ω. This group consists of matrices having the form

$$g = \begin{pmatrix} A & B \\ \bar{B} & \bar{A} \end{pmatrix} \quad : \quad AB^t = BA^t, \ AA^* - BB^* = I.$$

Denote by $\mathrm{Sp}_{\mathrm{res}}(H)$ a subgroup of $\mathrm{Sp}(H)$ cosisting of matrices g subject to the condition $\mathrm{tr}(BB^*) < \infty$. The group of linear transformations of H preserving ω and J is identified with the unitary group $\mathrm{U}(H)$ (it consists of matrices g with $B = 0$). Hence, $\mathrm{Sp}_{\mathrm{res}}(H)$ contains symplectic transfromations of H differing from a unitary by a Hilbert–Schmidt operator.

The homogeneous space $\mathrm{Sp}_{\mathrm{res}}(H)/\mathrm{U}(H)$, as in the finite-dimensional case, can be identified with the infinite-dimensional Siegel disc S^∞_{res} consisting of operators Z on W such that

$$Z^t = Z \quad , \quad I - Z^*Z \gg 0 \quad , \quad \mathrm{tr}(Z^*Z) < \infty.$$

On the other side, the manifold $S = \mathrm{Diff}(S^1)/S^1$ can be mapped holomorphically into the space S of univalent holomorphic functions in the unit disc $\Delta \subset \mathbb{C}$ which may be considered as a universal Teichmüller space in the sense that all finite-dimensional Teichmüller spaces T_g can be holomorphically embedded into S (cf. [11],[12]).

3 Geometric quantization

1 Prequantization

1.1 Quantization of a classical system

A classical (mechanical) system is given by its phase space, i.e. a symplectic manifold (M, ω) and an algebra of observables $\mathfrak{A}$, i.e. a Lie subalgebra of the Poisson algebra $C^\infty(M)$ of real smooth functions on M with the Poisson bracket as a Lie bracket.

To quantize this classical system means to construct an irreducible unitary Lie algebra representation

$$r : \quad \mathfrak{A} \longrightarrow \text{End}^* H$$

of the algebra $\mathfrak{A}$ in the algebra $\text{End}^* H$ of linear self-adjoint operators in a complex Hilbert space H called the quantization space. Here by a Lie algebra representation r we mean a linear mapping from $\mathfrak{A}$ into $\text{End}^* H$ such that

$$r(\{f, g\}) = i[\, r(f), r(g)\,]$$

where $[\, r(f), r(g)\,]$ is the usual commutator of linear operators $r(f), r(g)$. Unitary means that symplectomorphisms (canonical transformations) of M generated by observables from $\mathfrak{A}$ should correspond to unitary transformations of H. We shall impose on r additionally the normalization condition

$$r(1) = \text{id}\,,$$

i.e. the constant function equal to 1 corresponds under r to the identity operator on H.

A representation $r : \mathfrak{A} \to \text{End}^* H$ is called a prequantization if it satisfies to all of the above conditions besides irreducibility. In this case H will be called the prequantization space.

1.2 The Souriau–Kostant prequantization

If the phase space (M, ω) of a classical system satisfies to the quantization condition (cf. Sec.1, 2.3) it's always possible to construct its prequantization. For that consider the prequantum bundle L of M, i.e. an Hermitian line bundle $L \to M$ with an Hermitian connection ∇ such that $F_\nabla = \omega$. Take for the prequantization space H the space $\mathcal{L}^2(M, L; \omega)$ defined as the completion of the space of smooth sections of L over M with compact supports with respect to the inner product

$$(s, t) = \int_M < s, t > \omega^n$$

where $< \cdot, \cdot >$ is given by the Hermitian structure of L. Then we define the Souriau–Kostant prequantization

$$r : C^\infty(M) \longrightarrow \text{End}^* H$$

by setting

$$r(f) = f - i\nabla_{X_f} .$$

Here f in the right hand side is considered as an operator on sections $s \in \mathcal{L}^2(M, L; \omega)$ given by the multiplication by f; X_f is the Hamiltonian vector field corresponding to f and ∇_{X_f} is a component of ∇ in the direction of X_f.

Unfortunately, this prequantization cannot be a quantization as it follows, in fact, from the Heisenberg's uncertainty principle. According to this principle the true quantization space should be, roughly speaking, the "half" of the space $\mathcal{L}^2(M, L; \omega)$. Precise meaning to this "half" notion is given by the notion of polarization. We shall not pursue on that point here referring to general books on the geometric quantization [7]. In the string case the polarization will be provided by the restriction to spaces of holomorphic sections.

1.3 Prequantization of the bosonic string

In the string case we take for a phase space the loop space ΩG and for an algebra of observables the Poisson algebra $\mathfrak{A}$ generated by the Heisenberg algebra of ΩG and the algebra $\mathrm{Vect}(S^1)$ defined in the Sec.2, 2.1. To quantize the bosonic string means according to 1.1 to construct an irreducible Lie algebra representation $r : \mathfrak{A} \to \mathrm{End}^* H$. We concentrate here on the construction of a representation $r : \mathrm{Vect}(S^1) \to \mathrm{End}^* H$ of the Lie algebra $\mathrm{Vect}(S^1)$, it will generate a representation of the whole algebra $\mathfrak{A}$ in the usual way (cf. e.g. [14]). It is more convenient to consider a representation of the complexified algebra

$$r : \mathrm{Vect}^{\mathbb{C}}(S^1) \longrightarrow \mathrm{End}\, H$$

with the additional property that the complex conjugation in $\mathrm{Vect}^{\mathbb{C}}(S^1)$ corresponds under r to the Hermitian conjugation in $\mathrm{End}\, H$.

We want to construct such a representation following the same idea as in 1.2. To that end consider the prequantum bundle L over ΩG with the Hermitian connection ∇ such that F_∇ is equal to the symplectic form ω on ΩG. As we know, this bundle exists if a certain quantization condition is satisfied for ΩG. Referring to Pressley–Segal [8] for a general discussion of this condition we note only that they are satisfied if a compact group G is simple and simply connected. In the case of $G = \mathbb{R}^{d-1,1}$ the prequantum bundle L is trivial and flat.

To proceed as in 1.1 we should define the prequantization space H as the space of square integrable sections of L over ΩG with respect to the measure generated by ω and define a representation $r : \mathrm{Vect}^{\mathbb{C}}(S^1) \to \mathrm{End}\, H$ on base functions λ_n, $n = 0, \pm 1, \pm 2, \ldots$, of $\mathrm{Vect}^{\mathbb{C}}(S^1)$ by

$$r(\lambda_n) = \lambda_n - i\nabla_{L_n} . \tag{12}$$

But, as we know from 1.2, this prequantization has no chance to be a true quantization; for a true quantization we should introduce a polarization on ΩG. In our context it means that we should substitute H by a space of holomorphic sections of L over ΩG.

1.4 Polarization

We shall concentrate from now on the string case $G = \mathbb{R}^{d-1,1}$ because of its direct physical meaning. All the constructions extend to the case of compact G with some slight modifications (cf. below).

Points $\gamma \in \Omega = \Omega\mathbb{R}^{d-1,1}$ can be given by its Fourier decompositions

$$\gamma = \sum_{k \neq 0} \gamma_k z^k \quad , \quad z = e^{i\theta} \quad ,$$

so functions f on Ω can be considered as functions $f(\gamma) = f(\{\gamma_k\})$ depending upon infinite number of variables given by the Fourier coefficients $\{\gamma_k\}$.

Fix the complex structure J^0 on Ω (cf. Sec.2, 1.3). Then the symplectic structure ω on Ω has a Kähler potential K_0 with respect to J^0 given by the explicit formula

$$K_0(\gamma) = \sum_{k \neq 0} k < \gamma_k, \bar{\gamma}_k > \quad .$$

The prequantum bundle L in this case is trivial and sections of L over Ω are identified with functions on Ω, i.e. with functions $f(\{\gamma_k\})$. We define the polarized prequantization space

$$H_0 = \mathcal{O}_{J^0}(\Omega, L; \omega)$$

as the Fock space $F(\Omega, J^0)$ of functions $f(\{\gamma_k\})$ holomorphic w.r. to the complex structure J^0 with the inner product generated by the Kähler potential K_0:

$$(f, g) = \int f(\{\gamma_k\}) \bar{g}(\{\gamma_k\}) e^{-K_0(\{\gamma_k\})} \prod_{n=1}^{\infty} d\gamma_k d\bar{\gamma}_k \quad .$$

(Cf. [13] for the precise definition of the Fock space).

The operators $r(\lambda_n)$ from (12) are given in our case by the formulae

$$r(\lambda_n) = \sum_{k=1}^{\infty} \sqrt{k(n+k)} \langle \gamma_{n+k}, \frac{\partial}{\partial \gamma_k} \rangle + \frac{1}{2} \sum_{k=1}^{n-1} \sqrt{k(n-k)} \langle \gamma_k, \gamma_{n-k} \rangle \qquad (13a)$$

for $n > 0$ and

$$r(\lambda_0) = \sum_{k=1}^{\infty} k \langle \gamma_k, \frac{\partial}{\partial \gamma_k} \rangle \qquad (13b)$$

(cf. [14],[1]); analogous formula can be given for $r(\lambda_n)$ with $n < 0$. Using the creation and annihilation operators

$$a_k = \sqrt{k} \gamma_k \quad , \quad a_k^* = \sqrt{k} \frac{\partial}{\partial \gamma_k}$$

we generalize $(13a, b)$ to

$$r(\lambda_n) = \sum_{k=1}^{\infty} a_{k+n} a_k^* + \frac{1}{2} \sum_{k=1}^{n-1} a_k a_{n-k} \qquad (14a)$$

for $n > 0$ and

$$r(\lambda_0) = \sum_{k=1}^{\infty} a_k a_k^* + \beta \qquad (14b)$$

where we have added an arbitrary constant $\beta \in \mathbb{R}$ (called a normal ordering constant) which will be chosen later.

1.5 Projective representations in polarized prequantization spaces

The operators $r(\lambda_n)$ from (14) preserve the polarized prequantization space H_0 for $n \geq 0$ while those with $n < 0$ — not. So we need to mofidy the representation r by defining a new representation

$$
\begin{aligned}
r_0(\lambda_n) &: \ = r(\lambda_n) \quad \text{for} \quad n \geq 0 \\
r_0(\lambda_{-n}) &: \ = r^*(\lambda_n) \quad \text{for} \quad -n < 0 \,.
\end{aligned}
$$

Of course, for such a "rough" modification we should pay — the defined representation r_0 is not longer a true representation but only a projective one. More precisely, we have

$$[\, r_0(\lambda_n), r_0(\lambda_m)\,] - r_0(\{\lambda_n, \lambda_m\}) = \left(\frac{d}{12}m^3 - \beta m\right)\delta_{m,-n} \qquad (15)$$

so the right hand side (which should be zero for a true representation) is a diagonal operator. Thus we have constructed a projective representation

$$r_0 : \ \mathrm{Vect}^{\mathbb{C}}(S^1) \longrightarrow \mathrm{End}\, H_0 \,.$$

Using left translations by $\mathrm{Diff}(S^1)$ on Ω we can define in the same way polarized prequantization spaces

$$H_J = \mathcal{O}_J(\Omega, L; \omega)$$

with respect to any compatible complex structure J on Ω and corresponding projective representations

$$r_J : \ \mathrm{Vect}^{\mathbb{C}}(S^1) \longrightarrow \mathrm{End}\, H_J \,.$$

So far we have constructed only projective representations r_J instead of true representations. Even more unpleasant is the fact that our construction depends upon the reparametrization of the string generated by the action of $\mathrm{Diff}(S^1)$ on Ω.

As in the twistor theory of Riemannian manifolds [5] we propose to overcome these shortcomings by introducing a twistor space of all compatible polarizations (= complex structures) on Ω having a natural polarization (= complex structure) and defining a quantization directly in terms of this twistor space.

2 Twistor quantization

2.1 Twistor bundle of compatible pseudoKähler structures

We introduce now the twistor bundle $\pi : \mathcal{Z} \to \Omega$ of compatible pseudoKähler structures on Ω having the fiber $\pi^{-1}(\gamma)$ in a point $\gamma \in \Omega$ consisting of restrictions J_γ of compatible pseudoKähler structures J on Ω to $T_\gamma\Omega$; points $z \in \mathcal{Z}$ are the pairs $z = (\gamma, J_\gamma)$ where $\gamma \in \Omega$, J_γ is a complex structure on $T_\gamma\Omega$.

Compatible complex structures we are considering are invariant with respect to the LG–action on Ω so there is a natural action of LG on $\mathcal{Z}$ and the quotient of this action coincides with the manifold $\mathcal{S} = \mathrm{Diff}(S^1)/S^1$. Thus we have a double fibration

$$\begin{array}{ccc} \mathcal{Z} & \xrightarrow{\ p\ } & \mathcal{S} \\ {\scriptstyle \pi}\downarrow & & \\ \Omega & & \end{array} \qquad\qquad (16)$$

where p is the natural projection of $\mathcal{Z}$ to $\mathcal{S}$. The fiber $p^{-1}(J)$ in a point $J \in \mathcal{S}$ can be identified with the complex manifold $\Omega_J = (\Omega, J)$, i.e. with Ω provided with the complex structure corresponding to $J \in \mathcal{S}$. We shall denote a point $J \in \mathcal{S}$ and the corresponding complex structure on Ω by the same letter J.

2.2 Almost complex and real structures of $\mathcal{Z}$

We introduce now a natural almost complex structure on $\mathcal{Z}$. Consider the bundles $\pi^{-1}(T\Omega)$ and $p^{-1}(T\mathcal{S})$ over $\mathcal{Z}$ which are the pull-backs to $\mathcal{Z}$ of the tangent bundles $T\Omega$ and $T\mathcal{S}$ respectively. The projections π and p generate natural morphisms of bundles over $\mathcal{Z}$

$$d\pi : T\mathcal{Z} \longrightarrow \pi^{-1}(T\Omega) \quad , \quad dp : T\mathcal{Z} \longrightarrow p^{-1}(T\mathcal{S}) .$$

The kernel of $d\pi$ is the vertical subbundle V of $T\mathcal{Z}$ and the kernel of dp will be considered as a horizonthal subbundle H of $T\mathcal{Z}$. Note that the fiber V_z in a point $z \in \mathcal{Z}$ is identified by p with the tangent space $T_J\mathcal{S}$ in the point $J = p(z) \in \mathcal{S}$ and so has the complex structure $\mathcal{I}_J$ defined in the Sec.2, 2.4. Now we can define an almost complex structure $\mathcal{J}$ on $\mathcal{Z}$ exactly in the same way as in the Sec.1, 1.3,2.2. The projection $p : \mathcal{Z} \to \mathcal{S}$ will become a holomorphic mapping with respect to this structure $\mathcal{J}$.

We can define also a natural real structure on $\mathcal{Z}$. The space $\mathcal{S}$ has a real structure σ_0 generated by the mapping

$$f(e^{i\theta}) \longmapsto f(-e^{i\theta}) \quad , \ f \in \mathrm{Diff}(S^1), \ e^{i\theta} \in S^1 ,$$

of $\mathrm{Diff}(S^1)$ onto itself. The structure σ_0 has no fixed points on $\mathcal{S}$. A real structure σ on $\mathcal{Z}$ is generated by the action of σ_0 on fibers of π, i.e. $\sigma(\gamma, J_\gamma) := (\gamma, \sigma_0(J_\gamma))$. It transforms the almost complex structure $\mathcal{J}$ on $\mathcal{Z}$ into the conjugate almost complex structure $-\mathcal{J}$.

Hence, points of Ω by the twistor diagram (16) correspond to real holomorphic sections of $p : \mathcal{Z} \to \mathcal{S}$. It is similar to the case of hyperKähler manifolds (c.f. Sec.1, 1.4). In fact we can consider manifolds ΩG as symplectic infinite-dimensional analogues of hyperKähler manifolds. As in the hyperKähler case the twistor space $\mathcal{Z}$ of ΩG is a holomorphic bundle $p : \mathcal{Z} \to \mathcal{S}$ with a collection of real holomorphic sections parametrized by points of ΩG and having the normal bundle equal to the tensor product $\Omega\mathbf{g} \otimes K^{-1/2}$ where $\Omega\mathbf{g}$ is a trivial bundle over $\mathcal{S}$ with the fiber $\Omega\mathbf{g}$ and $K^{-1/2}$ is the square root of the anticanonical bundle of $\mathcal{S}$ (cf. below).

2.3 Ward's construction for ΩG

Let $L \to \Omega$ be the prequantum bundle over Ω with the connection ∇ having the curvature equal to ω. Applying the Ward's construction (Sec.1, 2.4) define the pull-back bundle $\tilde{L} = \pi^* L \to \mathcal{Z}$ and provide it with the natural complex structure.

The prequantum bundle $L \to \Omega$ is closely related to the Virasoro–Bott group Vir, the central extension of $\text{Diff}(S^1)$. There is a canonical action of Vir on $L \to \Omega$, namely Vir can be identified with the group of fiberwise mappings $L \to L$ covering transformations of Ω generated by elements of $\text{Diff}(S^1)$.

2.4 Fock bundle over S

We are going to reformulate now the problem of quantization in terms of the twistor space. Instead of a Fock space H we consider a Fock bundle

$$\tilde{H} \longrightarrow S$$

over S with the fiber $\tilde{H}_J$ in a point $J \in S$ consisting of holomorphic sections of $\tilde{L}$ over $p^{-1}(J) = \Omega_J$

$$\tilde{H}_J = \mathcal{O}(\Omega_J, \tilde{L}; \tilde{\omega})$$

where $\tilde{\omega}$ is the pull-back of ω to $\mathcal{Z}$.

Projective representations

$$r_J : \text{Vect}^{\mathbb{C}}(S^1) \longrightarrow \text{End } H_J$$

give rise to a connection D on the bundle $\tilde{H} \to S$. The covariant derivation on $\tilde{H}$ generated by D in the origin $o \in S$ is given in the base $\{L_n\}$ of $T_o S$ by the formula

$$D_{L_n} = \mathcal{L}_{L_n} + r_0(\lambda_n)$$

and extended to other points of S using the $\text{Diff}(S^1)$–action. The curvature F_D computed in the origin is equal to (cf. [1])

$$F_D(L_m, L_{-n}) = \left(\frac{d}{12} m^3 - \beta m \right) \delta_{m,n} \qquad (17)$$

and coincides with the cocycle (15) of the representation r_0.

Hence we obtain the following reformulation of the quantization problem in twistor terms — to quantize the string phase space Ω is equivalent to the construction of a flat unitary connection on a Fock bundle over S.

Unfortunately, the connection D on the Fock bundle $\tilde{H} \to S$ is never flat according to (17). The way how to overcome this obstacle is well known in the string theory and theory of twistors. Using the twistor terminology, we need to multiply tensorially our Fock bundle $\tilde{H}$ by the square root of the anticanonical bundle of S (this bundle has already appeared in the normal bundle of the fiber of $\mathcal{Z} \to \Omega$). In terms of string theory we should introduce "ghosts".

2.5 Quantization bundle

Consider the bundle $K^{-1/2} \to S$ which is the square root of the anticanonical bundle of S. (We refer to [8],[15] for the precise definition of the canonical bundle in the infinite-dimensional case). The bundle $K^{-1/2}$ is a complex line bundle which will be called following the physical tradition the ghost bundle of S. As is well known in the finite-dimensional case, $K^{-1/2}$ is a holomorphic bundle over S defining a spinor structure on S. The canonical Riemannian connection on S generates a spinorial connection $\tilde{\nabla}$ on $K^{-1/2}$ having the curvature equal to $-1/2$ Ric where Ric is the Ricci curvature of S. This curvature was computed in [1],[2]. According to 2.4 it coincides with the cocycle of a projective representation of the observables algebra $\mathrm{Vect}(S^1)$ in the space of holomorphic sections of $K^{-1/2}$ generated by the Clifford multiplication and can be computed also using the representation theory of $\mathrm{Vect}(S^1)$ (cf. [15]). In the origin $o \in S$ the curvature $F_{\tilde{\nabla}}$ in the base $\{L_n\}$ is equal to

$$F_{\tilde{\nabla}}(L_m, L_{-n}) = \left(-\frac{26}{12}m^3 + \frac{1}{6}m\right)\delta_{m,-n} . \tag{18}$$

Now we define the quantization bundle $\tilde{\mathcal{H}} \to S$ as the tensor product

$$\tilde{\mathcal{H}} = \tilde{H} \otimes K^{-1/2} \longrightarrow S$$

provided with the tensor-product connection

$$\tilde{D} = D \otimes 1 + 1 \otimes \tilde{\nabla} .$$

The curvature of $\tilde{D}$ in the origin, as it follows from $(17), (18)$, is equal to

$$F_{\tilde{D}}(L_m, L_{-n}) = \left[\frac{d-26}{12}m^3 + \left(\frac{1}{6} - \beta\right)\right]\delta_{m,n}$$

and so it's zero for $\beta = 1/6$ under the well known critical dimension condition

$$d = 26 .$$

In other words the quantization bundle $\tilde{\mathcal{H}} \to S$ under the condition $d = 26$ has a flat unitary connection $\tilde{D}$ thus providing a solution to the quantization problem.

According to our previous discussion in 2.4 this solution yields also a unitary representation of the algebra $\mathrm{Vect}(S^1)$ in the space of holomorphic sections of $\tilde{L} \otimes K^{-1/2}$ which can be constructed as follows. The Souriau–Kostant representation r defined in 1.4 does not preserve polarized prequantization spaces H_J identified with fibers of the Fock bundle $\tilde{\mathcal{H}} \to S$; in fact, some of operators $r(\lambda_n)$ map H_J into spaces $H_{J'}$ with different complex structures J'. The representation r can be easily extended (cf. [7]) to a representation $\tilde{r}$ in the polarized spaces $\tilde{\mathcal{H}}_J$ identified with fibers of the quantization bundle $\tilde{\mathcal{H}} \to S$. Operators $\tilde{r}(\lambda_n)$ of this extended representation will, generally speaking, map $\mathcal{H}_J$ into spaces $\mathcal{H}_{J'}$ with different J'. But assuming the condition $d = 26$ we can with the help of the flat unitary connection $\tilde{D}$ on $\tilde{\mathcal{H}}$ identify $\tilde{\mathcal{H}}_{J'}$ with $\tilde{\mathcal{H}}_J$ thus constructing a new representation of $\mathrm{Vect}(S^1)$ in $\tilde{\mathcal{H}}_J$ for an arbitrary but fixed polarization J.

2.6 Quantization in the case of a compact Lie group

For a compact Lie group G the above construction needs some modification (cf. Mickelsson [2]). In this case there is a family of prequantum bundles L^k over LG parametrized by a number $k \in \mathbb{Z}$. The bundle L^k has a connection D_k with the curvature equal to $k\omega$ where ω is a 2-form on LG given by

$$\omega(\xi, \eta) = \frac{\theta^2}{4\pi} \int_0^{2\pi} \operatorname{tr} \xi \cdot d\eta .$$

Here tr is the trace in the adjoint representation of the Lie algebra $\mathfrak{g}$ of G and θ^2 is the square length of the longest root of $\mathfrak{g}$. A section of L^k over LG can be considered as a map $s : \widehat{LG} \to \mathbb{C}$ of the central extension $\widehat{LG}$ of LG by S^1 such that $s(g \cdot h) = h^{-k} s(g)$ for $g \in \widehat{LG}$, $h \in S^1 \subset \widehat{LG}$.

Let k be a positive integer and T a maximal torus of G. Denote by λ an integral antidominant weight of G with respect to T, i.e. the longest weight in an irreducible representation of G. Introduce a line bundle $L^{k,\lambda} \to LG/T$ with the sections being the sections s of L^k over LG satisfying the condition

$$s(g \cdot t) = \lambda(t)^{-1} s(g)$$

for $t \in T, g \in \widehat{LG}$.

Considering again the twistor bundle $\mathcal{Z}_T \to LG/T$ of compatible Kähler structures on LG/T we construct Fock bundles $\tilde{H}^{k,\lambda} \to \mathcal{S}$ with fibers consisting of holomorphic sections of $L^{k,\lambda}$ over LG/T with a fixed Kähler structure. Using the projective representations of $\operatorname{Diff}(S^1)$ in $H_J^{k,\lambda}$ (cf. [16]) we define connections $D_{k,\lambda}$ on $\tilde{H}^{k,\lambda}$ with the curvature

$$F_{k,\lambda}(L_m, L_{-n}) = \left(\frac{c(\mathfrak{g})}{12} m^3 - \beta m \right) \delta_{m,n}$$

where

$$c(\mathfrak{g}) = \frac{k \dim \mathfrak{g}}{k + \kappa(\mathfrak{g})} .$$

Here $\kappa(\mathfrak{g})$ is the dual Coxeter number of $\mathfrak{g}$ which can be found for simple Lie algebras from the following table

$$\kappa(sl_n) = n, \quad \kappa(so_n) = n - 2, \quad \kappa(sp_{2n}) = n + 1,$$
$$\kappa(e_6) = 12, \quad \kappa(e_7) = 18, \quad \kappa(e_8) = 30, \quad \kappa(f_4) = 9, \quad \kappa(g_2) = 4 .$$

The quantization bundle $\tilde{\mathcal{H}}^{k,\lambda} \to \mathcal{S}$ is defined again as the tensor product

$$\tilde{\mathcal{H}}^{k,\lambda} = \tilde{H}^{k,\lambda} \otimes K^{-1/2} \longrightarrow \mathcal{S}$$

provided with the tensor-product connection $\tilde{D}_{k,\lambda} = D_{k,\lambda} \otimes 1 + 1 \otimes \tilde{\nabla}$. Its curvature $\mathcal{F}_{k,\lambda}$ in the origin $o \in \mathcal{S}$ is equal to

$$\mathcal{F}_{k,\lambda}(L_m, L_{-n}) = \left[\frac{c(\mathfrak{g}) - 26}{12} m^3 + \left(\frac{1}{6} - \beta \right) m \right] \delta_{m,n}$$

hence $\mathcal{F}_{k,\lambda} = 0$ for $\beta = 1/6$ under the well-known condition (cf. e.g. [17])

$$c(\mathfrak{g}) = \frac{k \dim \mathfrak{g}}{k + \kappa(\mathfrak{g})} = 26 .$$

References

[1]   Bowick M.J., Rajeev S.G. *The holomorphic geometry of closed bosonic string theory and* $Diff(S^1)/S^1$, Nucl. Phys. B293 (1987), 348–384; *Anomalies as curvature in complex geometry*, Nucl. Phys. B296 (1988), 1007–1033.

[2]   Pilch K., Warner N.P. *Holomorphic structure of superstring vacua*, Class. Quantum Grav. 4 (1987), 1183–1192; Mickelsson J. *String quantization on group manifolds and the holomorphic geometry of* $Diff(S^1)/S^1$, Commun. Math. Phys. 112 (1987), 653–661; Kirillov A.A., Juriev D.V. *Kähler geometry of the infinite-dimensional homogeneous space* $M = Diff(S^1)/Rot(S^1)$, Funkc. anal. i ego pril. 21 (1987), no. 4, 35–46 (Russian).

[3]   Penrose R. *Nonlinear gravitons and curved twistor theory*, Gen. Relat. Grav. 7 (1976), 31–52.

[4]   Hitchin N.J., Karlhede A., Lindstrøm U., Roček M. *Hyperkähler metrics and supersymmetry*, Commun. Math. Phys. 108 (1987), 535–589; Salamon S., *Quaternionic Kähler manifolds*, Invent. math. 67 (1982), 143–171; Mamone Capria M., Salamon S. *Yang–Mills fields on quaternionic spaces*, Nonlinearity 1 (1988), 517–530; Nitta T. *Vector bundles over quaternionic Kähler manifolds*, Tohôku Math. J. 40 (1988), 425–440; Topiwala P. *A new proof of the existence of Kähler –Einstein metrics on* $K3$, Invent. math. 89 (1987), 425–454.

[5]   Atiyah M.F., Hitchin N.J., Singer I.M. *Self-duality on four-dimensional Riemannian geometry*, Proc. Roy. Soc. London 362 (1978), 425–461; Bryant R. *Lie groups and twistor spaces*, Duke Math. J. 52 (1985), 223–261; Burstall F.E., Rawnsley J.H. *Twistor theory for Riemannian symmetric spaces with applications to harmonic maps of Riemann surfaces*, Springer Lect. Notes Math. 1424, Berlin–Heidelberg–New York, 1990; O'Brian N.R., Rawnsley J.H. *Twistor spaces*, Ann. Glob. Anal. Geom. 3 (1985), 29–58.

[6]   Vaisman I. *Symplectic twistor spaces*, J. Geom. Phys. 3 (1986), 507–524.

[7]   Kirillov A.A. *Geometric quantization*, Itogi nauki i tehn. Sovr. probl. matem. Fund. naprav. vol. 4, VINITI Moscow 1985, 141–178 (Russian); Sniatycki J. *Geometric quantization and quantum mechanics*, Springer, New York, 1980; Woodhouse N.J.M. *Geometric quantization*, 2nd ed., Clarendon Press, Oxford, 1992.

[8]   Pressley A., Segal G. *Loop groups*, Clarendon Press, Oxford, 1986.

[9]   Atiyah M.F. *Geometry of Yang–Mills fields*, Scuola norm. super., Pisa, 1979.

[10]   Segal G. *Unitary representations of some infinite dimensional groups*, Commun. Math. Phys. 80 (1981), 301–342.

[11]   Hong D.K., Rajeev S.G. *Universal Teichmüller space and* $Diff(S^1)/S^1$, Commun. Math. Phys. 135 (1991), 401–411; Nag S., Verjovsky A. $Diff(S^1)$ *and the Teichmüller spaces*, Commun. Math. Phys. 130 (1990), 123–128.

[12]   Lehto O. *Univalent functions and Teichmüller spaces*, Springer, Berlin–Heidelberg-New York, 1986.

[13]   Berezin F.A. *Method of second quantization*, Nauka, Moscow 1965 (Russian).

[14]   Neretin Ju.A. *Representations of Virasoro and affine algebras*, Itogi nauki i tehn. Sovr. probl. matem. Fund. naprav. vol. 22, VINITI, Moscow 1983, 163–224 (Russian).

[15]   Feigin B.L. *Semi-infinite cohomologies of Kac–Moody and Virasoro Lie algebras*, Uspehi mat. nauk 39, 2 (1984), 195–196 (Russian); Frenkel I.B., Garland H., Zuckerman G.J. *Semi-infinite cohomology and string theory*, Proc. Nat. Acad. Sci. USA 83 (1986), 8442–8446.

[16]   Goddard P., Olive D. *Kac–Moody and Virasoro algebras in relation to quantum physics*, Int. J. Mod. Phys. A1 (1986), 303–414.

[17]   Gepner D., Witten E. *String theory on group manifolds*, Nucl. Phys. B278 (1986), 493–549.

Compactifications of $\mathbb{C}^4$ of Index 3

Yuri G. Prokhorov

In his list of problems on complex and differentiable manifolds [H] F. Hirzebruch mentioned the classifications of all compactifications of $\mathbb{C}^n$ with second Betti number 1, i.e. the classifications of all compact complex manifolds V with $b_2(V) = 1$ and an analytic subset $A \subset V$ such that $V \setminus A \simeq \mathbb{C}^n$ (biholomorphically). This paper deals with compactifications of $\mathbb{C}^n$ (V, A) if $n = 4$, $b_2(V) = 1$ and V is projective.

First, let us make some remarks about dimension ≤ 3 : for $n = 1$ clearly $V \simeq \mathbb{P}^1$, for $n = 2$ Remmert and Van de Ven proved that $V \simeq \mathbb{P}^2$, $A = \mathbb{P}^1$ [RV]. In the case $n = 3$ there are 5 types of projective compactifications of $\mathbb{C}^3$ (see [PS], [P], [F1], [F2], [F3], [M], [Pr]).

Our main result is

Theorem 3.1. *There exist only four projective compactifications of $\mathbb{C}^4$ of index 3 with* $b_2(V) = 1$ (see §3 for more precise form of this theorem).

Note that the case of index $r \geq 4$ is an easy consequence of [KO] (see 1.2).

This theorem will be proved in §§3-4. Sect. 1 is preliminary. In Sect. 2 we describe a structure of the Fano 4-fold $V = V_5 \subset \mathbb{P}^7$ (a linear section of $\mathrm{Gr}(2, 5) \subset \mathbb{P}^9$).

1. Preliminaries

Here we collect some definitions and basic facts which are used in the sequel.

Definition. A pair (V, A) consisting of a compact complex manifold V and an analytic subset $A \subset V$ is called compactification of $\mathbb{C}^n$ if $V \setminus A \simeq \mathbb{C}^n$ (biholomorphically). Two compactifications (V, A) and (V', A') are called isomorphic if there is biholomorphism $\phi : V \longrightarrow V'$ such that $\phi(A) = A'$.

Proposition 1.1. [V], [PS], [K]. *Let (V, A) be a compactification of $\mathbb{C}^n$. Then*

(i) A is pure codimension 1;

(ii) $\chi_{\text{top}}(V) = \chi_{\text{top}}(A) + 1$;

(iii) if $b_2(V) = 1$ *and* V *is projective manifold, then* V *is Fano manifold (see below),* A *is irreducible and* $H^2(V, \mathbb{Z}) \simeq \text{Pic}\,(V) \simeq \mathbb{Z} \cdot A$.

We will be concerned only with compactifications of $\mathbb{C}^4$ (V, A) with $b_2(V) = 1$ (i.e. A is irreducible) and V is projective manifold. Now we collect the facts of Fano manifolds which are relevant for the subsequent sections.

Definitions. By Fano manifold we mean a smooth projective manifold V with ample $-K_V$. The index of a Fano manifold V is the largest positive integer $r = r(V)$ such that $-K_V \sim rH$ for some $H \in \text{Pic}\,(V)$. The selfintersection number $d = H^n$, where $n = \dim(V)$, is called the degree of V.

Theorem 1.2. [KO]. *Let* V *be a Fano n-fold of index* r. *Then*

$$(i)\ \ r \geq n + 1$$
$$(ii)\ \ r = n$$

$$\Longleftrightarrow \quad V \simeq \mathbb{P}^n;$$
$$\Longleftrightarrow \quad V \simeq Q \subset \mathbb{P}^{n+1}, \ \ \text{a smooth quadric.}$$

The classification of Fano n-folds of index $n - 1$ has been done by Iskovskikh [I] for $n = 3$ and Fujita [Fu] for $n \geq 4$. We state the classification of Fano 4-folds of index 3 with $b_2 = 1$.

Theorem 1.3. [Fu]. *Let* V *be a Fano 4-fold of index* 3 *and degree* d *with* $b_2(V) = 1$. *Then* $1 \leq d \leq 5$; *if* $d \geq 2$, $|H|$ *is free and if* $d \geq 3$, $|H|$ *is very ample. Furthermore* V *is one of the following*

(i) $d = 1$, $|H|$ *has only one base point,* $\chi_{\text{top}}(V) = 213$;

(ii) $d = 2$, *double covering of* $\mathbb{P}^4$ *branched along the divisor* $B \subset \mathbb{P}^4$ *of degree* 4, $\chi_{\text{top}}(V) = 66$;

(iii) $d = 3$, $V = V_3 \subset \mathbb{P}^5$ *is a cubic,* $\chi_{\text{top}}(V) = 27$;

(iv) $d = 4$, $V = V_4 \subset \mathbb{P}^6$ *is a complete intersection of two quadrics,* $\chi_{\text{top}}(V) = 12$;

(v) $d = 5$, $V = V_5 \subset \mathbb{P}^7$ *is a section of the Grassmannian* $G := \text{Gr}(2, 5) \subset \mathbb{P}^9$ *by a linear subspace of codimension* 2, *this Fano 4-fold is unique up to isomorphism,* $\chi_{\text{top}}(V) = 6$.

Remark 1.4. $|kH|$ gives an embedding of V for $d = 1$, $k = 3$ and for $d = 2$, $k = 2$. The above embeddings define projectively normal varieties which are intersections of quadrics (cf. [HW]).

2. Fano 4-fold $V = V_5^4 \subset \mathbb{P}^7$

Let $V = V_5^4 \subset \mathbb{P}^7$ be a Fano 4-fold of index 3 and degree 5 (see 1.3). This 4-fold may be represented as some 4-dimensional family of lines lying in $\mathbb{P}^4$. Every hyperplane section of the Grassmannian $G = \mathrm{Gr}(2,5) \subset \mathbb{P}^9$ is defined by $q \in \wedge^2 \mathbb{C}^{5*}$, $q \neq 0$, i.e. a skew-symmetric bilinear form on $\mathbb{C}^5$. We denote by $H_G(q)$ the hyperplane section of G corresponding to a form $q \in \wedge^2 \mathbb{C}^{5*}$, $q \neq 0$. By 1.3 (v) we have $V = H_G(q_1) \cap H_G(q_2)$ for some $q_i \in \wedge^2 \mathbb{C}^{5*}$, $q_i \neq 0$, $i = 1,2$.

It is well-known, that the integer homologies of the Grassmannian $G = \mathrm{Gr}(2,5)$ are freely generated by the following Schubert cycles ([GH, ch. 1, §5]):

σ	characterization of lines $l \subset \mathbb{P}^4$, $l \in \sigma$	$\dim \sigma$	$\deg \sigma$
$\sigma_{1,0}$	$l \cap \mathbb{P}^2 \neq \emptyset$	5	5
$\sigma_{2,0}$	$l \cap \mathbb{P}^1 \neq \emptyset$	4	3
$\sigma_{1,1}$	$l \subset \mathbb{P}^3$	4	2
$\sigma_{2,1}$	$l \subset \mathbb{P}^3$, $l \cap \mathbb{P}^1 \neq \emptyset$	3	2
$\sigma_{3,0}$	$l \ni p$	3	1
$\sigma_{2,2}$	$l \subset \mathbb{P}^2$	2	1
$\sigma_{3,1}$	$l \subset \mathbb{P}^3$, $l \ni p$	2	1
$\sigma_{3,2}$	$l \subset \mathbb{P}^2$, $l \ni p$	1	1

where $p \in \mathbb{P}^1 \subset \mathbb{P}^2 \subset \mathbb{P}^3 \subset \mathbb{P}^4$ is a flag in $\mathbb{P}^4$. In the last column we point out the degree of the Schubert variety σ relatively to the Plücker embedding. This number can be computed by the Pieri-formula ([GH, ch. 1, §5]).

Lemma 2.1. [T] *If $rk\,(q) = 4$, then $H_G(q)$ is smooth. If $rk\,(q) = 2$, then $H_G(q)$ is the Schubert variety $\sigma_{1,0}$ which is singular along a plane (Schubert variety $\sigma_{2,2}$).*

Hence the manifold V is defined by the pencil $\lambda_1 q_1 + \lambda_2 q_2 \in \wedge^2 \mathbb{C}^{5*}$ such that each form from this pencil has rank 4.

We shall need the description of linear subspaces of V. Since $\mathrm{Pic}\,(V) = \mathbb{Z} \cdot \mathcal{O}(1)$, then V does not contain 3-dimensional linear subspaces. Every plane on G is a Schubert variety $\sigma_{2,2}$ or $\sigma_{3,1}$. Every line on G is a Schubert variety $\sigma_{3,2}$.

Proposition 2.2. [T]. *(i) The manifold V contains exactly one $\sigma_{2,2}$-plane S and a 1-dimensional family of $\sigma_{3,1}$-planes.*

(ii) Each $\sigma_{3,1}$-plane P intersects S along a tangent line to the fixed conic $C \subset S$.

(iii) Any two $\sigma_{3,1}$-planes P_1 and P_2 intersect in a point $p \in S \setminus C$.

(iv) Let R be the union of all $\sigma_{3,1}$-planes on V. Then R is a hyperplane section of V, $\mathrm{Sing}(V) = S$.

(v) If $Z \subset R$ is a line, then $Z \cap S \neq \emptyset$ and Z is contained in S or in some $\sigma_{3,1}$-plane.

Sketch of the proof. Let $\lambda_1 q_1 + \lambda_2 q_2 \in \wedge^2 \mathbb{C}^{5*}$ be a pencil of skew-symmetric bilinear forms corresponding to the 4-fold $V \subset G$. Since $rk(\lambda_1 q_1 + \lambda_2 q_2) = 4$, each form $\lambda_1 q_1 + \lambda_2 q_2$ has the 1-dimensional kernel in $\mathbb{C}^5$ and defines a point in $\mathbb{P}^4 = P(\mathbb{C}^5)$. This point is called the center of $\{\lambda_1 q_1 + \lambda_2 q_2\}$.

Lemma 2.3. [T]. *The set of centers of forms from* $\{\lambda_1 q_1 + \lambda_2 q_2\} \subset \wedge^2 \mathbb{C}^{5*}$ *is a conic* $C_0 \subset \mathbb{P}^4 = P(\mathbb{C}^5)$.

Proof of lemma. The rational map

$$P(\wedge^2 \mathbb{C}^5) \longrightarrow \mathbb{P}^4, \quad \text{form} \longrightarrow \text{its center}$$

is defined by Pfaffians of 4×4-matrices, i.e. is quadratic. $\square$

Take a point $p \in C_0$, then each line $l \subset \mathbb{P}^4$ passing through p belongs to V (as a point). So $p \in C_0$ determines $\sigma_{3,1}$-plane $P \subset V$. If $\mathbb{P}^2 \subset \mathbb{P}^4$ is the linear span of C_0, then Schubert variety S of type $\sigma_{2,2}$ corresponding to the flag $\mathbb{P}^2 \subset \mathbb{P}^4$ lies in V. Points of the divisor R are represented by the lines in $\mathbb{P}^4$ meeting C_0. $\square$

We need formulas for Chern classes of V.

Lemma 2.4.

$c_1(G) \sim 5\sigma_{1,0}$,

$c_2(G) \sim 11\sigma_{2,0} + 12\sigma_{1,1}$,

$c_3(G) \sim 15\sigma_{3,0} + 30\sigma_{2,1}$,

$c_4(G) \sim 35\sigma_{3,1} + 25\sigma_{2,2}$.

Proof. Let $\mathcal{I}$ be a universal subbundle and Q be a universal factorbundle on G. Then $c_r(\mathcal{I}^*) \sim \sigma_{1,\dots,1}$, $c_r(Q) \sim \sigma_{r,0}$ [GH, ch. 3, §1] and $T_G \simeq \mathcal{I}^* \otimes Q$. Using the formula for Chern classes of tensor products (see for example [Ful, ex. 14.5.2]) we obtain 2.4. $\square$

Corollary 2.5. $c_2(V) \sim (4\sigma_{2,0} + 5\sigma_{1,1})\big|_V$, $\quad \chi_{top}(V) = 6$.

Corollary 2.6. *If* $S \subset V$ *is* $\sigma_{2,2}$-*plane, then* $c_2(N_{S/V}) = 2$, $c_1(N_{S/V}) = 0$. *If* $P \subset V$ *is* $\sigma_{3,1}$-*plane, then* $c_2(N_{P/V}) = 1$, $c_1(N_{P/V}) = 0$.

Now we consider the family of lines on V.

Proposition 2.7. *(i) The normal bundle* $N_{Z/V}$ *of* Z *in* V *is one of the following:*

$$N_{Z/V} \simeq \mathcal{O}_{\mathbb{P}^1} \oplus \mathcal{O}_{\mathbb{P}^1} \oplus \mathcal{O}_{\mathbb{P}^1}(1) \quad \text{or} \quad N_{Z/V} \simeq \mathcal{O}_{\mathbb{P}^1}(-1) \oplus \mathcal{O}_{\mathbb{P}^1}(1) \oplus \mathcal{O}_{\mathbb{P}^1}(1).$$

(ii) Let $\Gamma(V)$ *be the Hilbert scheme of lines on* V. *Then* $\Gamma(V)$ *is a smooth, irreducible and reduced variety of dimension* 4. *Furthermore* $\Gamma(V)$ *is isomorphic to the blow-up of a conic on* $\mathbb{P}^4$.

(iii) There is exactly one 1-dimensional family of lines passing through every point $p \in V$.

Proof. Similar to [I, ch.3, §1]. $\square$

According to 2.2 there exist only the following types of lines on V :

(a) $Z \not\subset R$, i.e. $Z \cap S = \emptyset$, $Z \cap R = \{point\}$;

(b) $Z \subset R, \quad Z \cap C = \emptyset$;

(c) $Z \subset R, \quad Z \cap S = Z \cap C = \{\text{point}\}$;

(d) $Z \subset S, \quad Z \cap C = \{2\text{points}\}$;

(e) $Z \subset S, \quad Z \cap C = \{\text{point}\}$, i.e. $Z = S \cap P$ where P is $\sigma_{3,1}$ − plane.

Proposition 2.8. *Let* $Z \subset V$ *be a line. Then*

(i) all lines meeting Z *cover up a divisor* $H(Z)$, *a hyperplane section of* V *which is singular along* Z;

(ii) $Z \subset S$ *(i.e. Z has a type (d) or (e))* $\iff H(Z) = R$;

(iii) if $H \subset V$ *is a hyperplane section and* $\dim(\mathrm{Sing}(H)) \geq 1$, *then* $H = H(Z)$ *for some line* $Z \subset V$. $\square$

3. Compactifications of $\mathbb{C}^4$

Now we give a precise form of the main theorem.

Theorem 3.1. *Let* (V, A) *be a compactification of* $\mathbb{C}^4$ *with* $b_2(V) = 1$. *Assume that V is a projective algebraic manifold and* $-K_V \sim 3 \cdot H$ *for some* $H \in \mathrm{Pic}\,(V)$. *Then V is a section of the Grassmannian* $G := \mathrm{Gr}(2, 5) \subset \mathbb{P}^9$ *by a linear subspace of codimension 2 and A is a singular hyperplane section of G. Furthermore A is one of the following*

(i) $A = R$ *and A is singular along* $\sigma_{2,2}$*-plane S (see 2.2 (iv))*;

(ii) $A = H(Z)$, *where Z is a line of type (a) (see 2.8 (i))*;

(iii) $A = H(Z)$, *where Z is a line of type (c)*;

(iv) A has only one singular point $p \notin R$, *ordinary double point.*

Each of these compactifications is unique up to isomorphism.

Remember that V is a Fano 4-fold of index 3 and A is a generator of $\mathrm{Pic}\,(V) = \mathbb{Z}$. In this section we assume that $V = V_5 \subset \mathbb{P}^7$. If A is non-singular then $\chi_{\mathrm{top}}(A) = 4$ [I]. By 1.1 (ii) we have $6 = \chi_{\mathrm{top}}(V) = \chi_{\mathrm{top}}(A) + 1 = 5$, contradiction. Hence $\mathrm{Sing}(A) \neq \emptyset$. First, we consider the case $\dim \mathrm{Sing}(A) \geq 1$.

Proposition 3.2 [Fu, II]. *Let* $S \subset V$ *be a* $\sigma_{2,2}$*-plane. Then*

(i) there exists a diagram below

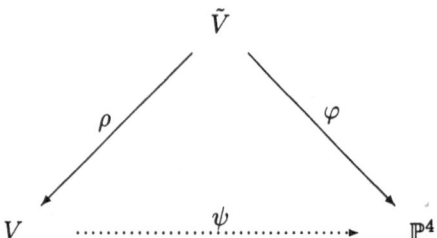

where $\rho : \tilde{V} \longrightarrow V$ is the blow-up of S, $\psi : V \cdots\!\!\rightarrow \mathbb{P}^4$ is the projection from S, $\varphi : \tilde{V} \longrightarrow \mathbb{P}^4$ is the blow-up of a rational normal cubic curve $Y \subset \mathbb{P}^4$;

(ii) the morphism $\varphi : \tilde{V} \longrightarrow \mathbb{P}^4$ is defined by $|H^* - E|$, where $H^* = \rho^* H$ and $E = \rho^{-1}(S)$ is an exceptional divisor;

(iii) $\varphi(E) = \mathbb{P}^3 = \langle Y \rangle$ is the linear span of Y;

(iv) the exceptional divisor $\tilde{R} = \varphi^{-1}(Y)$ of the morphism φ is the proper transform of R and $\tilde{R} \sim H^* - 2E$.

Sketch of the proof. Using 2.6 it is easy to compute $(H^* - E)^4 = 1$, $(H^* - E)^3 \cdot E = 1$, i.e. φ is a birational morphism, $\varphi(E) = \mathbb{P}^3$ and $\varphi(\tilde{V}) = \mathbb{P}^4$. Since $\tilde{R} \sim H^* - kE$, $k \geq 2$, we have $(H^* - E)^3 \cdot \tilde{R} = 2 - k \geq 0$. Thus $k = 2$ and $\dim \varphi(\tilde{R}) \leq 2$. $\square$

Corollary 3.3. *(V, R) is a compactification of $\mathbb{C}^4$.*

Proof. $V \setminus R \simeq \tilde{V} \setminus (\tilde{R} \cup E) \simeq \mathbb{P}^4 \setminus \varphi(E) \simeq \mathbb{C}^4$. $\square$

Corollary 3.4. *The group of biregular automorphisms of V is the following extension:*

$$1 \longrightarrow (\mathbb{C}^+)^5 \longrightarrow \mathrm{Aut}(V) \longrightarrow \mathrm{PSL}_2 \longrightarrow 0.$$

There exist only 4 $\mathrm{Aut}(V)$-orbits on $V : V \setminus R$, $R \setminus S$, $S \setminus C$, C.

Proof. $\mathrm{Aut}(V) \simeq \mathrm{St}_{\mathbb{P}^4}(Y)$, where $\mathrm{St}_{\mathbb{P}^4}(Y)$ is the stabilizer of $Y \subset \mathbb{P}^4$ in PGL_5. $\square$

Corollary 3.5. *There is a line of type (a) passing through every point $p \in V \setminus S$.* $\square$

Corollary 3.6. *Let $A \subset V$ be a hyperplane section of V and $p \in A$ be an isolated singular point such that $p \notin R$. Then p is an ordinary double point and it is the unique singularity of A on $A \setminus R$. The group $\mathrm{Aut}(V)$ transitively acts on the set of such hyperplane sections.*

Proof. Let $\tilde{A} \subset \tilde{V}$ be a strict transform of A. Then $\varphi(\tilde{A})$ is a quadric in $\mathbb{P}^4$ and $A \setminus R \simeq \varphi(\tilde{A}) \setminus \varphi(E)$. $\square$

Corollary 3.7. *The group $\mathrm{Aut}(V)$ transitively acts on the set of lines of types (a), (b), (c), (d) and (e).*

Proposition 3.8. (cf. [Fu, II])*Let $P \subset V$ be a $\sigma_{3,1}$-plane. Then*

(i) there exists a diagram below

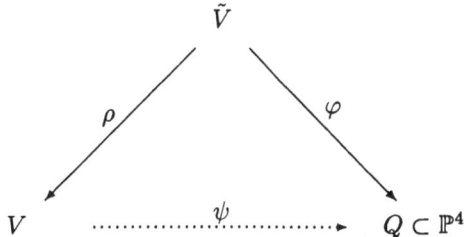

where $\rho : \tilde{V} \longrightarrow V$ is the blow-up of P, $Q \subset \mathbb{P}^4$ is a smooth quadric, $\psi : V \cdots\!\!\!\rightarrow Q \subset \mathbb{P}^4$ is the projection from P;

(ii) the morphism $\varphi : \tilde{V} \longrightarrow Q \subset \mathbb{P}^4$ is defined by $|H^* - E|$, where $H^* = \rho^* H$ and $E = \rho^{-1}(S)$ is the exceptional divisor;

(iii) a general fiber of φ is a strict transform of a line intersecting P, every 1-dimensional fiber of φ is isomorphic to $\mathbb{P}^1$;

(iv) there exists only one 2-dimensional fiber $\tilde{S}$ of φ, it is the strict transform of $\sigma_{2,2}$-plane S, $\tilde{S} \simeq \mathbb{P}^2$;

(v) the restriction $\varphi \big|_E \colon E \longrightarrow Q$ is the blow-up of a line $Y \subset Q$ and $\varphi(\tilde{S}) \in Y$.

Proposition 3.9. Let (V, A) be a compactification of $\mathbb{C}^4$ such that A contains $\sigma_{3,1}$-plane P. Using notations of 3.8 denote the strict transform of A on $\tilde{V}$ by $\tilde{A} \subset \tilde{V}$. Then $Q' := \varphi(\tilde{A})$ is a singular hyperplane section of Q, a cone with the vertex $p \in Q' = \varphi(\tilde{A})$. The compactification (V, A) is determined by the vertex p up to isomorphism. Furthermore for p and Q' we have one of the following cases:

(i) $p = \varphi(\tilde{S})$, $\quad Y \subset Q'$;

(ii) $p \neq \varphi(\tilde{S})$, $\quad p \in Y$, $\quad Y \subset Q'$;

(iii) $p \notin Y$, $\quad \varphi(\tilde{S}) \notin Q'$, $\quad Y \notin Q'$.

Remark 3.10. If $\mathrm{Sing}(A) \cap R \neq \emptyset$, then A contains a $\sigma_{3,1}$-plane. If $\dim \mathrm{Sing}(A) \geq 1$, then $\mathrm{Sing}(A) \cap R \neq \emptyset$.

Proof. It is easy to see that $(\tilde{V}, \tilde{A} \cup E)$ is a compactification of $\mathbb{C}^4$ with $b_2 = 2$. Since $\tilde{A} \sim H^* - E$, $Q' = \varphi(\tilde{A})$ is a hyperplane section of Q, restriction $\tilde{V} \setminus (E \cup \tilde{A}) \simeq \mathbb{C}^4 \longrightarrow Q \setminus Q'$ is surjective and has connected fibers. If Q' is smooth, then $\pi_1(Q \setminus Q') \simeq \mathbb{Z}/2\mathbb{Z}$, contradiction with simply-connectedness of $\mathbb{C}^4$. Hence Q' is a cone and $Q \setminus Q' \simeq \mathbb{C}^3$. The compactification (V, A) is defined by the vertex of Q'. Note that $p \in Y$ implies $Y \subset Q'$ and we have (i) and (ii) in this case. Now we assume $p \notin Y$. Then $Y \subset Q'$ and $\{p'\} := Y \cap Q'$ is a point. It remains to show that the case $\varphi(\tilde{S}) \in Q'$ (i.e. $\varphi(\tilde{S}) = p'$) is impossible. Indeed, in this case $\tilde{S} \subset \tilde{A}$. By 3.8 fiber over every point $p \in Q \setminus Q' \simeq \mathbb{C}^3$ is isomorphic to $\mathbb{P}^1$. Thus the morphism $\tilde{V} \setminus (\tilde{A} \cup E) \longrightarrow Q \setminus (Q' \cup Y)$ is $\mathbb{C}^1$-bundle. Hence $1 = \chi_{\mathrm{top}}(\mathbb{C}^4) = \chi_{\mathrm{top}}(\tilde{V} \setminus (\tilde{A} \cup E)) = \chi_{\mathrm{top}}(Q \setminus (Q' \cup Y)) \cdot \chi_{\mathrm{top}}(\mathbb{C}^1) = \chi_{\mathrm{top}}(Q) - \chi_{\mathrm{top}}(Q') - \chi_{\mathrm{top}}(Y) + 1 = 4 - 3 - 2 + 1 = 0$, contradiction. $\square$

Lemma 3.11. *If (V, A) is from 3.9 (i), then $A = R$. If (V, A) is from 3.9 (ii) or 3.9 (iii), then* $\dim \operatorname{Sing}(A) = 1$. $\square$

Lemma 3.12. *Let (V, A) be a compactification of $\mathbb{C}^4$ from 3.9. Then $A = H(Z)$, where Z is a line of type (a) in the case (iii) and Z is a line of type (c) in the case (ii).*

Proof. By 2.8 we have $A = H(Z)$. In the case (iii) $S \not\subset A = H(Z)$. Hence $Z \cap S = \emptyset$ and Z is of type (a). In the case (ii) $A = H(Z) \supset S$, but $Z \not\subset S$ (see 3.11, 2.8). Thus Z is of type (b) or (c). Consider the first case. Then $H(Z)$ contains $\sigma_{3,1}$-plane $P' \neq P$. It is easy to check that $\varphi(\tilde{P}')$ is a line on Q, where $\tilde{P}' \subset \tilde{V}$ is the strict transform of P'. So $\varphi(\tilde{S}') \in \varphi(\tilde{P}')$, $\varphi(\tilde{S}') \in Y$ and $\varphi(\tilde{P}') \subset Q'$, $Y \subset Q'$. Since Q' is 2-dimensional cone with vertex $p \neq \varphi(\tilde{S}')$, we have a contradiction. $\square$

Now we assume that A has only isolated singularities. By 3.10 we have that all singular points of A belong to $V \setminus R$. From 3.6 we conclude that A has only one singular point and the pair (V, A) is unique up to isomorphism.

Proposition 3.13. *Let $Z \subset V$ be a line of type (a). Then*

(i) there exists a diagram below

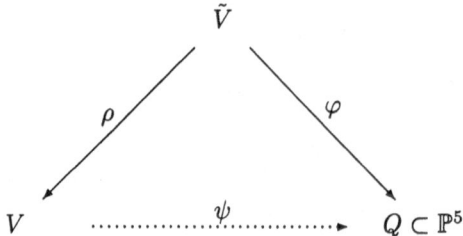

where $\rho : \tilde{V} \longrightarrow V$ is the blow-up of Z, $Q \subset \mathbb{P}^5$ is a smooth quadric, $\psi : V \dashrightarrow Q \subset \mathbb{P}^5$ is the projection from Z, $\varphi : \tilde{V} \longrightarrow Q$ is the blow-up of a smooth rational surface $F \subset Q$;

(ii) the morphism $\varphi : \tilde{V} \longrightarrow Q \subset \mathbb{P}^5$ is defined by $|H^* - E|$, where $H^* = \rho^* H$ and $E = \rho^{-1}(S)$ is the exceptional divisor;

(iii) $\varphi(E) = Q' = Q \cap \langle F \rangle$ is a 3-dimensional quadric cone;

(iv) the exceptional divisor $M = \varphi^{-1}(F)$ of the morphism φ is the proper transform of $H(Z)$ and $M \sim H^* - 2E$.

Remark 3.14. The proposition 3.13 gives another construction of the compactification from 3.1 (i).

Lemma 3.15. *There exists a compactification of $\mathbb{C}^4$ (V, A), where A has only isolated singularities.*

Proof. We use notations of the proposition 3.13. Assume that the quadric $Q \subset \mathbb{P}^5$ is defined by the equation: $x_2 x_3 + x_1 x_5 - x_0 x_4 = 0$. Take singular hyperplane sections of Q : $Q' := \{x_5 = 0\}$, $Q'' := \{x_0 = 0\}$ and a smooth surface in Q' of degree

$3 : F := \{x_5 = 0, \; x_2 x_3 = x_0 x_4, \; x_2 x_4 = x_1 x_3, \; x_0 x_1 = x_2^2\}$. Denote the strict transform of Q'' on $\tilde{V}$ by $\tilde{A}$ and put $A := \rho(\tilde{A})$. Lemma 3.15 is an easy consequence of the following

Claim. $\tilde{V} \setminus (M \cup E) \simeq \tilde{V} \setminus (\tilde{A} \cup E) \simeq \mathbb{C}^4$.

Proof of claim. Since Q' is a singular hyperplane section of Q and $F \subset Q'$, $\tilde{V} \setminus (M \cup E) \simeq Q \setminus Q' \simeq \mathbb{C}^4$. Take the coordinates in the affine part $Q^0 := Q \setminus Q' \simeq \mathbb{C}^4$:

$$(x_0, x_1, \dots, x_5) \longrightarrow (y_1 = x_1/x_0, y_2 = x_2/x_0, y_3 = x_3/x_0, y_4 = x_5/x_0).$$

Then $Q'^0 = \{y_4 = 0\}$, $F^0 = \{y_4 = 0, \; y_1 = y_2^2\}$, where $Q'^0 := Q' \setminus Q''$, $F^0 := F \setminus Q''$. By the coordinate transformation $z_1 = y_1 - y_2^2, \; z_2 = y_2, \; z_3 = y_3, \; z_4 = y_4$, we have $Q'^0 = \{z_4 = 0\}$, $F^0 = \{z_1 = z_4 = 0\}$. Let $\varphi^0 : \tilde{V}^0 \longrightarrow Q^0 \simeq \mathbb{C}^4$ be the blow-up of $F^0 \simeq \mathbb{P}^2$, M^0 be the exceptional divisor and E^0 be the strict transform of Q'^0. Then $\tilde{V} \setminus (\tilde{A} \cup E) \simeq \tilde{V}^0 \setminus E^0$. In $\mathbb{C}^4_{z_1, z_2, z_3, z_4} \times \mathbb{P}^1_{u_1, u_2}$ the variety $\tilde{V}^0$ is defined by the equation $z_1 u_2 = z_4 u_1$. So we can see that $\tilde{V} \setminus (\tilde{A} \cup E) \simeq \tilde{V}^0 \setminus E^0 \simeq \{z_1 u_2 = z_4 u_1, \; z_4 \neq 0\} = \{z_1 u_2 = z_4 u_1, \; z_4 \neq 0, \; u_2 \neq 0\}$. Taking the affine part $\mathbb{C}^5 \simeq \{u_2 = 1\}$ in $\mathbb{C}^4 \times \mathbb{P}^1$ we obtain $\tilde{V} \setminus (\tilde{A} \cup E) \simeq \{z_1 = z_4 u_1, \; z_4 \neq 0\}$. The projection $\mathbb{C}^5 \longrightarrow \mathbb{C}^4$, $(z_1, z_2, z_3, z_4, u_1) \longrightarrow (z_2, z_3, z_4, u_1)$ gives an isomorphism $\tilde{V} \setminus (\tilde{A} \cup E) \simeq \mathbb{C}^4$. The claim is proved. $\square$

4. The case $d \leq 4$

In this section we complete the proof of theorem 3.1.

Lemma 4.1. *Let V be a Fano n-fold of index $n - 1$ with $-K_V \sim (n-1)H$, $H \in \mathrm{Pic}\,(V)$ and $A \in |H|$ be a non-normal member. Then V has degree $d = 5$.*

Proof. Since A is a complete intersection in V, $\dim \mathrm{Sing}(A) = n - 2$. Taking a general section $H_1 \cap \dots \cap H_{n-3} \cap A$, for $H_i \in |H|$, we see that it is sufficient to consider only 3-dimensional case (here we keep in mind that for general $H_i \in |H|$ and $1 \leq k \leq n - 1$, the intersection $H_1 \cap \dots \cap H_k$ is smooth and irreducible [Fu]). Thus A is a non-normal surface. Let Z be an irreducible component of $\mathrm{Sing}(A)$ of dimensional 1. Then Z is a line on V, i.e. $Z \cdot H = 1$ and $Z \simeq \mathbb{P}^1$ (see, e.g., [PS, 2.1]). Let $\rho : \tilde{V} \longrightarrow V$ be the blow-up of Z, E be the exceptional divisor and $\tilde{A} \subset \tilde{V}$ be the strict transform of A. In the cases $d = 3, 4$ $|H|$ gives an embedding $V = V_d \subset \mathbb{P}^{d+1}$ [I]. Since $Z \subset V \subset \mathbb{P}^{d+1}$ is a line, Z is an intersection of divisors from $|H|$. Hence $|\rho^* H - E|$ is free and $(\rho^* H - E)^2 \tilde{A} \geq 0$. We have $\tilde{A} \sim \rho^* H - kE$, $k \geq 2$. But there are the following easy calculations: $(\rho^* H - E)^2 \tilde{A} = (\rho^* H - E)^2 (\rho^* H - kE) = d - 2k - 1 < 0$. In the case $d = 2$ $|2H|$ gives an embedding of V and the image of V is an intersection of quadrics (cf. [HW] and 1.4). Since the image of Z is a conic, Z is an intersection of divisors from $|2H|$. We have a contradiction analogous to the cases $d = 3, 4$. The proof in the case $d = 1$ is analogous, but instead of $|2H|$ we consider $|3H|$. $\square$

Now we assume that A is normal.

Lemma 4.2. *Let V be a Fano 4-fold of index 3 and degree d with $b_2(V) = 1$, $-K_V \sim 3H$, $H \in \text{Pic}\,(V)$ and $A \in |H|$ be a normal divisor. Then $\chi_{\text{top}}(A) \leq 2(12 - d)$.*

Proof. Take the general divisors H_1, $H_2 \in |H|$. By the Bertini theorem, $C := H_1 \cap H_2 \cap A$ is a smooth irreducible curve (in the case $d = 1$ we use also that the intersection $H_1 \cap H_2 \cap A$ is transverse in $Bs|H|$). Hence $C \cap \text{Sing}(A) = \emptyset$. Let $\rho : \tilde{A} \longrightarrow A$ be the blow-up of C and let $\mathcal{L}$ be the pencil on $\tilde{A}$ generated by the strict transforms of $H_1 \big|_A$ and $H_2 \big|_A$. Then $\mathcal{L}$ defines the morphism $\varphi : \tilde{A} \longrightarrow \mathbb{P}^1$. Every fiber F of φ is an irreducible normal surface with ample anticanonical divisor. By [HW] we have that F is a (possibly singular) Del Pezzo surface of degree d or a generalized cone over an elliptic curve. If F is a cone then $\chi_{\text{top}}(F) = 1$. If F is a Del Pezzo surface then $\chi_{\text{top}}(F) = 12 - d - k$, where $k = 10 - d - rk\,\text{Pic}\,(F) \geq 0$. So we have

$$\chi_{\text{top}}(A) = \chi_{\text{top}}(\tilde{A}) = \chi_{\text{top}}(\mathbb{P}^1) \cdot \chi_{\text{top}}(F_g) + \sum \big(\chi_{\text{top}}(F_i) - \chi_{\text{top}}(F_g)\big),$$

where F_g is a general and F_i is a special fibers of φ. Hence $\chi_{\text{top}}(A) \leq 2(12 - d)$. $\square$

Lemma 4.3. *If $d = 4$ and A is normal, then $\chi_{\text{top}}(A) \leq 10$.*

Proof. Let $p \in \text{Sing}(A)$. Projection from p gives the birational map $\psi : A \dashrightarrow Q_3$, where $Q_3 \subset \mathbb{P}^4$ is a quadric. This map contracts the cone with the vertex p to a (possibly reducible) curve Y of degree ≤ 4 lying on a hyperplane section $Q_2 \subset Q_3$. As in 4.2 we obtain $\chi_{\text{top}}(A) = \chi_{\text{top}}(Q_3) + \chi_{\text{top}}(Y) - \chi_{\text{top}}(Q_2) + 1 \leq 10$. $\square$

Now let (V, A) be a compactification of $\mathbb{C}^4$ with a normal divisor A and $d \leq 4$. Then 4.2 and 4.3 contradict to 1.1 (ii) and 2.6. The proof of the theorem is finished.

Remark. Note that lemma 4.1 follows also from results of the work [F4].

References

[F1]   M. FURUSHIMA. *Singular Del Pezzo surfaces and analytic compactifications of 3-dimensional complex affine space* $\mathbb{C}^3$, Nagoya Math. J. 104 (1986), 1-28.

[F2]   M. FURUSHIMA. *Complex analytic compactifications of* $\mathbb{C}^3$, Comp. Math. 76 (1990), 163-196.

[F3]   M. FURUSHIMA. *The structure of compactifications of* $\mathbb{C}^3$, Proc. Japan Acad. Ser. A. 68 (1992), 33-36.

[F4]   M. FURUSHIMA. *Non-normal Del Pezzo surfaces and Fano threefolds of the first kind*, J. Reine Angew. Math. 429(1992), 183-190.

[Fu]   T. FUJITA. *On the structure of polarized manifolds with total deficiency one, I, II and III*, J. Math. Soc. Japan. 32 (1980), 709-725, 33 (1981), 415-434, 36 (1984), 75-89.

[Ful]   W. FULTON. *Intersection theory*, Springer, 1984.

[GH]   P. GRIFFITHS, J. HARRIS. *Principles of algebraic geometry*, Wiley Interscience, 1978.

[H]   F. HIRZEBRUCH. *Some problems on differentiable and complex manifolds*, Ann. of Math. 60 (1954), 213-236.

[HW]   F. HIDAKA, K. WATANABE. *Normal Gorenstein surfaces with ample anticanonical divisor*, Tokyo J. Math. 4 (1981), 319-330.

[I]   V.A. ISKOVSKIKH. *Anticanonical models of three-dimensional algebraic varieties*, J. Soviet Math. 13-14 (1989), 745-814.

[K]   K. KODAIRA. *Holomorphic mappings of polydiscs into compact complex manifolds*, J. Differ. Geom. 6 (1971), 33-46.

[KO]   S. KOBAYASHI, T. OCHIAI. *Characterization of complex projective spaces and hyperquadrics*, J. Math. Kyoto Univ. 13 (1973), 31-47.

[M]   S. MUKAI. *On Fano 3-folds*, London Math. Soc. Lect. Note Ser. 179(1992), 255-263.

[P]   T. PETERNELL. *Compactifications of $\mathbb{C}^3$, II*, Math. Ann. 283 (1989), 121-137.

[Pr]   Yu.G. PROKHOROV. *Fano threefolds of genus 12 and compactifications of $\mathbb{C}^3$*, Algebra and analysis 3 (1991), 162-171 (Russian), English transl. in St. Petersburg Math. J. 3(1991), 855-864.

[PS]   T. PETERNELL, M. SCHNEIDER. *Compactifications of $\mathbb{C}^3$, I.* , Math. Ann. 280 (1988), 129-146.

[RV]   R. REMMERT, A. VAN DE VEN. *Zwei Sätze über die komplex projektive Ebene*, Nieuw Arch. Wisk. 8 (1960), 147-157.

[T]   J.P. TODD. *The locus representing the lines of four-dimensional space and its application to linear complexes in four dimensions*, Proc. London Math. Soc. Ser. 2. 30 (1930), 513-550.

[V]   A. VAN DE VEN. *Analytic compactifications of complex homology cells*, Math. Ann. 147 (1962), 189-204.

A Note on Cohomologies of Exceptional Bundles on a Quadric Surface

Alexei N. Rudakov

We prove that only one cohomology group could be different from zero for those exceptional vector bundles on a quadric surface which appear to be symmetric ones twisted by a line bundle, and that this is not true for ext-groups between two such bundles.

An algebraic vector bundle E or, strictly speaking, a coherent sheaf E on a proper smooth algebraic surface is called exceptional if

$$\text{Hom}(E, E) = k, \quad \text{Ext}^1(E, E) = 0, \quad \text{Ext}^2(E, E) = 0.$$

Really this definition implies that a canonical line bundle of the surface has no global sections because of Serre duality so it is reasonable to use it for Del Pezzo surfaces (there are weaker forms of the definition that we will not discuss here). Properties of these sheaves are closely related to a configuration of stable sheaves on a surface [D-LP, R2] and they were studied in great detail for the first two surfaces in the list of Del Pezzo surfaces, namely for the projective plane $\mathbb{P}^2$ and a quadric surface Q. Especially important for us is that there are fine rules to make many exact sequences for exceptional bundles which lead to helix theory [D, G-R].

It was proved for an exceptional sheaf E on $\mathbb{P}^2$, that only one of the cohomology groups $H^i(\mathbb{P}^2, E)$ could be different from zero [B-G in Sem] and as any line bundle is exceptional hence this gives a kind of a generalization of the well-known Bott-Borel-Weil theorem for line bundles on homogeneous varieties. And what is good for working with exceptional bundles on $\mathbb{P}^2$ that for any two such bundles E, F only one of the groups $\text{Ext}^i(E, F)$ could be different from zero [B-G in Sem].

In this paper we show that the second property is not valid for exceptional sheaves on a quadric surface Q, but the first one we prove under some restrictions on the sheaf E.

It is important to mention that if there exists an exceptional sheaf E on a quadric surface with prescribed rank $r(E) = r$ and first Chern class $c_1(E) = c$ then this E is unique and it is uniquely determined by the element $c/r \in \text{Pic } Q \otimes \mathbb{Q} \simeq \mathbb{Q}^2$ [G]. So the set of exceptional sheaves on Q corresponds to a set in a coordinate plane. The paper [R1] describes the structure of this point set and properties of the correspondence. The description is a little

simpler if the points lie on the line of equal coordinates or are of the kind $((p \pm 1)/r, \, p/r)$.
These are such points in the unit square related to sheaves with rank less than 17:

$$\left(\frac{0}{1}; \frac{0}{1}\right), \quad \left(\frac{1}{5}; \frac{2}{5}\right), \quad \left(\frac{2}{5}; \frac{1}{5}\right), \quad \left(\frac{1}{3}; \frac{1}{3}\right),$$

$$\left(\frac{4}{11}; \frac{4}{11}\right), \quad \left(\frac{0}{1}; \frac{1}{1}\right), \quad \left(\frac{1}{1}; \frac{0}{1}\right), \quad \left(\frac{5}{11}; \frac{5}{11}\right),$$

$$\left(\frac{2}{3}; \frac{2}{3}\right), \quad \left(\frac{3}{5}; \frac{4}{5}\right), \quad \left(\frac{4}{5}; \frac{3}{5}\right), \quad \left(\frac{1}{1}; \frac{1}{1}\right).$$

A sheaf $\mathcal{F}$ is called symmetrical exceptional if it is either an exceptional one corresponding to the point $\left(\frac{p}{r}, \frac{p}{r}\right)$ or direct sum of two exceptional sheaves related to the points $\left(\frac{p}{r}, \frac{p+1}{r}\right)$ and $\left(\frac{p+1}{r}, \frac{p}{r}\right)$ ([R1]). We will call semisymmetrical exceptional such a sheaf that is a symmetrical twisted by a line bundle. We remind that integer points correspond to line bundles and twisting by a line bundle correspond to translation on integer vector.

Our main result is the following theorem:

Theorem. *If E is a semisymmetrical exceptional sheaf on a quadric surface then only one of the groups $H^i(E)$ could be nonzero.*

Really by the very definitions of semisymmetrical sheaves and cohomology this theorem is equivalent to the following proposition.

Proposition 1. *Let L be a line bundle on a quadric surface. Then for a semisymmetrical exceptional sheaf E only one of the groups $\mathrm{Ext}^i(L, E)$ could be different from zero.*

As usual we will write line bundles on Q in the form $\mathcal{O}(m, n) = p_1^* \mathcal{O}(m) \otimes p_2^* \mathcal{O}(n)$, where p_i are two projections from Q on $\mathbb{P}^1$ related to isomorphism $Q \simeq \mathbb{P}^1 \times \mathbb{P}^1$.

Proposition 2. *Let M be an exceptional bundle that can be constructed as an extension via an exact sequence*

$$0 \longrightarrow \mathcal{O}(1, -1) \longrightarrow M \longrightarrow \mathcal{O}(0, 1) \oplus \mathcal{O}(0, 1) \longrightarrow 0. \qquad (1)$$

Then there is a semisymmetrical exceptional bundle E such that $\mathrm{Ext}^0(M, E) \neq 0$ and $\mathrm{Ext}^1(M, E) \neq 0$.

Proofs of both propositions are related to calculations based on helix theory for quadric surface ([R]). We will use it without further notice. We will use the notation $E\left(\frac{p}{r}; \frac{q}{r}\right)$ for an exceptional sheaf related to the point $\left(\frac{p}{r}; \frac{q}{r}\right)$ and the short notation $^i\langle A, B\rangle$ for $\mathrm{Ext}^i(A, B)$. So $^i\langle A, B\rangle = \mathrm{Ext}^i(A, B)$ by definition. And the sheaf M from proposition 2 is isomorphic to $E\left(\frac{1}{3}, \frac{1}{3}\right)$.

Proof of proposition 2. To prove the proposition we take $E = E\left(-\frac{2}{3}, \frac{7}{3}\right)$. It is the twist of M by $\mathcal{O}(-1, 2)$ so there is an exact sequence that is the twist of (1)

$$0 \longrightarrow \mathcal{O}(0,1) \longrightarrow E \longrightarrow \mathcal{O}(-1,3) \oplus \mathcal{O}(-1,3) \longrightarrow 0. \tag{2}$$

By Kunneth formula one can easily compute ext-groups between two line bundles. So we can compute the groups $^i\langle E\left(\frac{1}{3},\frac{1}{3}\right), \mathcal{O}(-1,3)\rangle$ via a long exact sequence and the only one nonzero is for $i = 1$.

At the same time there is another defining sequence for $E\left(\frac{1}{3},\frac{1}{3}\right)$:

$$0 \longrightarrow \mathcal{O}(-1,1) \longrightarrow E\left(\tfrac{1}{3},\tfrac{1}{3}\right) \longrightarrow \mathcal{O}(1,0) \oplus \mathcal{O}(1,0) \longrightarrow 0. \tag{3}$$

And from the related long exact sequence we see that the only one nonzero group from $^i\langle E\left(\frac{1}{3},\frac{1}{3}\right), \mathcal{O}(0,1)\rangle$ is for $i = 0$.

Now the long exact sequence related to (2) gives us the proposition. $\square$

Proof of proposition 1. We will specify $L = \mathcal{O}(1,1)$ and prove the following lemma.

Lemma 1. *The only possible nonzero group* $^i\langle L, E\rangle$, *where* $E = E(x, y)$ *is a semisymme-trical exceptional sheaf, could be*

$$for \ i = 0 \quad \Longleftrightarrow \quad x \geq 0, \quad y \geq 0, \quad x + y \geq 1,$$

$$for \ i = 1 \quad \Longleftrightarrow \quad either \ -1 \leq x \leq x + y \leq 1 \ or \ x \leq 0 \ or \ y \leq 0,$$

$$for \ i = 2 \quad \Longleftrightarrow \quad x \leq 0, \quad y \leq 0, \quad x + y \leq -1.$$

It may look strange that the areas have intersections but all the groups $^i\langle L, E\rangle$ are zero then (x, y) lies in intersection and this is so for an integer point with $x = 0$ or $y = 0$.

We will prove the lemma by induction on $\mathrm{rank}E$. The case $\mathrm{rank}E = 1$ is trivial by Kunneth formula. The general case is based on so called "symmetrical helices on a quadric". The main fact is that if $\mathrm{rank}E > 1$ then there is a pair of exact sequences of a type either

$$0 \longrightarrow A \longrightarrow V \otimes B \longrightarrow E \longrightarrow 0, \tag{4.1}$$
$$0 \longrightarrow E \longrightarrow W \otimes C \longrightarrow A \otimes \mathcal{O}(2,2) \longrightarrow 0, \tag{4.2}$$

or

$$0 \longrightarrow A \otimes \mathcal{O}(-2,-2) \longrightarrow V \otimes B \longrightarrow E \longrightarrow 0, \tag{5.1}$$
$$0 \longrightarrow E \longrightarrow W \otimes C \longrightarrow A \longrightarrow 0, \tag{5.2}$$

where V and W are trivial bundles and A, B, C are semisymmetrical exceptional sheaves of a smaller rank and such that their points are in the same unit square as the point (x, y). Moreover if $\mathrm{rank}E > 3$ then all the points for A, B, C are either below or above a diagonal of the square parallel to the line $x + y = 0$.

The case $\mathrm{rank}E = 3$ which is the least possible after 1 needs a separate investigation. And it is important to mention that such E is either a twist of $E\left(\frac{1}{3},\frac{1}{3}\right)$ by a line bundle or a twist of a dual $E\left(\frac{1}{3},\frac{1}{3}\right)^*$. So we will get short exact sequences connecting E with line bundles from (1) and (2). Hence we prove the statement in this case.

To proceed the proof of a general case we will look at each zone separately. Let $x \geq 0$, $y \geq 0$, $x + y \geq 1$ first. Then either (4.1) or (5.1) is valid for E and the points related to A, B are in the same zone. So by induction $^1\langle L, B \rangle = {}^2\langle L, B \rangle = 0$, and by the same reasoning $^2\langle L, A \otimes \mathcal{O}(-2, -2) \rangle = {}^2\langle L, A \rangle = 0$.

Let us write a long exact sequence:

$$
\begin{array}{ccccccc}
0 & \longrightarrow & {}^0\langle L, * \rangle & \longrightarrow & {}^0\langle L, V \otimes B \rangle & \longrightarrow & {}^0\langle L, E \rangle & \longrightarrow \\
& \longrightarrow & {}^1\langle L, * \rangle & \longrightarrow & 0 & \longrightarrow & {}^1\langle L, E \rangle & \longrightarrow \\
& \longrightarrow & 0 & \longrightarrow & 0 & \longrightarrow & {}^2\langle L, E \rangle & \longrightarrow 0,
\end{array}
$$

where $*$ denotes A or $A \otimes \mathcal{O}(-2, -2)$. And so only $^0\langle L, E \rangle$ could be nonzero. Quite similar is the case when $x \leq 0$, $y \leq 0$, $x + y \leq -1$. Here we use (4.2) or (5.2). And by induction
$$
{}^0\langle L, A \otimes \mathcal{O}(2, 2) \rangle = {}^0\langle L, A \rangle = 0, \quad {}^0\langle L, C \rangle = {}^1\langle L, C \rangle = 0.
$$
So the long exact sequence will have the structure

$$
\begin{array}{ccccccc}
0 & \longrightarrow & {}^0\langle L, E \rangle & \longrightarrow & 0 & \longrightarrow & 0 & \longrightarrow \\
& \longrightarrow & {}^1\langle L, E \rangle & \longrightarrow & 0 & \longrightarrow & {}^1\langle L, * \rangle & \longrightarrow \\
& \longrightarrow & {}^2\langle L, E \rangle & \longrightarrow & {}^2\langle L, W \otimes C \rangle & \longrightarrow & {}^2\langle L, * \rangle & \longrightarrow 0,
\end{array}
$$

hence we get the statement.

Coming to the last case $i = 1$, I mention first that here either $-1 < x + y < 1$ or $x < 0$ or $y < 0$ because if $\mathrm{rank} E > 1$ then its point lies inside a unit square. Now using induction and long exact sequences we have from (4.1) or (5.1):

$$
\begin{array}{ccccccc}
0 & \longrightarrow & {}^0\langle L, * \rangle & \longrightarrow & 0 & \longrightarrow & {}^0\langle L, E \rangle & \longrightarrow \\
& \longrightarrow & {}^1\langle L, * \rangle & \longrightarrow & {}^1\langle L, W \otimes B \rangle & \longrightarrow & {}^1\langle L, E \rangle & \longrightarrow \\
& \longrightarrow & {}^2\langle L, * \rangle & \longrightarrow & 0 & \longrightarrow & {}^2\langle L, E \rangle & \longrightarrow 0.
\end{array}
$$

So $^2\langle L, E \rangle = 0$. And the same diagram for (4.2) or (5.2) is:

$$
\begin{array}{ccccccc}
0 & \longrightarrow & {}^0\langle L, E \rangle & \longrightarrow & 0 & \longrightarrow & {}^0\langle L, * \rangle & \longrightarrow \\
& \longrightarrow & {}^1\langle L, E \rangle & \longrightarrow & {}^1\langle L, W \otimes C \rangle & \longrightarrow & {}^1\langle L, * \rangle & \longrightarrow \\
& \longrightarrow & {}^2\langle L, E \rangle & \longrightarrow & 0 & \longrightarrow & {}^2\langle L, * \rangle & \longrightarrow 0.
\end{array}
$$

Hence $^0\langle L, E\rangle = 0$, so we get the statement. $\square$

Remarks: 1. Riemann-Roch theorem permits us to compute $\chi(L, E) = \sum(-1)^i \dim {}^i\langle L, E\rangle$ and the result is (for $L = \mathcal{O}(1, 1),\ r = r(E)$)

$$r\left(xy - \frac{1}{2} + \frac{1}{2r^2}\right)$$

and as a result of the lemma this is a dimension of the nonzero ext-group up to sign.

2. It would be interesting to know if the statement of the theorem is valid without semi-symmetricity condition. Some computations that support that it is true were made by my student S. Zyuzina.

3. For symmetrical exceptional sheaves on a quadric surface the uniqueness of nonzero cohomology group and the uniqueness of nonzero ext-group are valid as it was proved by my student K. Americ (unpublished).

References

[B-G]   A.I. BONDAL, A.L. GORODENTSEV. *On the Functors* Ext *Applied to Exceptional Bundles on* $\mathbb{P}^2$, in "Helices and Vector Bundles: Seminaire Rudakov", Lond. Math. Soc., Lect. Notes Ser. 148(1990), p.39-44.

[D]   J.-M. DREZET. *Fibrés exceptionnels et suite spectrale de Beilinson généralisée sur* $\mathbb{P}_2(\mathbb{C})$, Math. Ann., B.275, h.1, (1986), p.25-48.

[D-LeP]   J.-M. DREZET, J.LE POTIER. *Fibrés stables et Fibrés exceptionnels sur* $\mathbb{P}_2$, Ann.sciet., ENS, 4 serie, t.18, (1985), p.193-243.

[G]   A.L. GORODENTSEV. *Exceptional vector bundles on surfaces with moveable anticanonical divisor*, Math. USSR Izv. ser. math. 1988, v.52 N4, p.740-750 (Russian).

[G-R]   A.L. GORODENTSEV, A.N. RUDAKOV. *Exceptional vector bundles on the projective spaces*, Duke Math. J. 1987, v.54, N1, p.115-130.

[R1]   A.N. RUDAKOV. *Exceptional vector bundles on a quadric*, Math. USSR Izv., ser. math. 1988, v.52, N4, p.782-812.

[R2]   A.N. RUDAKOV. *A description of Chern classes of semistable sheaves on a quadric surface*, Preprint Erlangen 1990.

Exceptional Vector Bundles on a Del Pezzo Surface

Alexei N. Rudakov

An exceptional bundle is basically a bundle whose analytical or algebraic structure is uniquely determined by its topological one. This phenomenon was discovered in the mid eighties while studying vector bundles over complex algebraic surfaces or real 4-manifolds. Since then there were several attempts to investigate exceptional vector bundles and to use them in vector bundle theory and in the classification of 4-manifolds. Main idea of this article is to combine results and conjectures for bundles on a Del Pezzo surface. The author feels strongly that the context of Del Pezzo surfaces has several advantages and understanding here could develop really rapidly. Our aim is to influence such a development.

In the first three sections of the text we briefly remind basic facts and definitions, then we state new results and conjectures.

I wish to express my gratitude to Sergey Kuleshov, Boris Karpov and Andrey Tyurin and other members of our vector bundle seminar for helpful discussions and encouragement. I am thankful to Alexander Tikhomirov whose kind assistance in preparation the manuscript for publishing was a great help.

1 Del Pezzo surfaces

A smooth projective surface S is called Del Pezzo surface if its anticanonical sheaf is ample. There are $h^1(\mathcal{O}_S) = h^2(\mathcal{O}_S) = 0$ for such a surface and it would be a rational surface. We will work over an algebraically closed field k of a characteristic zero and then there is a classification of Del Pezzo surfaces. They are the projective plane $\mathbb{P}^2$, a quadric surface Q, and a surface made out of $\mathbb{P}^2$ by several blows up. In any case its Neron-Severy group coincides with Picard group $\operatorname{Pic} S$ and is isomorphic to $\mathbb{Z}^r$.

The rank r of the group $\operatorname{Pic} S$ is an important invariant of Del Pezzo surface. If $r = 1$ then $S \simeq \mathbb{P}^2$, if $r = 2$ then S is quadric Q or $\mathbb{F}_1$ that is blowing up of a point in a plane. And in general $r \leq 9$.

We will write $\mathcal{O}$ for a structure sheaf of a surface in question and K for canonical sheaf and K' for anticanonical sheaf, so $K' = K^{-1}$. It is also convenient to use shortened notations for Ext-vector spaces as follows ${}^i\langle A, B\rangle$ instead of $\mathrm{Ext}^i_{\mathcal{O}_S}(A, B)$. As usual $\chi(A, B) = \sum(-1)^i \dim {}^i\langle A, B\rangle$ is a biadditive function on sheaves and so a bilinear "scalar product" on a Grothendick group $K_0(S)$.

For Del Pezzo surfaces the form χ is nondegenerate on K_0, but it is not symmetric. Because of Serre duality

$${}^i\langle A, B\rangle \simeq {}^{2-i}\langle B, K \otimes A\rangle^*,$$

we would have

$$\chi(A, B) = \chi(B, A \otimes K).$$

The Riemann-Roch theorem states

$$\chi(A, B) = r_A r_B - c_1(A)c_1(B) + r_A s_B + r_B s_A + \frac{1}{2}\left(r_A c_1(B) - r_B c_1(A)\right) \cdot K',$$

where r_F is a rank of a sheaf F, $c_1(F)$ is a first Chern class and

$$s_F = \frac{1}{2}c_1(F)^2 - c_2(F).$$

Functions r, c_1, s are additive and χ looks just as bilinear form need to be.

2 Rigid and exceptional sheaves

In many classification problems we are faced up with a dilemma: are the objects separate or do they appear to be depending on continuous parameters. Looking at coherent sheaves on a surface we could ask a closed question: is there such a sheaf F that $H^1(S, \mathrm{End}F) = 0$ or the same ${}^1\langle F, F\rangle = 0$? In other words, do there exist sheaves without nontrivial deformations? What are these sheaves like?

The condition ${}^1\langle F, F\rangle = 0$ is usually called rigidity and there is a beautiful result about rigid sheaves on Del Pezzo.

Theorem 2.1. (Kuleshov [3]) *Any rigid sheaf without torsion on a Del Pezzo surface is a finite direct sum of indecomposable rigid sheaves.*

For an indecomposable rigid sheaf F without torsion there would be valid statements:

(1) F is (Mumford-Takemoto) stable relative to K';

(2) ${}^0\langle F, F\rangle \simeq k$ (ground field), ${}^2\langle F, F\rangle = 0$;

(3) the discriminant $\triangle_F = \frac{1}{r_F}\left(c_2(F) - \frac{r_F - 1}{2r_F}c_1(F)^2\right)$ is less than $\frac{1}{2}$.

And one can reformulate this in a little different manner.

Proposition 2.2. *The following conditions are equivalent for a sheaf F without torsion on a Del Pezzo surface:*

(1) F is K'-stable and rigid;

(2) $^0\langle F, F \rangle \simeq k$ and F is rigid;

(3) $^0\langle F, F \rangle \simeq k$, $^1\langle F, F \rangle = 0$, $^2\langle F, F \rangle = 0$;

(4) F is K'-stable and $\chi(F, F) = 0$;

(5) F is K'-stable and $\Delta_F < \frac{1}{2}$;

(6) F is rigid and indecomposable.

Usually we take condition (3) as a definition for an exceptional sheaf. And there is another theorem

Theorem 2.3. ([2], [4]) *An exceptional sheaf on Del Pezzo surface S is either torsion sheaf or without torsion. If first then it is zero out of an exceptional line on S and its restriction on this line would be an invertible sheaf (a sheaf of sections of a line bundle or simply line bundle).*

3 Main properties of exceptional sheaves on Del Pezzo

So proposition 2.2 and theorem 2.3 give us several properties of exceptional sheaves. We would like to mention some more.

First for an exceptional F we have $\chi(F, F) = 1$ so an image $[F]$ of F in $K_0(S)$ lies on a quadric $\chi(x, x) = 1$. And if $r_F \neq 0$ then we can compute $c_2(F)$ from r_F, $c_1(F)$. Really we have

$$\Delta_F = \frac{1}{2}\left(1 - \frac{1}{r_F^2}\right).$$

And if $r_F = 0$ then $\chi(F, F) = 1 \iff c_1(F)^2 = -1$.

Proposition 3.1. *If E and F are exceptional and $r_E = r_F \neq 0$, $c_1(E) = c_1(F)$ then $E \simeq F$.*

This is mostly a consequence of stability. And it has an important implication. Nontorsion exceptional sheaves on Del Pezzo we could discriminate by their numerical data (r, c_1). Really one can make a little stronger statement.

Corollary. *If $r_E \neq 0$, $r_F \neq 0$ and $\dfrac{c_1(F)}{r_F} = \dfrac{c_1(E)}{r_E}$, then $E \simeq F$.*

One only needs to mention that the Riemann-Roch theorem and $\chi(E, E) = 1$ imply that $c_1(E)$ and r_E have no common multiple.

We introduce a notation $\varepsilon_E = \dfrac{c_1(E)}{r_E}$ and we can say that one could describe the set of nontorsion exceptional sheaves on Del Pezzo having found the set of their points ε_E in $\text{Pic } S \otimes \mathbb{Q}$. So we have an extended Picard space $\text{Pic }_\mathbb{Q} S = \text{Pic } S \otimes \mathbb{Q}$ with a scalar product which is an extension of an intersection form on $\text{Pic } S$. The point in $\text{Pic }_\mathbb{Q}(S)$ of a type ε_E where E is exceptional will be called exceptional point. The form χ is related to this scalar product for exceptional A, B as

$$\chi(A, B) = r_A r_B \left(\frac{1}{2} \left[(\varepsilon_B - \varepsilon_A + \rho)^2 - \rho^2 \right] + \delta_A + \delta_B \right),$$

where A, B – exceptional nontorsion sheaves, $\rho = \frac{1}{2} K'$ and $\delta = \frac{1}{2r^2}$. Also we will fix a notation for K'-slope $\mu_E = \frac{1}{r_E} c_1(E) \cdot K'$. Then for

$$\chi(A, B) = r_A r_B \left(\frac{1}{2}(\varepsilon_A^2 + \varepsilon_B^2) + \frac{1}{2}(\mu_B - \mu_A) + \delta_A + \delta_B \right).$$

And the antisymmetric part of χ is for exceptional A, B

$$a(A, B) = \chi(A, B) - \chi(B, A) = r_A r_B (\mu_B - \mu_A).$$

4 Semistable sheaves on Del Pezzo

I want to announce here an important result on relative Chern classes or K_0-images of K'-semistable sheaves on Del Pezzo. It was first established for $\mathbb{P}^2$ ([1]) and then for Q ([6]) and now I can prove it for any Del Pezzo.

Let us denote Exc subset of classes of exceptional sheaves in K_0 and use the same notation μ for natural extension of $\mu_E = \frac{c_1(E)}{r_E}$ on K_0. This extension is quite natural for Del Pezzo because here one can use coordinates in $K_0(S)(r, c_1, s)$, $r \in \mathbb{Z}$, $s \in \frac{1}{2}\mathbb{Z}$, $c_1 \in \text{Pic } S$ such that an image of a sheaf F is $(r_F, c_1(F), s_F)$. Then for $x = (r, c_1, s)$ it would be $\mu(x) = \frac{1}{r} c_1 K'$.

Theorem 4.1. *Let $x = (r, c_1, s) \in K_0(S) \backslash \text{Exc and } r > 0$. Then $x = (r(F), c_1(F), s(F))$ for a K'-semistable sheaf F iff:*
for $z \in \text{Exc such that } r(z) \leq r(x) \text{ and } \mu(x) \leq \mu(z) \leq \mu(x) + K^2 \text{ there is } \chi(z, x) \leq 0$.

So we see that the description of "semistable cone" in $K_0(S)$ depends on Exc.

The condition $x \in K_0(S) \setminus \text{Exc}$ one can replace by

$$\Delta(x) \geq \frac{1}{2},$$

so if $\Delta(x) < \frac{1}{2}$, $r > 0$, then the class x contains semistable sheaf iff $x \in \mathbb{N} \cdot \text{Exc}$ that means x is a multiple of exceptional class and for $\Delta(x) \geq \frac{1}{2}$, $r > 0$ one needs to use the theorem.

Quite probably one could prove stronger results.

Conjecture 4.2. *If* $\Delta(x) > \frac{1}{2}$ *and* x *satisfies the conditions of the theorem then there is a* K'-*stable sheaf* F *in the class* x. *If* $\Delta(x) = \frac{1}{2}$ *then all the semistable sheaves in the class* x *are not stable.*

5 Exceptional tuples and braid group

The main tool to study exceptional sheaves is a braid group action on exceptional tuples. Let us remind principal definitions.

A tuple $(E_1, \ldots, E_p)$ of exceptional sheaves is called exceptional if $^k\langle E_j, E_i \rangle = 0$ for $i < j$ and any k.

An important result about sheaves on a Del Pezzo surface is that there is a braid group Br_p action on exceptional p-tuples.

Really this is the conclusion of several theorems and an important part is a description of the action for $p = 2$, the action on exceptional pairs. Here $\mathrm{Br}_2 \simeq \mathbb{Z}$ so we have an action of a generator of the group that sometimes is called mutation of a pair in question. And an orbit of the group $\mathrm{Br}_2 \simeq \mathbb{Z}$ is called a chain of pairs or sometimes a tendril.

We have now a kind of classification of mutations and tendrils for Del Pezzo surfaces.

Proposition 5.1. *Any chain (or tendril) of exceptional pairs on Del Pezzo surface looks like this:*
$$\ldots (A_{-2}, A_{-1}); (A_{-1}, A_0); (A_0, A_1); (A_1, A_2) \ldots$$
where either

(a) $A_{2n} = A$, $\quad A_{2n+1} = B$; $\quad$ *or*

(b) $A_{3n} = A$, $\quad A_{3n+1} = B$, $\quad A_{3n+2} = C$; $\quad$ *or*

(c) A_i *are all different and*

for (a) $\chi(A_i, A_{i+1}) = 0$,
for (b) $\chi(A_i, A_{i+1}) = \pm 1$,
for (c) $\chi(A_i, A_{i+1}) = \pm d$, $\quad d \geq 2$.
And more: for (c) there are two possibilities:

(c1) $\chi(A_i, A_{i+1}) = d > 2$ $\quad$ *for any* i *or*

(c2) $\chi(A_{i_0}, A_{i_0+1}) = -d$ *and* $\chi(A_i, A_{i+1}) = d > 2$ *for* $i \neq i_0$.

For (c1) chain we have $\ldots \mu(A_i) < \mu(A_{i+1}) < \ldots$ and for (c2) chain it is that
$$\mu(A_{i_0+1}) < \mu(A_{i_0+2}) \ldots < c < \ldots \mu(A_{i_0-1}) < \mu(A_{i_0}).$$
And there are exact sequences those connect members of the chain:

if $\mu(A_i) < \mu(A_{i+1}) < \mu(A_{i+2})$ and $\chi(A_i, A_{i+1}) = d$ then it is

$$0 \longrightarrow A_i \longrightarrow A_{i+1}^{\oplus d} \longrightarrow A_{i+2} \longrightarrow 0.$$

If $\mu(A_{i+2}) < \mu(A_i) < \mu(A_{i+1})$ then it would be like

$$0 \longrightarrow A_{i+2} \longrightarrow A_i \longrightarrow A_{i+1}^{\oplus d} \longrightarrow 0.$$

If $\mu(A_{i+1}) < \mu(A_{i+2}) < \mu(A_i)$ then

$$0 \longrightarrow A_{i+1}^{\oplus d} \longrightarrow A_{i+2} \longrightarrow A_i \longrightarrow 0.$$

So whenever one has an exceptional pair of exceptional bundles (E, F) and $\chi(E, F) \neq 0$ one can make a new exceptional sheaf via braid group action and for $\chi(E, F) \neq 0$, 1, -1, not unique new one.

Conjecture 5.2. *Any exceptional sheaf on Del Pezzo could be put in an exceptional pair.*

Conjecture 5.3. *An exceptional p-tuple on a Del Pezzo surface S with $\mathrm{rk}\,\mathrm{Pic}\,S = r$ could be put into an exceptional $(p+1)$-tuple iff $p \leq r + 1$.*

An evidence for these conjectures are $S = \mathbb{P}^2$, Q. And in both cases it appears that exceptional sheaves, exceptional pairs and exceptional tuples are forming together a special kind of structure which is not well understood yet.

Maximal possible length of an exceptional tuple is $p_{max} = r + 2$ and there is a nice conjecture.

Conjecture 5.4. *A group $\mathrm{Br}_{p_{max}}$ acts on exceptional p_{max}-tuples transitively (that is p_{max}-tuples constitute one orbit for $\mathrm{Br}_{p_{max}}$).*

It is known for $p_{max} \leq 4$ or $r \leq 2$ and probably would be more difficult as p_{max} grows.

Also it would be interesting to know possible finite orbits for Br_p, $2 < p < p_{max}$. It is in some sense analogous to tame quivers and probably there would be finite amount of types.

Another important question is to find all exceptional sheaves of small rank on a given Del Pezzo surface and exceptional tuples among them.

References

[1]   J.-M. DREZET, J. LE POTIER. *Fibrés stables et fibrés exceptionnels sur* $\mathbb{P}_2$, Ann. scient. ENS (1985), t.18, 193-243.

[2]   A.L. GORODENTSEV. *Exceptional bundles on a surface with a moveable anticanonical class*, Math. USSR Izv. 52(1988), N4, 740-757 (Russian).

[3]   S.A. KULESHOV. *Rigid sheaves on surfaces*, Present collection.

[4]   D.O. ORLOV. *Derived categories of the coherent sheaves of the monoidal transforms*, thesis, Yaroslavl', 1991 (Russian).

[5]   A.N. RUDAKOV. *Exceptional vector bundles on a quadric*, Math. USSR Izv. (1989), 33, 115-138 (Russian).

[6]   A.N. RUDAKOV. *A description of Chern classes of semistable sheaves on a quadric surface*, (preprint) Erlangen Schriftenreihe N88 (1990).

Standard Bundles on a Hilbert Scheme of Points on a Surface

A.S.Tikhomirov

Let S be a smooth irreducible algebraic surface over $\mathbb{C}$, H_d a Hilbert scheme of 0-dimensional subschemes of length d in S, $\dim H_d = 2d$, and $Z_d \subset S \times H_d$ a universal family with natural projections $S \xleftarrow{\tau_d} Z_d \xrightarrow{\pi_d} H_d$. Fix an arbitrary divisor D on S and denote $\mathcal{E}_D^d = \pi_{d*}\tau_d^* \mathcal{O}_S(D)$. Since π_d is a flat finite morphism of degree d, the sheaf $\mathcal{E}_D^d$ is in fact the vector bundle of rank d over H_d. We call $\mathcal{E}_D^d$ the standard vector bundle over H_d. The problem of computation of its Segre classes is connected with a number of questions of enumerative geometry. In recent times it has got applications to the description of the smooth structure of the 4-manifold underlying S – see [10]. In this paper we consider the question of computation of the degree $\delta_d = \int\limits_{H_d} s_{2d}(\mathcal{E}_D^d)$ of the top Segre class $s_{2d}(\mathcal{E}_D^d)$ of $\mathcal{E}_D^d$ as the function of numerical invariants x, y, z, w of S and D, where

$$x = (D^2), \quad y = (D \cdot K_S), \quad z = (K_S^2) - \chi_{top}(S) = s_2(\Omega_S), \quad w = (K_S^2), \qquad (1)$$

(here K_S is the canonical class of S). The main result of the paper is

Theorem 1. δ_d *is the polynomial with rational coefficients of* x, y, z, w.

The proof of this theorem is first given in §§1–2 for the very ample divisor D. For this purpose δ_d is interpreted in §1 in terms of the projective embedding of S into the space $\mathbb{P}^{3d-2}$ by an appropriate linear subseries of the complete linear series $|D|$ (proposition 1), and using the multiple-point formulas of Kleiman (proposition 3), and then is extended to the case of arbitrary D in §3. One technical result used in the proof (proposition 2) is put into the Appendix (§5).

The explicit formula for the polynomial δ_d is known for $d = 2$ – this is essentially the classic formula of double points of the surface in $\mathbb{P}^4$: $\delta_2 = \frac{1}{2}(x^2 - 10x - 5y - z)$. In the proof of theorem 1 we obtain the method of finding the polynomial δ_3, and the second main result of the paper (see §4) is

Theorem 2.

$$\delta_3 = \frac{1}{3!}x^3 - 5x^2 - \frac{5}{2}xy - \frac{1}{2}xz + \frac{112}{3}x + 32y + \frac{20}{3}z + \frac{8}{3}w.$$

Remark that we take $k = \mathbb{C}$ as the base field for technical reasons, because our construction uses one result of Beltrametti and Sommese [1] proved over $k = \mathbb{C}$ (see §1, lemma 1). Since this result is evidently true for any algebraically closed field of characteristic 0, all the assertions of this paper are valid over an arbitrary $k = \bar{k}$, $\mathrm{char}\,k = 0$.

1. Top Segre class of the standard bundle $\mathcal{E}_D^d$ in the case of very ample D

First introduce some notations. Let D_1 be a very ample and D_2 an arbitrary divisors on S, $D = nD_1 + D_2$, $n \geq 1$, $\mathcal{E} = \mathcal{E}_D^d$, $d \geq 3$, $\Pi = P(\mathcal{E}^\vee)$ and $\rho : \Pi \longrightarrow H_d$ be the natural projection. Next, let n be large enough, so that:

1) $N = h^0(D) \geq 3d$,

2) the map $i_{n,1} : S \longrightarrow \mathbb{P}^{N-1}$ by the complete linear series $|D|$ is the embedding, and

3) there exists $(N - 3d)$-dimensional linear subspace L in $\mathbb{P}^{N-1}$ such that $i_L = pr_L \circ i_{n,1} :$ $S \longrightarrow \mathbb{P}^{3d-2}$ is the embedding, where $pr_L : \mathbb{P}^{N-1} \longrightarrow \mathbb{P}^{3d-2}$ is the (rational) linear projection.
(One easily sees, that for $n \gg 1$ the conditions 1) – 3) are satisfied.)

For an arbitrary closed point $\xi \in H_d$ by Ξ we shall denote the corresponding 0-dimensional subscheme of length d in S, i.e. $\Xi = (\tau_d \circ \pi_d^{-1})(\xi)$. Besides, for any subscheme $\mathcal{X}$ in $\mathbb{P}^{3d-2}$ (respectively, in $\mathbb{P}^{N-1}$) by $\langle \mathcal{X} \rangle$ we denote its linear span in $\mathbb{P}^{3d-2}$ (respectively, in $\mathbb{P}^{N-1}$).

We shall use in the sequel the following result of Beltrametti and Sommese [1], lemma 0.0.1 and theorem 3.1.

Lemma 1. *1) For any $r \geq 2$ there exists $\psi(r) \in \mathbb{N}$ such, that for every $n \geq \psi(r)$*

$$\dim \langle i_{n,1}(\Xi) \rangle = r - 1 \tag{2}$$

for any closed point $\xi \in H_r$, so that there is defined a morphism $i_{n,r} : H_r \longrightarrow G_{(r-1)} :$ $\xi \longmapsto \langle i_{n,1}(\Xi) \rangle$, where $G_{(r-1)}$ is the grassmanian of $(r - 1)$-dimensional subspaces of $\mathbb{P}^{N-1}$, and

2) $i_{n,r}$ is the embedding.

From this lemma immediately outcomes

Corollary. *For $n \geq \psi(d)$ the embedding $i_{n,d} : H_d \hookrightarrow G_{(d-1)}$ induces the embedding $j_{n,d} :$ $\Pi = P(\mathcal{E}^\vee) \hookrightarrow \mathbb{P}^{N-1} \times H_d$ (namely, $\Pi \simeq \Gamma_{(d-1)} \times H_d$, where $\Gamma_{(d-1)} \subset \mathbb{P}^{N-1} \times G_{(d-1)}$ is the flag variety) such that*

$$\mathcal{O}_{P(\mathcal{E}^\vee)}(1) = j_{n,d}^* \left(\mathcal{O}_{\mathbb{P}^{N-1}}(1) \boxtimes \mathcal{O}_{H_d} \right) \tag{3}$$

and one has the commutative diagram

$$
\begin{array}{ccccc}
W & \xleftarrow{\;w_{n,d}\;} & \Pi & \xrightarrow{\;\rho\;} & H_d \\
\downarrow & & \downarrow{\scriptstyle j_{n,d}} & & \downarrow{\scriptstyle i_{n,d}} \\
\mathbb{P}^{N-1} & \xleftarrow{\;p_1\;} & \mathbb{P}^{N-1} \times H_d & \xrightarrow{\;p_2\;} & G_{(d-1)}
\end{array}
\qquad (4)
$$

where ρ is the natural projection, p_1 and p_2 are projections onto the factors, $w_{n,d} = p_1 \circ j_{n,d}$ and $W = w_{n,d}(\Pi)$.

Now denote $\Delta_d = \{\xi \in H_d \mid \#\Xi < d\}$. As it is well-known [3], Δ_d is the proper closed subset of H_d.

Lemma 2. *For $n \gg 1$ the morphism $w_{n,d} : \Pi \longrightarrow W$ in the diagram* (4) *is birational.*

Proof. We prove the assertion of lemma for $n \geq \psi(2d)$. Everywhere in the proof we identify the subschemes Ξ in S with their images under the embedding $i_{n,1} : S \hookrightarrow \mathbb{P}^{N-1}$. First, according to assertion 1) of lemma 1

$$
\dim\langle\Xi\rangle = \ell(\Xi) - 1 = r - 1, \quad \xi \in H_r, \quad r \leq 2d. \qquad (5)
$$

Suppose that $\dim W < \dim \Pi$. Then from the definition of W it follows immediately that for almost any closed point $x \in W$ there are satisfied the following conditions:

1) there exist two closed points $\xi_1, \xi_2 \in H_d - \Delta_d$, $\xi_1 \neq \xi_2$, such that

$$
x \in \langle\Xi_1\rangle \cap \langle\Xi_2\rangle, \qquad (6)
$$

2) $\Xi_a = \Xi_1 \cup \Xi_2$ and $\Xi_b = \Xi_1 \cap \Xi_2$ are reduced subschemes, such that

$$
\Xi_b \neq \Xi_1, \qquad (7)
$$

3) $x \notin \langle\Xi'\rangle$ for any $\Xi' \subsetneq \Xi_1$.

Now from (5) and (7) we have: $\dim\langle\Xi_a\rangle + \dim\langle\Xi_b\rangle = \dim\langle\Xi_1\rangle + \dim\langle\Xi_2\rangle$. Besides, since Ξ_a is reduced, $\dim\langle\Xi_a\rangle \leq \dim\langle\langle\Xi_1\rangle, \langle\Xi_2\rangle\rangle$, wherefrom by the previous equality $\dim\langle\Xi_b\rangle \geq \dim(\langle\Xi_1\rangle \cap \langle\Xi_2\rangle)$. But it is evident that $\langle\Xi_b\rangle \subset \langle\Xi_1\rangle \cap \langle\Xi_2\rangle$. Thus, $\langle\Xi_1\rangle \cap \langle\Xi_2\rangle = \langle\Xi_b\rangle = \langle\Xi_1 \cap \Xi_2\rangle$. But this contradicts to the condition 3) above in view of (6) and (7).

Hence, $\dim W = \dim \Pi$, and the fibre of $w_{n,d}$ is finite over almost any closed point of W. Now if the degree of $w_{n,d}$ is more than 1, then again find $\xi_1, \xi_2 \in H_d - \Delta_d$, $\xi_1 \neq \xi_2$, with the properties 1) – 3) above and come to a contradiction, q.e.d.

Lemma 3. *Let $n \gg 1$. Then for almost any $(N - 3d)$-dimensional linear subspace $L \subset \mathbb{P}^{N-1}$ the scheme $L \cap W$ is 0-dimensional and reduced, and such that:*

1) $\rho \circ w_{n,d}^{-1}(L \cap W) = \{\xi_1, \ldots, \xi_{\delta_d}\} \subset H_d - \Delta_d$, *where* $\delta_d = \#L \cap W$, *i.e. the schemes* $\Xi_1, \ldots, \Xi_{\delta_d} \subset S$ *are reduced subschemes of length d:*

$$
\Xi_k = (\Xi_k)_{\mathrm{red}}, \quad \#\Xi_k = \ell(\Xi_k) = d, \quad 1 \leq k \leq \delta_d,
$$

2) $\dim\langle i_L(\Xi_k)\rangle = d - 2$, $1 \leq k \leq \delta_d$,

3) *for any closed point* $\xi \in H_d - \{\xi_1, \ldots, \xi_{\delta_d}\}$

$$\dim\langle i_L(\Xi)\rangle = \dim\langle i_{n,1}(\Xi)\rangle = d - 1$$

(in other words, the set of $(d-2)$-dimensional d-secant spaces of the surface $i_L(S)$ in $\mathbb{P}^{3d-2}$ is finite and consists of the spaces, corresponding to the points $i_{n,d}(\xi_k)$, $1 \le k \le \delta_d$),

4) *for any vector* $0 \ne v \in T_{\xi_k} H_d$, $1 \le k \le \delta_d$, *considered as a subscheme in H_d, there is correctly defined subscheme $\Xi_v = (\tau_d \circ \pi_d^{-1})(v)$ (by the definition, its ideal sheaf $\mathcal{I}_{\Xi_v,S}$ is $\tau_{d*}\pi_d^*(\mathcal{I}_{v,H_d})$), such that $\dim(\langle i_{n,1}(\Xi_v)\rangle \cap L) = 0$, so that $\dim\langle i_L(\Xi_v)\rangle = \ell(\Xi_v) - 2$.*

Proof. Consider the diagram

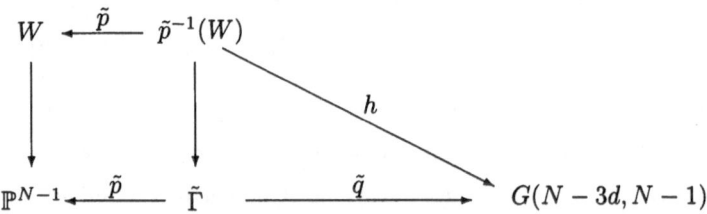

in which $G(N - 3d, N - 1)$ is the grassmanian of $(N - 3d)$-dimensional subspaces of $\mathbb{P}^{N-1}$, $\tilde{\Gamma} \subset \mathbb{P}^{N-1} \times G(N - 3d, N - 1)$ is the flag variety with projections $\tilde{p}$ and $\tilde{q}$, and $h = \tilde{q}\,|_{\tilde{p}^{-1}(W)}$. Since by lemma 2 dim $W = \dim \Pi = 3d - 1$, then by the Bezout theorem for any $\{L\} \in G(N - 3d, N - 1)$ the intersection $L \cap W$ is nonempty, i.e. h is surjective. On the other hand, dim $\tilde{p}^{-1}(W) = \dim W + \dim(\text{fibre of } \tilde{p}) = 3Nd - 9d^2 + 6d - N - 1 = \dim G(N - 3d, N - 1)$. Thus, by the property of generic smoothness [5], III, corollary 10.7, there exists nonempty open subset $U \subset G(N - 3d, N - 1)$ such that for any closed point $\{L\} \in G(N - 3d, N - 1)$ the fibre $h^{-1}(\{L\})$ is a reduced 0-dimensional scheme, isomorphic to $L \cap W$ in view of lemma 2. Whence we immediately obtain assertions 1) – 3) of lemma.

To prove the assertion 4), consider the diagram (4) and take an arbitrary vector $0 \ne v \in T_{\xi_k} H_d$, $1 \le k \le \delta_d$. Then $P_v = (w_{n,d} \circ \rho^{-1})(v)$ is a subscheme of W, defined by the ideal sheaf $\mathcal{I}_{P_v,W} = (p_1 \circ j_{n,d})_* \rho^* \left(\mathcal{I}_{v,H_d} \underset{\mathcal{O}_{\mathbb{P}^{N-1}}}{\otimes} \mathcal{O}_W \right)$, and such that:

a) $P_v \cap i_{n,1}(S) \supset i_{n,1}(\Xi_v) \supset i_{n,1}(\Xi_k)$, where $\Xi_k = \tau_d \circ \pi_d^{-1}(\xi_k)$,

b) $P_v \supset \langle i_{n,1}(\Xi_k)\rangle$,

c) $\mathrm{Supp}P_v = \langle i_{n,1}(\Xi_k)\rangle$.

From a) and b) it follows that

$$W \cap L \supset P_v \cap L = \langle P_v\rangle \cap L \supset \langle i_{n,1}(\Xi_v)\rangle \cap L \supset \langle i_{n,1}(\Xi_k)\rangle \cap L \qquad (8)$$

(here the equality $P_v \cap L = \langle P_v\rangle \cap L$ outcomes from the linearity of L). Besides, since $L \cap W$ is 0-dimensional, from the condition c) above we get:

$$x := \mathrm{Supp}(P_v \cap L) = \langle i_{n,1}(\Xi_k) \rangle \cap L = \{pt\}. \tag{9}$$

Moreover, since $L \cap W$ is reduced, from (8) and (9) it follows that

$$P_v \cap L = \langle i_{n,1}(\Xi_k) \rangle \cap L = x. \tag{10}$$

Now, taking $n \geq \psi(2d)$, by lemma 1 we have $\dim \langle i_{n,1}(\Xi_v) \rangle = \ell(\Xi_v) - 1$, since $\ell(\Xi_v) \leq 2d$ by its construction. Whence by (10) we get $\dim \langle i_L(\Xi_v) \rangle = \dim \langle i_{n,1}(\Xi_v) \rangle - 1 = \ell(\Xi_v) - 2$, q.e.d.

Now we come to the main result of this section

Proposition 1. *Under the conditions of lemma 3 the degree of the top Segre class $s_{2d}(\mathcal{E})$ of the vector bundle $\mathcal{E} = \mathcal{E}_D^d$, $D = nD_1 + D_2$, equals to the number δ_d of $(d-2)$-dimensional d-chords of the surface $i_L(S)$ in $\mathbb{P}^{3d-2}$:*

$$\int_{H_d} s_{2d}(\mathcal{E}) = \delta_d = \#L \cap W.$$

Proof. Consider the diagram (4). In view of lemmas 2 and 3 one has the transversal intersection $\kappa = P(\mathcal{E}^\vee) \cap (L \times H_d) \simeq W \cap L$ in $\mathbb{P}^{N-1} \times H_d$. This transversality means that $\kappa = (s)_0$, where $s \in H^0 \left(\bigoplus_1^{3d-1} \mathcal{O}_{P(\mathcal{E}^\vee)}(1) \right)$, so that

$$\delta_d = \#\kappa = \int_\Pi [(s)_0] = \int_{P(\mathcal{E}^\vee)} c_{3d-1} \left(\bigoplus_1^{3d-1} \mathcal{O}_{P(\mathcal{E}^\vee)}(1) \right) =$$

$$= \int_{P(\mathcal{E}^\vee)} c_1 \left(\mathcal{O}_{P(\mathcal{E}^\vee)}(1) \right)^{3d-1} = \int_{H_d} \rho_* c_1 \left(\mathcal{O}_{P(\mathcal{E}^\vee)}(1) \right)^{3d-1} =$$

$$= \int_{H_d} s_{2d}(\mathcal{E}^\vee) = \int_{H_d} s_{2d}(\mathcal{E})$$

(in the third equality of this chain we use the property of the localized top Chern class of a bundle – see [4], proposition 14.1), q.e.d.

2. Multiple-point formulas

In this section we use the theory of multiple points of a morphism in the following setup. Let $f : X \longrightarrow Y$ be the morphism of smooth projective varieties. Let $d \geq 2$. According to the construction of Kleiman [7], §4.1, there is defined the sequence of schemes $X_0(= Y), X_1(= X), X_2, \ldots, X_d$ and morphisms $f_0 = f, f_r : X_{r+1} \longrightarrow X_r$, $1 \leq r \leq d - 1$ (f_r is called the r-th derived map of f), such that for $1 \leq r \leq d$ $N_r(f) = (f_0 \circ f_1 \circ \ldots \circ f_{r-1})(X_r)$ is the set of r-ple points of the map f. (In particular, $N_r(f)$ contains the closure in Y of the set $\{y \in Y \mid \#f^{-1}(y) \geq r\}$, $2 \leq r \leq d$.) The set $M_r = (f_1 \circ \ldots \circ f_{r-1})(X_r) \subset X$ (respectively, the cycle $[M_r] = (f_1 \circ \ldots \circ f_{r-1})_*([X_r]) \in Z_*X$) is called the set (respectively, the cycle) of r-ple points of the morphism f.

Next, let the following conditions be satisfied for f:

(i) the morphism $f : X \longrightarrow Y$ and its derived morphisms $f_r : X_{r+1} \longrightarrow X_r$, $1 \leq r \leq d - 1$, are quasifinite, and $\dim Y = dm$, $\dim X = (d-1)m$ for some $m \geq 1$,

(ii) the scheme X_d is 0-dimensional and reduced and coincides as a set with $(f_0 \circ \ldots \circ f_{d-1})^{-1}\{y \in Y \mid \#f^{-1}(y) = d\}$,

(iii) $\operatorname{codim} f_r = \operatorname{codim} f = m$, $1 \leq r \leq d - 1$.

Now denote by m_1 the class $[X]$ and respectively by m_r the class of the cycle $[M_r]$ in A_*X, $2 \leq r \leq d$, and let $[N_d(f)]$ be the class in A_*Y of the scheme $N_d(f)$ as a reduced 0-dimensional subscheme in Y. From the construction of X_d [7], §4, and the condition (ii) it follows immediately that $f_*m_d = d![N_d(f)]$, so that

$$\int_Y f_*m_d = d! \, (\#N_d(f)) . \tag{11}$$

Next, the condition (iii) means that f is d-generic (in the sense of Kleiman [7], proposition 4.4) locally complete intersection morphism. Under these conditions the theorem 1.1 of [6] is true, which states that the classes m_r are given by the recursive formulas of the type:

$$m_{r+1} = f^* f_* m_r - \sum_{i=0}^{r-1}(-1)^{i+1}\frac{r!}{(r-i)!}P_i \cdot m_r, \quad 1 \leq r \leq d-1, \tag{12}$$

where $P_i \in \mathbb{Q}[c_0, c_1, \ldots, c_{m(d-1)}]$, and $c_i = c_i(\nu_f)$ are the Chern classes of the normal sheaf $\nu_f = f^*T_Y/T_X$ of the morphism f. Iterating the formula (12), we get the following expression for m_d:

$$m_d = \sum_{p=1}^{d-1} \sum_{|I_p| \leq d-1} (f^* f_*)^{i_1}\left(P_{d,I_p,1}(f^* f_*)^{i_2}\left(P_{d,I_p,2} \cdots (f^* f_*)^{i_p} P_{d,I_p,p}\right)\cdots\right), \tag{13}$$

where

$$(f^* f_*)^k := \underbrace{f^* f_* \ldots f^* f_*}_{k \; times},$$

by I_p is denoted the multiindex $(i_1, \ldots, i_p) \in \mathbb{N}^p$, $|I_p| = i_1 + \ldots + i_p$ is its length, and $P_{d,I_p,k}$ are some polynomials in $\mathbb{Q}[c_0, \ldots, c_{m(d-1)}]$.

Now remark, that by the projection formula for any $a \in A_*X$ we have $f_*(f^* f_*)^k a = f_*\left(f^* f_*(f^* f_*)^{k-1}a \cdot [X]\right) = f_*(f^* f_*)^{k-1}a \cdot f_*[X] = \ldots = f_*a \cdot (f_*[X])^{k-1}$. From here and (12) we have

Lemma 4. *Let the morphism $f : X \longrightarrow Y$ satisfy the conditions (i)-(iii) above. Then the class $m_d \in A_*X$ of the cycle of d-ple points of the morphism f satisfies the equality*

$$f_*m_d = \sum_{p=1}^{d-1} \sum_{|I_p| \leq d-1} (f_*[X])^{|I_p|-(p-1)}\left(f_*P_{d,I_p,1}\right) \cdots \left(f_*P_{d,I_p,p}\right), \tag{14}$$

where $P_{d,I_p,k}$, $1 \leq k \leq p$, $1 \leq p \leq d-1$, are some polynomials with rational coefficients of the Chern classes $c_i = c_i(f^*T_Y/T_X)$ of the normal sheaf of the map f.

Now consider the embedding $i_L : S \hookrightarrow P = \mathbb{P}^{3d-2}$ introduced in §1. Let $G_k = G(k, 3d-2)$ be the grassmanian of k-dimensional subspaces in P, and $\Gamma_k \subset P \times G_k$ be the flag variety with the projections $P \xleftarrow{p_k} \Gamma_k \xrightarrow{q_k} G_k$, $0 \le k \le 3d - 3$. Consider the diagram

$$
\begin{array}{ccccc}
P & \xleftarrow{\quad p \quad} & \Gamma & \xrightarrow{\quad q \quad} & G \\
\Big\uparrow\scriptstyle i & & \Big\uparrow\scriptstyle j & & \Big\| \\
S & \xleftarrow{\quad g \quad} & X & \xrightarrow{\quad f \quad} & Y
\end{array}
\tag{15}
$$

in which $i = i_L$, $\Gamma = \Gamma_{d-2}$, $p = p_{d-2}$, $q = q_{d-2}$, $X = p^{-1}(i_L(S))$, $f = q\big|_X$, $g = i_L^{-1} \circ (p\big|_X)$, and j is the natural embedding.

Proposition 2. *In notations of §1 let $n \gg 1$ and L be the sufficiently general $(N - 3d)$-dimensional subspace in $\mathbb{P}^{N-1}$. Then the morphism $f : X \longrightarrow Y$ in the diagram (15) satisfies the conditions (i)-(iii) of lemma 4 for $m = 2(d - 1)$.*

Proof. See §5.

Now we prove the theorem 1 in the case when $D = nD_1 + D_2$, where D_1 is ample on S. This is given by the following

Proposition 3. *For any $d \ge 1$ there exists such a polynomial $\delta_d \in \mathbb{Q}[x_1, x_2, x_3, x_4]$, that for any ample divisor D_1 on S and arbitrary divisor D_2 on S*

$$
\int_{H_d} s_{2d}\left(\mathcal{E}_{nD_1+D_2}^d\right) = \delta_d(x, y, z, w) \quad for\ n \gg 1,
$$

where x, y, z, w are given by (1) for $D = nD_1 + D_2$.

Proof. For $d = 1$ and 2 this result is classic. Suppose that $d \ge 3$. Denote $\mathcal{E} = \mathcal{E}_{nD_1+D_2}^d$ and let $n \gg 1$. From propositions 1 and 2 and lemmas 3 and 4 it follows that

$$
\int_{H_d} s_{2d}(\mathcal{E}) = \frac{1}{d!} \int_G f_* m_d,
\tag{16}
$$

where $f : X \longrightarrow Y$ is taken from (15), and $f_* m_d$ is given by the formula (14). To perform the computations, consider the diagram (15) and introduce the following notations:

$$
R = c_1\left(\mathcal{O}_P(1)\right), \quad R_S = i^* R, \quad K_S = c_1(\Omega_S), \quad W_S = s_2(\Omega_S), \quad \mathcal{R} = \mathbb{Z}[R_S, K_S, W_S].
$$

From (1) it is evident now that

$$
\begin{aligned}
i_*[S] &= x R^{3d-4}, \\
i_* R_S &= x R^{3d-3}, \\
i_* K_S &= y R^{3d-3}, \\
i_* R_S^2 &= x R^{3d-2},
\end{aligned}
$$

$$i_* R_S K_S = y R^{3d-2},$$
$$i_* K_S^2 = w R^{3d-2}, \qquad (17)$$
$$i_* W_S = z R^{3d-2}.$$

Now consider the exact triple

$$0 \longrightarrow \mathcal{O}_X \longrightarrow f^*Q \otimes \mathcal{O}_X(g^*R_S) \longrightarrow j^*T_{\Gamma/G} \longrightarrow 0,$$

where Q is the tautological rank $(d-1)$-bundle on G. From this triple we find

$$s_t(j^*T_{\Gamma/G}) = s_t\big(f^*Q(g^*R_S)\big) = \sum_{p \geq 0} t^p \sum_{i=0}^{p}(-1)^{p+i}\binom{d-1+p}{d-1+i}s_i(f^*Q)g^*R_S^{p-i}.$$
$$(18)$$

Besides, since $\Gamma = P(Q)$ and $p^*R = \mathcal{O}_{\Gamma/G}(1)$, then

$$s_i(f^*Q) = f^*s_i(Q) = f^*q_*R^{d-1+i}, \quad i \geq 0. \qquad (19)$$

Next, from the exact triple

$$0 \longrightarrow T_S \longrightarrow i^*T_P \longrightarrow N_{S/P} \longrightarrow 0$$

and the equality $N_{X/\Gamma} = g^*N_{S/P}$ we find:

$$c_t(N_{X/\Gamma}) = c_t(g^*i^*T_P)s_t(g^*T_S) = (1+g^*R_S t)^{3d-1}(1+g^*K_S t+g^*W_S t^2). \quad (20)$$

Now from the exact triples

$$0 \longrightarrow j^*T_{\Gamma/G} \longrightarrow j^*T_\Gamma \longrightarrow f^*T_G \longrightarrow 0,$$
$$0 \longrightarrow T_X \longrightarrow j^*T_\Gamma \longrightarrow N_{X/\Gamma} \longrightarrow 0$$

we find $c_t(\nu_f) = c_t(f^*T_G/T_X) = c_t(f^*T_G)s_t(T_X) = c_t(j^*T_\Gamma)s_t(j^*T_{\Gamma/G})s_t(T_X) = c_t(N_{X/\Gamma})s_t(j^*T_{\Gamma/G})$. From here and (18)–(20) one has:

$$c_i = c_i(\nu_f) = \sum_j g^*a_{ij}f^*q_*R^{b_{ij}}, \quad a_{ij} \in \mathcal{R}, \quad b_{ij} \geq 0, \quad i \geq 0. \qquad (21)$$

Next, since the left square in (15) is cartesian, then in view of (17) we get:

$$f_*g^*[S] = q_*j_*g^*[S] = q_*p^*i_*[S] = x q_* p^* R^{3d-4}, \qquad (22)$$

and similarly

$$f_*g^*R_S = x q_* p^* R^{3d-3}, f_*g^*R_S K_S = y q_* p^* R^{3d-2},$$
$$f_*g^*K_S = y q_* p^* R^{3d-3}, f_*g^*K_S^2 = w q_* p^* R^{3d-2}, \qquad (23)$$
$$f_*g^*R_S^2 = x q_* p^* R^{3d-2}, f_*g^*W_S = z q_* p^* R^{3d-2}.$$

Now substituting (21) into the polynomials $P_{d,I_p,k}$ in (14) and using (22) and (23), we obtain:

$$f_*[X], \quad f_*P_{d,I_p,k} \in A[x,y,z,w], \quad 1 \leq k \leq p, \quad |I_p| \leq d-1, \qquad (24)$$

where by A we denote the ring $\mathbb{Q}[q_*p^*R^{d-1}, \ldots, q_*p^*R^{\dim \Gamma}]$. Now from (14) and (24) we get f_*m_d as the polynomial in $A[x,y,z,w]$, from where in view of (16) we find $\delta_d = \int_{H_d} s_{2d}(\mathcal{E}) = \frac{1}{d!}\int_G f_*m_d$ as the polynomial $\delta_d(x,y,z,w) \in \mathbb{Q}[x,y,z,w]$, q.e.d.

3. Proof of theorem 1

In this section we prove theorem 1 in full generality, i.e. for any divisor D on S. For this, we shall need the following

Proposition 4. *For any given $d \geq 1$ and fixed divisors D_1 and D_2 on S the number*

$$\delta_{d,D_1,D_2}(n) = \int_{H_d} \left(s(\mathcal{E}_{nD_1}^d)\right)^3 \left(c(\mathcal{E}_{nD_1+D_2}^d)\right)^3 s(\mathcal{E}_{nD_1+2D_2}^d),$$

considered as a function of $n \in \mathbb{Z}$, is the polynomial with rational coefficients.

Proof. Consider the rational map $\varphi_{d+1}' : S \times H_d \cdots\!\!\rightarrow H_{d+1} : (x, \Xi) \longmapsto (x \cup \Xi)$, evidently regular outside Z_d. (As above, $Z_d \subset S \times H_d$ and $S \xleftarrow{\tau_d} Z_d \xrightarrow{\pi_d} H_d$ are natural projections.) As it is shown by Ellingsrud [2], this map is resolved by the diagram

where σ_{z_d} is the blowing-up of Z_d in $S \times H_d$ and φ_{d+1} is regular birational morphism. Denote by $\widetilde{Z}_d$ the (Cartier) divisor $\sigma_{z_d}^{-1}(Z_d)$ on $S \widetilde{\times} H_d$ and let $S \xleftarrow{\theta_d} S \widetilde{\times} H_d \xrightarrow{\psi_d} H_d$ be the natural projections. We have a commutative diagram of morphisms:

(25)

Consider the bundles $\varphi_{d+1*}\pi_{d+1}^* \mathcal{E}_D^{d+1}$, $\psi_d^* \mathcal{E}_D^d$ and $\theta_d^* \mathcal{O}_S(D) = \mathcal{O}_{S \widetilde{\times} H_d}(\theta_d^* D)$ on $S \widetilde{\times} H_d$. These bundles are included into the exact triple:

$$0 \longrightarrow \mathcal{O}_{S \widetilde{\times} H_d}(\theta_d^* D - \widetilde{Z}_d) \longrightarrow \varphi_{d+1*}\pi_{d+1}^* \mathcal{E}_D^{d+1} \longrightarrow \psi_d^* \mathcal{E}_D^d \longrightarrow 0. \quad (26)$$

(It is sufficient to consider the diagram

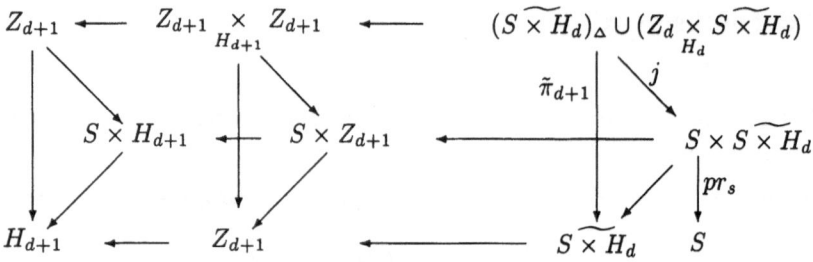

in which all the squares are cartesian, $\tilde{\pi}_{d+1}\left((S \widetilde{\times} H_d)_\Delta \cap (Z_d \underset{H_d}{\times} S \widetilde{\times} H_d)\right)$ is isomorphic to $\widetilde{Z}_d$ and $\rho_d := \tilde{\pi}_{d+1}\big|_{(S\widetilde{\times}H_d)_\Delta} : (S \widetilde{\times} H_d)_\Delta \longrightarrow S \widetilde{\times} H_d$ is the isomorphism, so that one has the exact triple

$$0 \longrightarrow \mathcal{O}_{(S\widetilde{\times}H_d)_\Delta}(-\rho_d^*\widetilde{Z}_d) \longrightarrow \mathcal{O}_{(S\widetilde{\times}H_d)_\Delta \cup (Z_d \underset{H_d}{\times} S\widetilde{\times}H_d)} \longrightarrow \mathcal{O}_{Z_d \underset{H_d}{\times} (S\widetilde{\times}H_d)} \longrightarrow 0,$$

and apply to this triple, twisted by $j^* pr_s^* \mathcal{O}_S(D)$, the functor π_{d+1*}.)

From (26) we have:

$$\varphi_{d\,|\,1*}\pi_{d+1}^* c(\mathcal{E}_D^{d+1}) = \psi_d^* c(\mathcal{E}_D^d)\big(1 + (\theta_d^* D - \widetilde{Z}_d)\big), \tag{27}$$

$$\varphi_{d+1*}\pi_{d+1}^* s(\mathcal{E}_D^{d+1}) = \psi_d^* s(\mathcal{E}_D^d)\Big[\sum_{k=0}^{2d+2} \big(\widetilde{Z}_d - \theta_d^* D\big)^k\Big]. \tag{28}$$

Now using the diagram (25), we construct the diagram of natural maps, in which all the squares are cartesian:

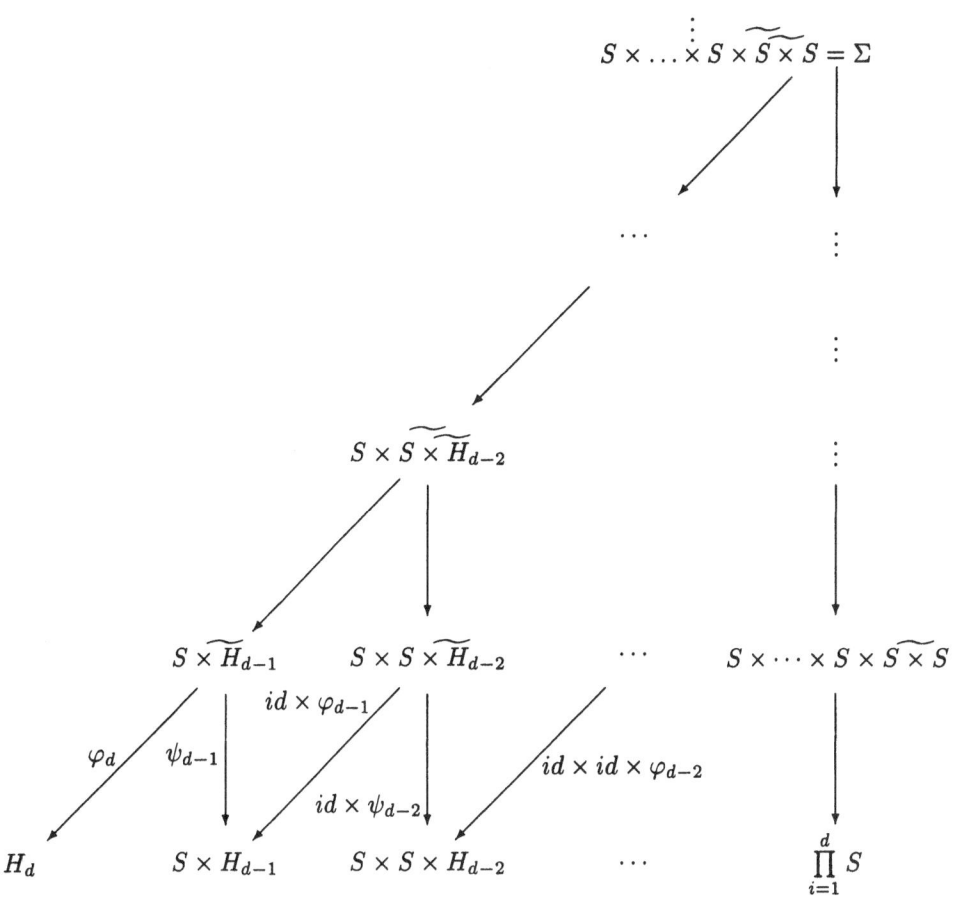

Introduce in this diagram the following notations for maps:

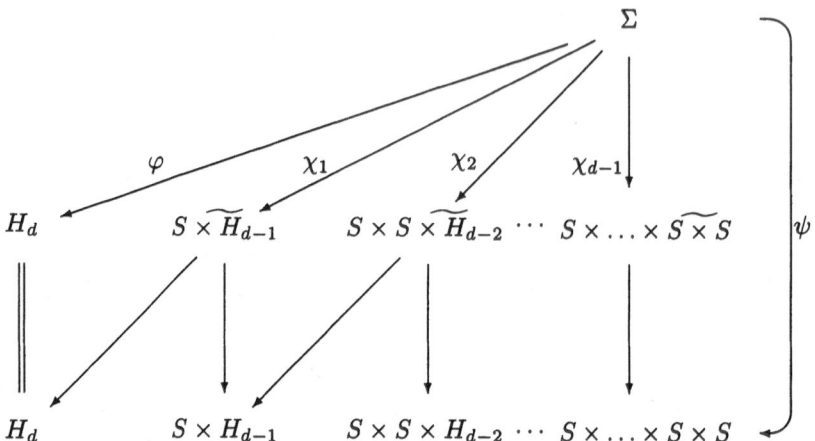

and denote

$$W_i = \chi_i^* \left(\prod_{k=1}^{i-1} S \times Z_{d-i} \right), \quad 1 \le i \le d-1, \quad W_d := 0,$$
$$D_{(i)} = \psi^* pr_i^* D, \quad D_{j,i} = \psi^* pr_i^* D_j, \quad j = 1, 2,$$

where $pr_i : \prod_{k=1}^{d} S \longrightarrow S$ are the projections onto the factors, $1 \le i \le d$. Then the formulas (27) and (28) and the induction on d give:

$$\varphi^* c(\mathcal{E}_D^d) = \prod_{i=1}^{d} (1 + D_{(i)} - W_i), \quad \varphi^* s(\mathcal{E}_D^d) = \prod_{i=1}^{d} \left(\sum_{k=0}^{2d} (W_i - D_{(i)})^k \right). \tag{29}$$

Since φ is generically finite morphism of degree $d!$, then

$$\delta_{d,D_1,D_2}(n) = \frac{1}{d!} \int_{\Sigma} \varphi^* \left(s(\mathcal{E}_{nD_1}^d) \right)^3 \varphi^* \left(c(\mathcal{E}_{nD_1+D_2}^d) \right)^3 \varphi^* s(\mathcal{E}_{nD_1+2D_2}^d).$$

Substituting here the formulas (29), one obtains:

$$\delta_{d,D_1,D_2}(n) = \frac{1}{d!} \sum_{\Sigma(k_i+l_i+m_i)=2d} a_{k_1...k_d l_1...l_d m_1...m_d} n^{\Sigma l_i} \times$$
$$\times \int_{\Sigma} W_1^{k_1} \dots W_d^{k_d} D_{1,1}^{l_1} \dots D_{1,d}^{l_d} D_{2,1}^{m_1} \dots D_{2,d}^{m_d},$$

where $a_{k_1...m_d}$ are integers, as well as the corresponding intersection indices of Cartier divisors $\int_{\Sigma} W_1^{k_1} \dots D_{2,d}^{m_d}$. Whence the assertion follows, q.e.d.

Now we pass to the proof of theorem 1.

Let D be an arbitrary divisor on S. Represent D in the form $D = D_1 - D_2$, where D_1 and D_2 are very ample divisors on S (evidently, such D_1 and D_2 exist). Thus for $n \gg 1$ and divisors nD_1, $nD_1 + D_2$, $nD_1 + 2D_2$ the proposition 3 is true. Since D_2 is very ample, the following sequence is exact:

$$0 \longrightarrow \mathcal{O}_S(nD_1 - D_2) \longrightarrow 3\mathcal{O}_S(nD_1) \longrightarrow 3\mathcal{O}_S(nD_1 + D_2) \longrightarrow \mathcal{O}_S(nD_1 + 2D_2) \longrightarrow 0. \tag{30}$$

Now since $\pi_d : Z_d \longrightarrow H_d$ is the finite morphism, then applying to (30) the functor $R^i \pi_{d*} \tau_d^*$, we get the exact triple of vector bundles on H_d:

$$0 \longrightarrow \mathcal{E}^d_{nD_1 - D_2} \longrightarrow \overset{3}{\underset{1}{\bigoplus}} \mathcal{E}^d_{nD_1} \longrightarrow \overset{3}{\underset{1}{\bigoplus}} \mathcal{E}^d_{nD_1 + D_2} \longrightarrow \mathcal{E}^d_{nD_1 + 2D_2} \longrightarrow 0,$$

wherefrom in view of proposition 4

$$\int_{H_d} s_{2d}(\mathcal{E}^d_{nD_1 - D_2}) = \int_{H_d} \left[(s(\mathcal{E}^d_{nD_1}))^3 (c(\mathcal{E}^d_{nD_1 + D_2}))^3 s(\mathcal{E}^d_{nD_1 + 2D_2}) \right]_{2d}$$

$$= \delta_{d, D_1, D_2}(n), \quad n \in \mathbb{Z}. \tag{31}$$

Next, by the proposition 3, one has:

$$\int_{H_d} s_{2d}(\mathcal{E}^d_{nD_1 - D_2}) = \delta_d \Big((nD_1 - D_2)^2, ((nD_1 - D_2) \cdot K_S), z, w \Big) = \varphi(n), \quad n \gg 1,$$

$$\tag{32}$$

where $\varphi \in \mathbb{Q}[t]$. From here and (31) we get: $\varphi(n) = \delta_{d, D_1, D_2}(n)$ for $n \gg 1$, hence $\varphi = \delta_{d, D_1, D_2}$ in $\mathbb{Q}[t]$. In particular, $\varphi(1) = \delta_{d, D_1, D_2}(1)$, i.e. in view of (31) and (32) $\int_{H_d} s_{2d}(\mathcal{E}^d_{D_1 - D_2}) = \delta_d \Big((D_1 - D_2)^2, ((D_1 - D_2) \cdot K_S), z, w \Big)$, i.e., in view of (1) and notation $D = D_1 - D_2$, $\int_{H_d} s_{2d}(\mathcal{E}^d_D) = \delta_d(x, y, z, w)$, q.e.d.

4. Computation of the polynomial $\delta_3(x, y, z, w)$

In this section we prove the theorem 2. For this purpose we use the method developed in §2. First, remark that for $d = 3$ the formula (14) becomes the well-known triple-point formula of Kleiman [7]:

$$m_3 = f^* f_* m_2 - 2c_4 m_2 + (2c_3 c_5 + 4c_2 c_6 + 8c_1 c_7 + 16c_8) m_1,$$

which together with the double-point formula $m_2 = f^* f_* m_1 - c_4 m_1$, $m_1 = [X]$, gives:

$$m_3 = f^* f_* f^* f_* [X] - f^* f_* c_4 - 2c_4 f^* f_* [X] + (2c_4^2 + 2c_3 c_5 + 4c_2 c_6 + 8c_1 c_7 + 16c_8)[X],$$

whence we obtain the following case of formula (14) for $d = 3$:

$$f_* m_3 = (f_*[X])^3 - 3(f_* c_4)(f_*[X])^2 + 2f_*\big((c_4^2 + c_3 c_5 + 2c_2 c_6 + 4c_1 c_7 + 8c_8)[X]\big).$$

$$(33)$$

Now as in the proof of the proposition 3 (see (20) and the subsequent formulas of §2) we have in the case $d = 3$:

$$c_t(\nu_f) = c_t(N_{X/\Gamma})/c_t(j^* T_{\Gamma/G}) = \frac{1 + g^*(8R_S + K_S)t + g^*(28R_S^2 + 8R_S K_S + W_S)t^2}{1 + (2g^* R_S - f^* h)t},$$

where $h = q_* R^2$. Developing this expression as a series in t, we get $c_i = c_i(\nu_f)$:

$$c_1 = g^*(6R_S + K_S) + f^* h,$$

$$c_2 = g^*(16R_S^2 + 6R_S K_S + W_S) + g^*(4R_S + K_S)f^* h + f^* h^2,$$

$$c_3 = g^*(8R_S^2 + 4R_S K_S + W_S)f^* h + g^*(2R_S + K_S)f^* h^2 + f^* h^3,$$

$$c_4 = g^*(4R_S^2 + 2R_S K_S + W_S)f^* h^2 + g^* K_S f^* h^3 + f^* h^4,$$

$$c_5 = g^*(R_S^2 + W_S)f^* h^3 + g^*(-2R_S + K_S)f^* h^4 + f^* h^5,$$

$$c_6 = g^*(8R_S^2 + 2R_S K_S + W_S)f^* h^4 + g^*(-4R_S + K_S)f^* h^5 + f^* h^6,$$

$$c_7 = g^*(16R_S^2 + 4R_S K_S + W_S)f^* h^5 + g^*(-6R_S + K_S)f^* h^6 + f^* h^7,$$

$$c_8 = g^*(28R_S^2 - 6R_S K_S + W_S)f^* h^6 + g^*(-8R_S + K_S)f^* h^7 + f^* h^8.$$

$$(34)$$

Denote by $\sigma_{a,b}$ the class in $A_* G$ of the Schubert cycle $\{l \in G \mid \dim(\mathbb{P}^a \cap l) \geq 0, l \subset \mathbb{P}^b\}$. One easily has for $R = c_1(\mathcal{O}_P(1))$:

$$f_* R^5 = \sigma_{2,7}, \quad f_* R^6 = \sigma_{1,7}, \quad f_* R^7 = \sigma_{0,7},$$

wherefrom by (23),

$$f_*[X] = x\sigma_{2,7}, \qquad f_* g^* R_S K_S = y\sigma_{0,7},$$

$$f_* g^* R_S = x\sigma_{1,7}, \qquad f_* g^* K_S^2 = w\sigma_{0,7},$$

$$f_* g^* K_S = y\sigma_{1,7}, \qquad f_* g^* W_S = z\sigma_{0,7}.$$

$$f_* g^* R_S^2 = x\sigma_{0,7},$$

$$(35)$$

Now from (34), (35) and projection formula we find:

$$(f_*[X])^3 = x^3 \sigma_{2,7}^3,$$

$$(f_* c_4)(f_*[X])^2 = x^2 \sigma_{2,7}^2 h^4 + xy\sigma_{1,7}\sigma_{2,7} h^3 + (4x^2 + 2xy + xz)\sigma_{0,7}\sigma_{2,7} h^2,$$

$$f_*(c_4^2[X]) = x\sigma_{2,7}h^8 + 2y\sigma_{1,7}h^7 + (8x + 4y + 2z + w)\sigma_{0,7}h^6 =$$

$$= f_*(c_3c_5[X]) = f_*(c_2c_6[X]),$$

(36)

$$f_*(c_1c_7[X]) = x\sigma_{2,7}h^8 + 2y\sigma_{1,7}h^7 + (-20x - 4y + z + w)\sigma_{0,7}h^6,$$

$$f_*(c_8[X]) = x\sigma_{2,7}h^8 + (-8x + y)\sigma_{1,7}h^7 + (28x - 6y + z)\sigma_{0,7}h^6.$$

Next, in A_*G one has by the Schubert's calculus:

$$\int_G \sigma_{2,7}^3 = 1, \quad \int_G \sigma_{2,7}h^8 = 20, \quad \int_G \sigma_{1,7}h^7 = 6, \quad \int_G \sigma_{0,7}h^6 = 1, \qquad (37)$$

$$\int_G \sigma_{2,7}^2 h^4 = 6, \quad \int_G \sigma_{1,7}\sigma_{2,7}h^3 = 3, \quad \int_G \sigma_{0,7}\sigma_{2,7}h^2 = 1.$$

Now substituting (36) and (37) into (33), we get the desired formula for $\delta_3(x, y, z, w)$.

Remark 1. P.Le Barz in [8], [9] obtained the following formula for the number of trisecant lines of a surface S in $\mathbb{P}^7$:

$$\delta_3 = \frac{n}{3}(7n^2 - 48n + 56) + 8t - 8d(n - 5) + \delta(n - 8),$$

where n is the degree of S in $\mathbb{P}^7$, d is the degree of the double curve C of the projection of S into $\mathbb{P}^3$, t is the number of triple points on C and δ is the number of double points of the projection of S into $\mathbb{P}^4$. One can check that this formula coincides with our formula for $\delta_3(x, y, z, w)$ if we translate the variables n, d, t, δ into x, y, z and w from (1).

Remark 2. It is curious to check the formula for δ_3 for well-known classic examples. The first such example is the case of $S = \mathbb{P}^1 \times \mathbb{P}^1$ embedded into $\mathbb{P}^7$ by the general 7-dimensional linear subseries of $|\mathcal{O}_{\mathbb{P}^1}(2) \boxtimes \mathcal{O}_{\mathbb{P}^1}(2)|$. It is geometrically evident that in this case S has no trisecant lines, i.e. $\delta_3 = 0$. Now one has $x = 8$, $y = -8$, $z = 4$, $w = 8$. Substituting this to the formula of theorem 2, we get the desired result $\delta_3 = 0$.

Another example is of the intersection S of 4 hyperquadrics and a hypersurface of degree $n \geq 2$ in $\mathbb{P}^7$. It is well-known that there are 512 lines on the intersection of 4 hyperquadrics. Since each of these lines is naturally counted $\binom{n}{3}$ times as a trisecant line for S, one should expect the answer $512\binom{n}{3} = \frac{256n(n-1)(n-2)}{3}$ for δ_3. Substituting now the corresponding values of x, y, z, w(say, here $x = 16n$, $y = 16n^2$, etc.) into the formula of theorem 2, one obtaines the desired result $\delta_3 = \frac{256}{3}n(n - 1)(n - 2)$. In particular, we have $\delta_3 = 0$, when $n = 2$, i.e. there are no trisecant lines for the intersection of 5 hyperquadrics in $\mathbb{P}^7$, as one can expect by the evident geometric reasons.

5. Appendix: proof of proposition 2

(i) Consider the diagram (15). Since $S \simeq i(S)$ and $p : \Gamma \longrightarrow P$ is the smooth projective morphism of relative dimension equal to $\dim G(d-3, 3d-3) = d(d-2)$, then $f : X \longrightarrow Y$ is the morphism of smooth projective varieties, and $m := \operatorname{codim} f = \dim Y - \dim X = \dim G_{d-2} - \dim X = 2d(d-1) - 2d(d-2) - 2 = 2(d-1)$. Thus,

$$\dim Y = md, \quad \dim X = m(d-1), \quad m = 2(d-1) = \operatorname{codim} f.$$

Next, the quasifiniteness of f easily follows from lemma 3. In fact, suppose that for some closed point $\{\mathbb{P}^{d-2}\} \in Y$ $\dim f^{-1}(\{\mathbb{P}^{d-2}\}) \geq 1$ and let $C \subset S$ be some irreducible 1-dimensional component of the set $(g \circ f^{-1})(\{\mathbb{P}^{d-2}\})$, so that $\langle i_L(C) \rangle \subset \mathbb{P}^{d-2}$, i.e. $\dim \langle i_L(C) \rangle \leq d - 2$. Then for any subscheme $\Xi \subset C$ of length d (i.e. $\xi \in Hilb^d C \subset H_d$) $\dim \langle i_L(\Xi) \rangle \leq d - 2$. Thus by the assertion 3 of lemma 3 $Hilb^d C$ is a finite subset in H_d, which is contained in $\{\xi_1, \ldots, \xi_{\delta_d}\}$. But this contradicts to the fact that $\dim Hilb^d C = d \geq 1$, since $k = \bar{k}$. Next, the quasifiniteness of f_r for $r \geq 1$ follows immediately from its construction and the quasifiniteness of f.

(ii) Now for the set $\{\xi_1, \ldots, \xi_{\delta_d}\} \subset H_d$ from lemma 3 consider the points $y_i = \langle i_L(\Xi_i) \rangle \in Y = G_{d-2}, \ 1 \leq i \leq \delta_d$, and denote $N_d = \{y_1, \ldots, y_{\delta_d}\}$. From the assertion 3 of lemma 3 it follows that for $y \in Y - N_d$ $\ell(f^{-1}(y)) < d$, so that $M_d \subset f^{-1}(N_d)$. On the other hand, the assertions 1 and 2 of lemma 3 imply that $f^{-1}(y_i) \in M_d$ for $y_i \in N_d$, and

$$\#f^{-1}(y_i) = d = \ell(f^{-1}(y_i)), \quad y_i \in N_d. \tag{38}$$

Thus, $M_d = f^{-1}(N_d)$, so that X_d is 0-dimensional and coincides as a set with $(f_0 \circ \ldots \circ f_{d-1})^{-1}(N_d)$:

$$X_d = (f_0 \circ \ldots \circ f_{d-1})^{-1}(N_d). \tag{39}$$

Now in view of (38) the assertion that X is reduced is equivalent to the fact that the equality (39) is true in the scheme-theoretic sense, where N_d is understood as a reduced 0-dimensional scheme of length δ_d. This again in view of (38) means that for any $y \in N_d$ and any vector $0 \neq v \in T_y Y$, considered as a subscheme in Y, the length of 0-dimensional subscheme $f^{-1}(v)$ of X is less than $2\ell(f^{-1}(y)) = 2d$:

$$\ell(f^{-1}(v)) < 2d. \tag{40}$$

Let $\Xi = (g \circ f^{-1})(y) = x_1 \cup \ldots \cup x_d$, where by (38) all the points $x_1, \ldots, x_d$ are distinct. Remark, that the simple dimension counting (see, e.g., the equality (47) below) shows, that for sufficiently general choice of L $\langle i_{n,1}(\Xi') \rangle \cap L = \emptyset$ for any subscheme $\Xi' \underset{\neq}{\subset} \Xi$, so that

$$\dim \langle i_L(\Xi') \rangle = \ell(\Xi') - 1, \quad \Xi' \underset{\neq}{\subset} \Xi. \tag{41}$$

Now let $P_v = (p \circ q^{-1})(v)$, where, as in the proof of the assertion 4 of lemma 3, the scheme P_v is correctly defined as a subscheme of P with the support $\mathbb{P}_y^{d-2} = (p \circ q^{-1})(y)$. (Namely, the ideal sheaf $\mathcal{I}_{P_v,P}$ is defined as $p_*q^*\mathcal{I}_{v,Y}$.) Consider the set $\Sigma_v = \{z \in q^{-1}(y) \mid$ there exists $w \in T_z q^{-1}(v)$ such that $dp_* w = 0\}$. One easily verifies that the following equality is true for P_v and Σ_v:

$$\dim\langle P_v \rangle = 2d - 4 - \dim\langle p(\Sigma_v) \rangle. \tag{42}$$

Since $v \neq 0$, one evidently has $\dim\langle P_v \rangle > \dim \mathbb{P}_y^{d-2} = d - 2$, hence from (42) we get:

$$\dim\langle p(\Sigma_v) \rangle < d - 2. \tag{43}$$

Now suppose that (40) is false, i.e. $\ell(f^{-1}(v)) = 2d$. This means that for every point $z_i \in f^{-1}(y)$, where $g(z_i) = x_i$, there exists nonzero vector $v_i \in T_{z_i} f^{-1}(y)$ such that $df_*(v_i) = v$, $1 \leq i \leq d$. Considering v_i as a 0-dimensional subscheme of $f^{-1}(v)$, let $u_i = g(v_i)$, $1 \leq i \leq d$. Renumerate the points $z_1, \ldots, z_d \in f^{-1}(v)$ in such a way that $dg_*(v_i) = 0$, $1 \leq i \leq k$, and $dg_*(v_i) = u_i \neq 0$ for $k + 1 \leq i \leq d$. Take in S the 0-dimensional subscheme $\Xi_v = (g \circ f^{-1})(v) = g\left(\bigcup_{i=1}^{d} v_i\right) = x_1 \cup \ldots \cup x_k \cup u_{k+1} \cup \ldots \cup u_d$, $\ell(\Xi_v) = 2d - k$ (here we identify $u_i \in T_{x_i} S$ with the corresponding subschemes of length 2 in S). The conditions $dg_*(v_i) = 0$, $1 \leq i \leq k$, mean that

$$\{x_1, \ldots, x_k\} \subset p(\Sigma_v). \tag{44}$$

But from (41) it follows immediately that

$$\dim\langle x_1, \ldots, x_k \rangle = \left\{ \begin{array}{ll} k - 1, & k \leq d - 1, \\ d - 2, & k = d. \end{array} \right\}$$

From here and (43) and (44) we obtain:

$$k \leq d - 2, \quad \dim\langle x_1, \ldots, x_k \rangle = k - 1 \leq \dim\langle p(\Sigma_v) \rangle. \tag{45}$$

Next, since by the construction $i_L(\Xi_v) \subset P_v$, then $\langle i_L(\Xi_v) \rangle \subset \langle P_v \rangle$, wherefrom by (42) and (45) we have:

$$\dim\langle i_L(\Xi_v) \rangle \leq 2d - 4 - \dim\langle x_1, \ldots, x_k \rangle = 2d - 3 - k = \ell(\Xi_v) - 3. \tag{46}$$

But this contradicts to the assertion 4) of lemma 3.

(iii) Let $1 \leq r \leq d - 1$ and $U_r = H_r - \Delta_r = \{\xi \in H_r \mid \#\Xi = r\}$, $\bar{U}_r = H_r$, $\dim U_r = 2r$. Next, for any multiindex $k = (k_1, \ldots, k_r)$, $0 \leq k_1 < \ldots < k_r \leq 3d - 2$ denote by $\Gamma_k \subset G_{k_1} \times \ldots \times G_{k_r}$ the flag variety with natural projections $p_{k'}^k : \Gamma_k \longrightarrow \Gamma_{k'}$ for any subset k' in k. Consider the diagrams

$$G_{r-1} \xleftarrow{\ p_{r-1}\ } \Gamma_{r-1,d-2} \xrightarrow{\ q_{d-2}\ } G_{d-2},$$

where $p_{r-1} = p_{r-1}^{r-1,d-2}$, $q_{d-2} = p_{d-2}^{r-1,d-2}$, and

$$\mathbb{P}^{N-1} \xleftarrow{\ p_{(r-1)}\ } \Gamma_{(r-1)} \xrightarrow{\ q_{(r-1)}\ } G_{(r-1)},$$

where $G_{(r-1)}$ is the grassmanian of $(r-1)$-dimensional subspaces in $\mathbb{P}^{N-1}$, $\Gamma_{(r-1)} \subset \mathbb{P}^{N-1} \times G_{(r-1)}$ – the flag variety, and $p_{(r-1)}$ and $q_{(r-1)}$ – the natural projections. Let $W_r = p_{(r-1)} \left(q_{(r-1)}^{-1} (i_{n,r}(H_r)) \right)$. One has: $\dim W_r \leq \dim q_{(r-1)}^{-1} (i_{n,r}(H_r)) = \dim H_r + \dim(\text{fibre of } q_{(r-1)}) = 2r + r - 1 = 3r - 1$. Thus, under the general choice of the $(N-3d)$-dimensional subspace L in $\mathbb{P}^{N-1}$,

$$L \cap W_r = \emptyset, \quad 1 \leq r \leq d - 1. \tag{47}$$

The condition (47) means that for $r \leq d - 1$ there are defined the morphisms $i_r :$ $H_r \longrightarrow G_r : \xi \longmapsto \langle i_L(\Xi) \rangle$, such that $i_r : H_r \longrightarrow i_r(H_r)$ are the bijections on the sets of closed points and, as a corollary, generically the embeddings. From here we have for $U'_r = i_r(U_r)$:

$$\dim U'_r = 2r. \tag{48}$$

Besides, by the definition of the set M_r of r-ple points of f,

$$(q_{d-2} \circ p_{r-1}^{-1})(U'_r) \subset M_r. \tag{49}$$

Next, from lemma 3 it follows immediately that $q_{d-2} \big|_{p_{r-1}^{-1}(U'_r)}$ is the quasifinite morphism, hence in view of (48) $\dim(q_{d-2} \circ p_{r-1}^{-1})(U'_r) = \dim p_{r-1}^{-1}(U'_r) = 2r + \dim(\text{fibre of } p_{r-1}) = 2r + 2d(d - 1 - r) = 2(d - r)(d - 1)$. From here and the quasifiniteness of $f_1, \ldots, f_r$ (the property (i) above) we find:

$$\dim X_r = \dim M_r \geq 2(d - r)(d - 1). \tag{50}$$

Now denote $H_1 = S$, $G_0 = P$, $H_{(r)} = \prod_{k=1}^{r} H_k$, $G_{(r)} = \prod_{k=0}^{r-1} G_k$, $i_{(1)} = i_L$, $i_{(r)} = i_L \times i_2 \times \ldots \times i_r$, $r \geq 2$, $\Gamma_{(r)} = \Gamma_{0,1,\ldots,r-1}$, and let $Z_{(r)} = \{(\xi_1, \ldots, \xi_r) \in H_{(r)} \mid \Xi_1 \subset \ldots \subset \Xi_r\}$ is the full graph of incidence. One has the cartesian square

$$\begin{array}{ccc}
H_{(r)} & \xrightarrow{\ i_{(r)}\ } & G_{(r)} \\
\uparrow & & \uparrow \\
& & \\
Z_{(r)} & \xrightarrow{\ i_{(r)}\ } & \Gamma_{(r)}
\end{array} \tag{51}$$

Since $i_r : H_r \longrightarrow i_r(H_r)$ are bijections, $i_{(r)} : H_{(r)} \longrightarrow i_{(r)}(H_{(r)})$ are also bijections, hence

$$\dim i_{(r)}(Z_{(r)}) = \dim Z_{(r)} = 2r, \quad 1 \le r \le d-1. \tag{52}$$

Now for $1 \le r \le d-1$ we construct the number of maps $e_r : X_r \longrightarrow \Pi_r$, where $\Pi_r = \Gamma_{0,1,\dots,r-1,d-2}$, $e_1 = j$, such that the following diagram is commutative:

$$
\begin{array}{ccccccccccc}
G_{d-2} = X_0 & \xrightarrow{f \,=\, f_0} & X = X_1 & \xleftarrow{f_1} & X_2 & \longleftarrow & \cdots & \xleftarrow{f_{d-3}} & X_{d-2} & \xleftarrow{f_{d-2}} & X_{d-1} \\
e_0 = id \downarrow \cong & & \downarrow e_1 = j & & \downarrow e_2 & & & & \downarrow e_{d-2} & & \downarrow e_{d-1} \\
\Pi_0 & \xleftarrow{p^0} & \Pi_1 & \xleftarrow{p^1} & \Pi_2 & \longleftarrow & \cdots & \xleftarrow{p^{d-3}} & \Pi_{d-2} & \xleftarrow{p^{d-2}} & \Pi_{d-1}
\end{array}
$$

$$\tag{53}$$

in which we put $\Pi_0 = G_{d-2}$, $e_0 = id$, $p^0 = q$, $p^r = p^{0,1,\dots,r,d-2}_{0,1,\dots,r-1,d-2}$, $r \ge 1$. We construct this diagram by the induction on r, beginning from $r = 1$ (the left square in (53)). Suppose that $r \le d-2$ and that the embedding e_r is already constructed. Consider the cartesian square

$$
\begin{array}{ccc}
X_r & \xrightarrow{\quad e_r \quad} & \Pi_r \\
\Delta_x \downarrow & & \downarrow \Delta_\Pi \\
X_r \underset{X_{r-1}}{\times} X_r = X_r \underset{\Pi_{r-1}}{\times} X_r & \xrightarrow{\quad e_r \underset{\Pi_{r-1}}{\times} e_r \quad} & \Pi_r \underset{\Pi_{r-1}}{\times} \Pi_r
\end{array}
$$

in which the vertical embeddings are the diagonal maps, and the bottom left equality outcomes from the injectivity of e_{r-1}. This square induces the cartesian square

$$
\begin{array}{ccc}
X_{r+1} = X_r \underset{X_{r-1}}{\widetilde{\times}} X_r & \xrightarrow{\quad h_r \quad} & \Pi_r \underset{\Pi_{r-1}}{\widetilde{\times}} \Pi_r \\
\sigma_x \downarrow & & \downarrow \sigma_\Pi \\
X_r \underset{X_{r-1}}{\times} X_r & \longrightarrow & \Pi_r \underset{\Pi_{r-1}}{\times} \Pi_r
\end{array}
\tag{54}
$$

in which σ_x and σ_Π are the blowing-ups of the diagonals $\Delta_x(X_r)$ and $\Delta_\Pi(\Pi_r)$ respectively, and the top left equality outcomes from the definition of X_r by Kleiman (see [7], §4). Next, from the construction of flag varieties Π_r it follows immediately that there is uniquely defined a morphism $\varphi_r : \Pi_r \underset{\Pi_{r-1}}{\times} \Pi_r \longrightarrow \Pi_{r+1}$ such that the following diagram is commutative:

$$
\begin{array}{ccc}
\Pi_r \underset{\Pi_{r-1}}{\widetilde{\times}} \Pi_r & \xrightarrow{\;\varphi_r\;} & \Pi_{r+1} \\
\Big\downarrow{\sigma_\Pi} & & \Big\downarrow{p^r} \\
\Pi_r \underset{\Pi_{r-1}}{\times} \Pi_r & \xrightarrow{\;pr_{1,r}\;} & \Pi_r
\end{array}
\tag{55}
$$

where $pr_{1,r}$ is the projection of the fibred square onto the first factor. Let $e_{r+1} = \varphi_r \circ h_r :$ $X_{r+1} \longrightarrow \Pi_{r+1}$ be the composition of top horizontal morphisms in (54) and (55). By the construction one has the commutative diagram:

$$
\begin{array}{ccc}
X_{r+1} & \xrightarrow{\;e_{r+1}\;} & \Pi_{r+1} \\
\Big\downarrow{f_r} & & \Big\downarrow{p^r} \\
X_r & \xrightarrow{\;e_r\;} & \Pi_r
\end{array}
$$

(here we take into account that by the definition of the morphism f_r (see [7], §4) we have $f_r = pr_{1,x} \circ \sigma_x$, where $pr_{1,x} : X_r \underset{X_{r-1}}{\times} X_r \longrightarrow X_r$ is the projection onto the first factor).

This diagram gives us the desired step of induction. We have only to check that e_{r+1} is the embedding for $r \le d - 3$.

For this purpose remark, that in the diagram (55) the fibre of the projection φ_r over the given closed point $(\varphi_r \circ h_r)(y)$ is by the construction the projective line m in the factorspace $P(V/W)$, where $P(V) = P$ and $P(W) = \langle i_L(\Xi) \rangle$ for some point $\xi \in H_{r-1}$, defined by the point y. Hence the noninjectivity of e_{r+1} at the point y would mean that on the line m there lie at least two (possibly, infinitely near) points from $h_r(X_{r+1})$. This means that there would exist a subscheme $\Xi' \supset \Xi$ of S, $\ell(\Xi') = \ell(\Xi) + 3 = r + 2$, such that

$$
\dim \langle i_L(\Xi') \rangle = \dim \langle i_L(\Xi) \rangle + 2 \le r
\tag{56}
$$

for a point $\xi' \in H_{r+2}$, corresponding to the scheme Ξ', $r \le d - 3$. But from (47) and lemma 1 it follows that (under the general choice of L) $\dim \langle i_L(\Xi) \rangle = k - 1$ for any point $\xi' \in H_k$, $k \le d-1$, which contradicts to (56). Hence, e_{r+1} is the embedding for $r \le d-3$, and the diagram (53) is constructed.

To finish the proof, remark that by the construction for $0 \le k \le r - 1$ the image of any closed point $x \in X_r$ under the composition $p_{0,k}^{0,1,\dots,r-1,d-2} \circ e_r$ lies in $j_k(Z_k)$, where $j_k : Z_k \longrightarrow \Gamma_{0,k}$ is defined as the composition of the embedding $Z_k \hookrightarrow S \times H_k$ and the map $i_L \times i_k : S \times H_k \longrightarrow P \times G_k$. From here by the universal property of the full flag variety $Z_{(r)}$ from (51) it follows that

$$
\left(p_{0,1,\dots,r-1}^{0,1,\dots,r-1,d-2} \circ e_r \right) (X_r) \subset i_{(r)}(Z_{(r)}),
$$

so that $\dim X_r \leq \dim i_{(r)}(Z_{(r)}) + \dim(\text{fibre of } p_{0,1,\ldots,r}^{0,1,\ldots,r-1,d-2})$, i.e. in view of (52), $\dim X_r \leq 2r + 2d(d-r-1) \leq 2(d-r)(d-1)$. Thus, the inequality in (50) becomes the equality, wherefrom (iii) follows for $m = 2(d-1)$.

References

[1]   M.BELTRAMETTI, A.J.SOMMESE. *Zero cycles and k-th order embeddings of smooth projective surfaces*, Symposia Mathematica, vol.XXXII (1991), 33-48.

[2]   G.ELLINGSRUD. *Another proof of the irreducibility of the punctual Hilbert scheme of a smooth surface*, Manuscript, Bergen, 1991.

[3]   J.FOGARTY. *Algebraic families on an algebraic surface: II, the Picard scheme of the punctual Hilbert scheme*, Amer.J.Math., 95(1973), 660-687.

[4]   W.FULTON. *Intersection theory*, Springer-Verlag, 1984.

[5]   R.HARTSHORNE. *Algebraic geometry*, Springer-Verlag, 1977.

[6]   S.KATZ. *Iteration of multiple point formulas and applications to conics*, Lecture Notes in Math. 1311(1988), 147-155.

[7]   S.KLEIMAN. *Multiple-point formulas. I: Iteration*, Acta Math. 147(1981), 13-49.

[8]   P.LE BARZ. *Formules pour les trisécants des surfaces algébriques*, L'Enseingement Mathematique, t.33(1987), 1-66.

[9]   P.LE BARZ. *Formules pour les multisécants des surfaces*, C.R.Acad.Sci.Paris, t.292 (1981), Ser.I, 797-800.

[10]   V.Ya.PIDSTRIGACH, A.N.TYURIN. *Invariants of the smooth structures of an algebraic surfaces arising from Dirac operator*, Izv. RAN, Ser. math. 56, 2(1992), 279-371 (Russian).

Top Segre Class of a Standard Vector Bundle $\mathcal{E}_D^4$ on the Hilbert Scheme $Hilb^4 S$ of a Surface S

A.S. Tikhomirov, T.L. Troshina

This work is a continuation of the paper [4] of the present collection. Using the results of [4] (and keeping the notations introduced there), we compute here the degree $\delta_4 = \int_{H_4} s_8(\mathcal{E}_D^4)$ of the top Segre class of a standard rank-4 vector bundle $\mathcal{E}_D^4$ on the Hilbert scheme $H_4 = Hilb^4 S$ of 0-dimensional subschemes of length 4 on a smooth projective surface S. The main result of this paper is

Theorem 1. *Let* $Z_4 \subset S \times H_4$ *be the universal flat family of 0-dimensional subschemes of length 4 in S with natural projections* $S \xleftarrow{\tau_4} Z_4 \xrightarrow{\pi_4} H_4$. *Let D be an arbitrary divisor on S and $\mathcal{E}_D^4 = \pi_{4*}\tau_4^* \mathcal{O}_S(D)$ be a standard rank-4 vector bundle on H_4, corresponding to D. Then the degree δ_4 of the top Segre class $s_8(\mathcal{E}_D^4)$ of $\mathcal{E}_D^4$ is the following rational polynomial of the variables* $x = (D^2), y = (K_S \cdot D), z = (K_S^2) - \chi_{top}(S), w = (K_S^2)$:

$$\delta_4(x,y,z,w) = \frac{1}{4!}(x^4 - 60x^3 - 30x^2 y - 6x^2 z + 1196x^2 + 1068xy + 220xz +$$

$$+ 64xw + 75y^2 + 30yz + 3z^2 - 7920x - 9042y - 1944z - 1356w).$$

In the proof of this theorem we use the following geometric construction. Let D be the ample divisor in S and $n \geq 1$ is such that:

a) $N = h^0(\mathcal{O}_S(nD)) \geq 12$, the linear series $|nD|$ defines the embedding $i_n : S \hookrightarrow \mathbb{P}^{N-1}$ and there exists $(N-12)$-dimensional linear subspace L in $\mathbb{P}^{N-1}$ such that $i_L = pr_L \circ i_n : S \longrightarrow \mathbb{P}^{10}$ is the embedding, where $pr_L : \mathbb{P}^{N-1} \dashrightarrow \mathbb{P}^{10}$ is the (rational) linear projection with centre L,

b) the induced morphism $i_{L,2} : H_2 = Hilb^2 S \longrightarrow G(1,10) : \xi \longmapsto \langle i_L(\Xi) \rangle$ is the embedding, where $G(1,10)$ is the grassmanian of lines in $\mathbb{P}^{10}$,

c) for any point $s \in S$ the projective plane $\mathcal{P}T_s S$ in $\mathbb{P}^{10}$, tangent to the surface $S \simeq i_L(S)$, intersects S in a subscheme $s^{(2)}$, given by the ideal $\mathcal{I}_{s^{(2)},S} = \mathcal{I}_{s,S}^2$. (The existence of this number $n \geq 1$ and subspace L in $\mathbb{P}^{N-1}$ is guaranteed by lemma 1 below.)

Let $Ch \subset \mathbb{P}^{10}$ be the variety filled by chordal lines of the surface S in $\mathbb{P}^{10}$ (here and everywhere below we identify S with $i_L(S)$). Then $S \subset \mathrm{Sing}Ch$ and, under the generic choice of L, $\mathrm{Sing}(Ch - S)$ consists of a finite set of points $x_1, \ldots, x_M$ such that through each of the points x_i, $1 \le i \le M$ there pass two distinct chords l_{1i}, l_{2i} of the surface S, intersecting it in pairs of distinct points, i.e. $P_i^2 = \langle l_{1i}, l_{2i} \rangle$ is the 4-secant plane of S in $\mathbb{P}^{10}$. Since in each such 4-secant plane of S lie by the construction 3 of the points $x_1, \ldots, x_M$, the number of these 4-secant planes equals $\frac{1}{3}M$. Now according to [4], lemma 3 and proposition 1, this number $\frac{1}{3}M$ is exactly equal to $\delta_4 = \int_{H_4} s_8(\mathcal{E}_{nD}^4)$:

$$\delta_4(x, y, z, w) = \frac{1}{3}M. \tag{1}$$

Thus the problem of computation of δ_4 is reduced to finding the number M. For this purpose consider the blowing-up $\sigma : P \longrightarrow \mathbb{P}^{10}$ of the surface S in $\mathbb{P}^{10}$ and let $\widetilde{Ch}$ be the proper inverse image via σ of the chordal variety Ch in $\mathbb{P}^{10}$. The condition c) above means that $\widetilde{Ch}$ is smooth along $\widetilde{Ch} \cap \sigma^{-1}(S)$ (see the proof of the proposition 1 below), so that $\widetilde{Ch}$ has only a finite number M of isolated double points $\tilde{x}_i = \sigma^{-1}(x_i)$, $1 \le i \le M$, vanishing under the normalization $\nu : X \longrightarrow \widetilde{Ch}$ of the variety $\widetilde{Ch}$. Moreover, one can assume (see proposition 1) that

1) the morphism ν has nonvanishing differential, so that in the diagram

$$\tag{2}$$

where i_p is the natural embedding, the composition $f = i_p \circ \nu$ is the immersion and

2) $2M = \deg M_2$, where M_2 is the cycle of double points of the immersion, and the degree of M_2 is calculated by the double point formula $\deg M_2 = \int_X (f^* f_*[X] - c_5(f^* T_P / T_X))$.

Thus by (1)

$$\delta_4(x, y, z, w) = \frac{1}{6} \int_X (f^* f_*[X] - c_5(f^* T_P / T_X)). \tag{3}$$

For the effective computation of the right side of (3) we need to know the class $f_*[X] = [\widetilde{Ch}]$ in $A_*(P)$ and also the Chern polynomial $c_t(T_X)$ of the sheaf T_X. The computation of $[\widetilde{Ch}]$ is done in §2. Since the standard blow-up-formula of Fulton [2], theorem 6.7, does not work effectively in our situation because $S \subset \mathrm{Sing}Ch$, we develop here another attitude, based on some arguments from the jet theory (see lemma 2 and §2).

In §3 we compute $c_t(T_X)$ and then $c_t(f^* T_P / T_X)$, using the isomorphism $X \simeq q^{-1} i_{L,2}(H_2)$, where $q : \Gamma \longrightarrow G(1,10)$ is the natural projection of the flag variety $\Gamma \subset \mathbb{P}^{10} \times G(1,10)$.

§1 is preliminary and is devoted to the justification of the described construction (see proposition 1 and its corollary).

Everywhere in this paper we work over the base field $k = \mathbb{C}$. By the word "point" we mean the closed point of a scheme.

1. Preliminary constructions

Let, as above, S be a smooth projective surface, $H_k = Hilb^k S$ be the Hilbert scheme of 0-dimensional subschemes of length k on S, $k \geq 2$, D be the ample divisor on S, and let $n \geq 1$ be large enough, so that the map $i_n : S \longrightarrow \mathbb{P}^{N-1}$ by the complete linear series $|nD|$, where $N = h^0(\mathcal{O}_S(nD)) \geq 12$, be the embedding.

Lemma 1. *Let $n \gg 1$ and $pr_L : \mathbb{P}^{N-1} \cdots\cdots\twoheadrightarrow \mathbb{P}^{10}$ be the (rational) linear projection from the general $(N-12)$-dimensional linear subspace L in $\mathbb{P}^{N-1}$. Then*

1) the map $i_L = pr_L \circ i_n : S \longrightarrow \mathbb{P}^{10}$ is the embedding, and the induced map $i_{L,2} : H_2 \longrightarrow G(1,10) : \xi \longmapsto \langle i_L(\Xi) \rangle$ is also the embedding,

2) for any point $s \in S$ the ideal sheaf of the subscheme $\mathcal{P}T_s S \cap S$ of S equals $\mathcal{I}^2_{s,S}$, where we identify S with the surface $i_L(S)$ in $\mathbb{P}^{10}$ and $\mathcal{P}T_s S$ denotes the projective plane in $\mathbb{P}^{10}$, tangent to the surface S at the point s.

Proof. 1) One easily sees that for $n \gg 1$ i_L is the embedding. Next, let $G(1, N-1)$ be the grassmanian of lines of $\mathbb{P}^{N-1}$. It is known (see [4], lemma 1) that for $n \gg 1$ the map $i_{n,2} : H_2 \longrightarrow G(1, N-1) : \xi \longmapsto \langle i_n(\Xi) \rangle$ is the embedding. Now the property of having the nondegenerate differential for the map $i_{L,2} : H_2 \longrightarrow G(1,10) :$ $\xi \longmapsto \langle i_L(\Xi) \rangle$ is equivalent to the fact that the closure $\bar{\mathcal{Y}}$ in $\mathbb{P}^{N-1}$ of the variety $\mathcal{Y} = \bigcup_{\substack{x_1, x_2 \in S \\ x_1 \neq x_2}} \langle \bigcup_{i=1,2} \mathcal{P}T_{i_n(x_i)} i_n(S) \rangle$ does not intersect L. The last is evidently fulfilled since

$$\dim \mathcal{Y} \leq 2 \dim S + \dim \langle \bigcup_{i=1,2} \mathcal{P}T_{i_n(x_i)} i_n(S) \rangle = 4 + 5 = 9 < 11 = \text{codim}(L, \mathbb{P}^{N-1}).$$

At last, the injectivity of $i_{L,2}$ as the map of sets is equivalent to the fact that for any two points $\xi_1, \xi_2 \in H_2$, $\xi_1 \neq \xi_2$, $i_L(\Xi_1) \neq i_L(\Xi_2)$). Since $i_L : S \longrightarrow \mathbb{P}^{10}$ is the embedding, the above inequality evidently means that $i_L(S)$ does not have trisecant lines in $\mathbb{P}^{10}$. The last is equivalent to the fact that L does not intersect the variety $\mathcal{W} = \bigcup_{\xi \in H_3} \langle i_n(\Xi) \rangle$ in $\mathbb{P}^{N-1}$. But this is evident for L chosen generically by the easy dimension counting. Hence $i_{L,2}$ is the embedding.

2) This property is equivalent to the fact that for any point $\xi \in H_4$ a) $\dim \langle i_n(\Xi) \rangle = 3$ and

b) $L \cap (\bigcup_{\xi \in H_4} \langle i_n(\Xi) \rangle) = \emptyset$. The condition a) is again fulfilled by [4], lemma 1, and b) is checked by simple dimension counting, q.e.d.

Now consider the flag variety $\Gamma \subset \mathbb{P}^{10} \times G(1,10)$ with natural projections $\mathbb{P}^{10} \xleftarrow{pr} \Gamma \xrightarrow{qr} G(1,10)$, identify S with its image under the embedding $i = i_L : S \hookrightarrow \mathbb{P}^{10}$, respectively H_2 with its image under the embedding $i_{L,2} : H_2 \hookrightarrow G(1,10)$ (the assertion 1) of lemma 1 above), and let $X = q_\Gamma^{-1}(H_2)$, $p = pr \big|_X$, $q = q_\Gamma \big|_X$ and $Ch = p(X)$ be the chordal variety of S (i.e. the variety filled by chordal lines of S). Let, as above, $\sigma : P \longrightarrow \mathbb{P}^{10}$ be the blowing-up of the surface S (as a reduced subscheme) in $\mathbb{P}^{10}$, $D = \sigma^{-1}(S)$ be the exceptional divisor and $\widetilde{Ch}$ be the proper inverse image of Ch under σ. Here $\widetilde{Ch}$ is also a variety and $\sigma_{Ch} = \sigma \big|_{\widetilde{Ch}} : \widetilde{Ch} \longrightarrow Ch$ is the blowing-up of S (as a reduced subscheme) in Ch. Hence by the well-known functorial property of blowing-ups (see, e.g., [2], appendix B.6.9.) $Z = \sigma_{Ch}^{-1}(S)$ coincides with the scheme-theoretic intersection $\widetilde{Ch} \cap D$ in P:

$$Z = \sigma_{Ch}^{-1}(S) = \widetilde{Ch} \cap D. \tag{4}$$

Thus we have a diagram of natural maps:

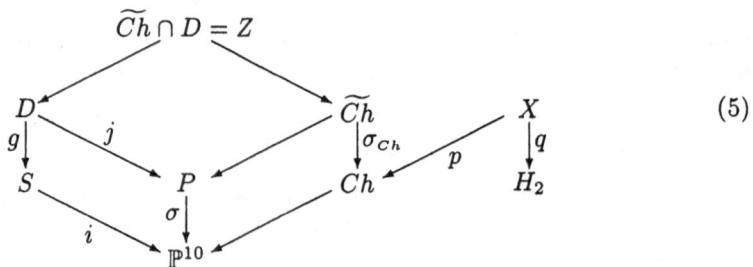

$$\tag{5}$$

Lemma 2. *Under the conditions of lemma 1 Z_{red} is nonsingular and isomorphic to the blow-up $\widetilde{S \times S}$ of the variety $S \times S$ along the diagonal $S \xhookrightarrow{\Delta} S \times S$.*

Proof. Consider the vector bundle $J_1 T_S$ of 1-jets for the embedding $i : S \hookrightarrow \mathbb{P}^{10}$, i.e. the vector bundle defined as the extension:

$$0 \longrightarrow \mathcal{O}_S \xrightarrow{i_0} J_1 T_S \longrightarrow T_S \longrightarrow 0, \tag{6}$$

given by the element $\xi \in \text{Ext}^1(T_S, \mathcal{O}_S) \simeq H^1(\Omega_S) \simeq H^{1,1}(S) \hookrightarrow H^2(S, \mathbb{C})$ coinciding with the image of the positive generator (the class of hyperplane section) in $H^2(\mathbb{P}^{10}, \mathbb{Z}) \simeq \mathbb{Z}$ under the natural map $H^2(\mathbb{P}^{10}, \mathbb{Z}) \xrightarrow{i^*} H^2(S, \mathbb{Z})/\text{Tors} \hookrightarrow H^2(S, \mathbb{C})$ (see, e.g., [5]). Denote $Y = P(J_1 T_S)$, $\theta : Y \longrightarrow S$ the structure morphism, and let $y : Y \hookrightarrow \mathbb{P}^{10} \times S$ be the embedding uniquely defined by the property that the composition $\psi : Y \xrightarrow{y} \mathbb{P}^{10} \times S \xrightarrow{pr_{\mathbb{P}^{10}}} \mathbb{P}^{10}$ coincides with the map by the subseries $\mathbb{P}^{10\vee} \subset \big| \mathcal{O}_{Y/S}(1) \otimes \theta^* i^* \mathcal{O}_{\mathbb{P}^{10}}(1) \big|$ corresponding to the rational linear projection $i_L : \mathbb{P}^{N-1} \dashrightarrow \mathbb{P}^{10}$, so that

$$\psi(Y) = \bigcup_{s \in S} PT_s S. \tag{7}$$

Let $S_\Delta^{(2)}$ be the "first infinitesimal neighbourhood" of the diagonal S_Δ in $S \times S$, i.e. the subscheme in $S \times S$ defined by the ideal sheaf $\mathcal{I}_{S_\Delta^{(2)}, S \times S} = \mathcal{I}_{S_\Delta, S \times S}^2$. Let also $s_0 : S \simeq P(\mathcal{O}_S) \hookrightarrow P(J_1 T_S) = Y$ be the section of θ, corresponding to the injection i_0 in (6). Since by the definition of Y $\ N_{s_0(S)/Y} \simeq T_S \simeq N_{S_\Delta/S \times S}$, there is naturally defined the embedding $S_\Delta^{(2)} \overset{s_0^{(2)}}{\hookrightarrow} Y$ extending the embedding $S_\Delta \simeq S \overset{s_0}{\hookrightarrow} Y$. Moreover, from (7) and the assertion 2) of lemma 1 follows the cartesian square

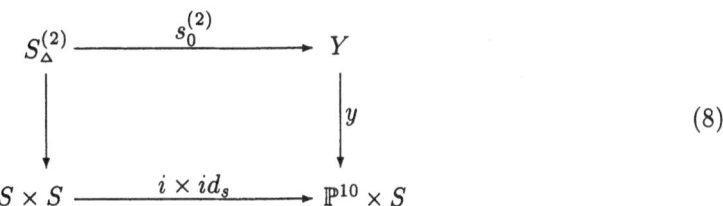

$$(8)$$

Now since for any point $s \in S \simeq i(S)$ one has the coincidence of normal spaces:

$$N_{S/\mathbb{P}^{10}}\big|_s = N_{\mathcal{P}T_s S/\mathbb{P}^{10}}\big|_s,$$

the following diagram is commutative:

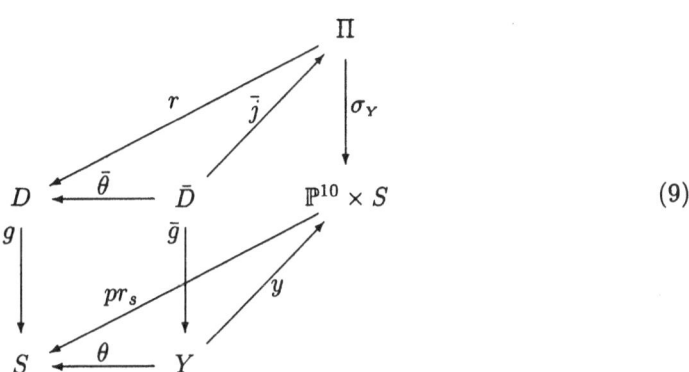

$$(9)$$

in which $\sigma_Y : \Pi \longrightarrow \mathbb{P}^{10} \times S$ is the blowing-up of the subvariety Y in $\mathbb{P}^{10} \times S$, $\bar{D}$ is the exceptional divisor of σ_Y, r, $\bar{\theta}$, $\bar{g}$ and $\bar{j}$ are natural projections, and the left square is cartesian, so that

$$\bar{D} = Y \underset{S}{\times} D. \tag{10}$$

Moreover,

1) from (8) and the functorial properties of blowing-ups outcomes the commutativity of the diagram:

$$
\begin{array}{ccccc}
\Delta & \longrightarrow & \widetilde{S \times S} & \xrightarrow{\;\tilde{i}\;} & \Pi \\
\downarrow{\scriptstyle \rho} & & \downarrow{\scriptstyle \sigma_\Delta} & & \downarrow{\scriptstyle \sigma_Y} \\
S_\Delta & \longrightarrow & S \times S & \xrightarrow{i \times id_s} & \mathbb{P}^{10} \times S
\end{array}
\tag{11}
$$

where $\sigma_\Delta : \widetilde{S \times S} \longrightarrow S \times S$ is the blowing-up of the diagonal S_Δ in $S \times S$, Δ is the exceptional divisor of σ_Δ and $\tilde{i}$ and ρ are the induced maps; besides,

$$
\tilde{i}^* \bar{D} = 2\Delta,
\tag{12}
$$

2) the image of $\widetilde{S \times S}$ under the composition $r \circ \tilde{i} : \widetilde{S \times S} \longrightarrow D$ coincides with Z_{red} as a set:

$$
(r \circ \tilde{i})(\widetilde{S \times S}) \stackrel{\text{sets}}{=} Z_{red}.
\tag{13}
$$

Now show that $r \circ \tilde{i}$ is the embedding. Let $S \xleftarrow{p_1} \widetilde{S \times S} \xrightarrow{p_2} S$ be the natural projections. If the map $r \circ \tilde{i}$ is not injective at the point $x \in \widetilde{S \times S}$, i.e. $(r \circ \tilde{i})(x) = (r \circ \tilde{i})(x')$ for $x' \neq x$, then by the construction of $r \circ \tilde{i}$ we have: $p_1(x) = p_1(x') =: s_1, \; s_2 := p_2(x) \neq p_2(x') =: s'_2, \; PT_{s_1} S \cap \langle i(s_2), i(s'_2) \rangle \neq \emptyset$.

However, the last is impossible for the general choice of L in $\mathbb{P}^{10}$ (simple dimension counting). Similarly one checks that the differential of the morphism $r \circ \tilde{i}$ is nondegenerate. Hence, $r \circ \tilde{i}$ is the embedding. From here and (13) the assertion of lemma follows, q.e.d.

Lemma 3. *Under the conditions of lemma 1 there exists a morphism* $\nu : X \longrightarrow \widetilde{Ch}$ *such that the following diagram is commutative*

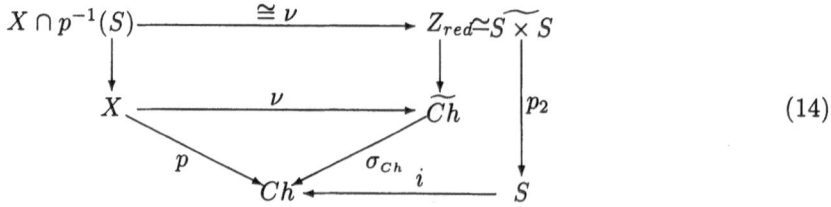

$$
\tag{14}
$$

Proof. From the results of [1] it follows that the following diagram is commutative, in which $\tilde{\Gamma} \subset \mathbb{P}^{N-1} \times G(1, N-1)$ is the flag variety with natural projections $\tilde{p}_r$ and $\tilde{q}_r$ and $i_{n,2} : H_2 \hookrightarrow G(1, N-1)$ is the natural embedding (see [4], lemma 1):

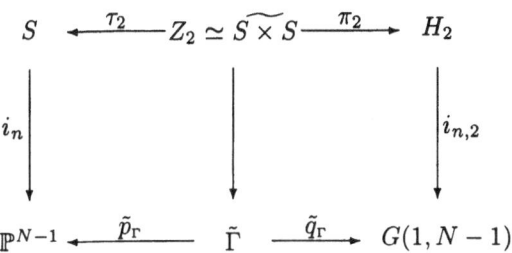

Denote $\Gamma_{H_2} = \tilde{q}_r^{-1} i_{n,2}(H_2)$ and let $\tilde{q} = \tilde{q}_r \mid_{\Gamma_{H_2}}$. Then from the diagram above and the definition of X it follows that $X = (pr_L \times id_{H_2})(\Gamma_{H_2})$, i.e. there is a commutative diagram

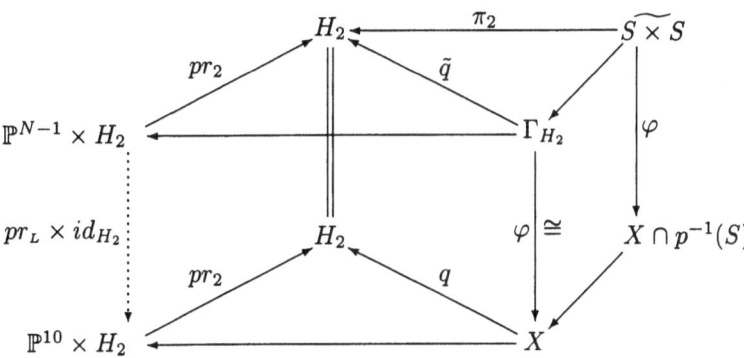

Thus, $X \cap p^{-1}(S) \simeq \widetilde{S \times S}$ is a Cartier divisor in X, so by the universal property of the blowing-up there exists a morphism $\nu : X \longrightarrow \widetilde{Ch}$, such that the diagram (14) is commutative, where the upper isomorphism $X \cap p^{-1}(S) \xrightarrow[\simeq]{\nu} Z_{red}$ follows from lemma 2, q.e.d.

Now we prove the main result of this section.

Proposition 1. *Under the conditions and notations of lemmas* $1 - 3$

1) $\nu : X \longrightarrow \widetilde{Ch}$ *is the normalization map and* $Z = Z_{red} \simeq \widetilde{S \times S}$,

2) the composition map $f : X \xrightarrow{\nu} \widetilde{Ch} \xrightarrow{i_p} P$ *is the immersion and* $\widetilde{Ch} = f(X)$ *has only a finite set* $\{x_1, \ldots, x_M\}$ *of isolated double points, such that for* $1 \le i \le M$

$$f^{-1}(x_i) = \{y_{i1}, y_{i2}\}, \qquad y_{i1} \ne y_{i2}; \qquad (*)$$

3) the cycle M_2 *of double points of* f *equals* $\sum_{i=1}^{M} (y_{i1} + y_{i2})$, *hence its degree* $\deg M_2 = 2M$ *is given by the double-point formula:*

$$2M = \int_X \left(f^* f_*[X] - c_5(f^* T_P / T_X) \right).$$

Proof. Take any chordal line l of S in $\mathbb{P}^{10}$ such that $\Xi = l \cap S = x_1 \cup x_2$, where $x_1 \neq x_2$. Let $\tilde{l} \subset \widetilde{Ch}$ be the proper inverse image of l under σ and $\tilde{x}_1 \cup \tilde{x}_2 = (\sigma \mid_{\tilde{l}})^{-1}(x_1 \cup x_2)$. Now one easily checks that $\tilde{l}$ intersects D transversally in P, i.e., in particular,

$$\dim \left(T_{\tilde{x}_1}(Z_{red}) + T_{\tilde{x}_1}\tilde{l} \right) = 5. \tag{15}$$

Next, let $\xi \in H_2$ be the point corresponding to Ξ. By the construction, for $l_\xi = q^{-1}(\xi) \subset X$ one has the isomorphism $\nu \mid_{l_\xi} : l_\xi \xrightarrow{\simeq} \tilde{l}$, hence, in particular, the isomorphism:

$$d\nu_* : T_{y_1} l_\xi \xrightarrow{\simeq} T_{\tilde{x}_1}\tilde{l}, \tag{16}$$

where $y_1 := (\nu \mid_{l_\xi})^{-1}(\tilde{x}_1)$. On the other hand, in view of lemma 3 there is the isomorphism

$$d\nu_* : T_{y_i} \widetilde{S \times S} \xrightarrow{\simeq} T_{\tilde{x}_i}(Z_{red}). \tag{17}$$

From (15)–(17) follows the isomorphism

$$d\nu_* : T_{y_1} X = T_{y_1} \widetilde{S \times S} + T_{y_1} l_\xi \xrightarrow{\simeq} T_{\tilde{x}_1}(Z_{red}) + T_{\tilde{x}_1}\tilde{l},$$

i.e. $d\nu_*$ is nondegenerate at y_1. Hence, $\widetilde{Ch}$ intersects D transversally at $\tilde{x}_1$. This means, in view of (4) and the irreducibility of Z_{red}, that $Z = Z_{red}$, i.e. $\widetilde{Ch} \cap D$ is scheme-theoretically isomorphic to $\widetilde{S \times S}$:

$$Z = \widetilde{Ch} \cap D \simeq \widetilde{S \times S}. \tag{18}$$

Whence, since $\widetilde{S \times S}$ is smooth and D is a smooth divisor in P, it follows that $\widetilde{Ch}$ is smooth along Z.

Now the property of $\nu \mid_{X - \widetilde{S \times S}} : X - \widetilde{S \times S} \longrightarrow \widetilde{Ch} - Z$ to be the immersion and the isomorphism outside the finite set $\{x_1, \ldots, x_M\}$ of points satisfying (*) is proved by the reproduction of the argument from the proof of lemma 3 of [4]. Moreover, the simple dimension counting shows that, for the general choice of L in $\mathbb{P}^{N-1}$, for $1 \leq i \leq M$ and each $y_{ij} \in f^{-1}(x_i)$, $1 \leq j \leq 2$,

(i) $\{u_{ij}, v_{ij}\} = p(q^{-1}q(y_{ij}) \cap \widetilde{S \times S})$ is the pair of distinct points, different from $x_{i0} = \sigma(x_i)$, and

(ii) $\langle \mathcal{P}T_{u_{i1}}S, \mathcal{P}T_{v_{i1}}S \rangle \cap \langle \mathcal{P}T_{u_{i2}}S, \mathcal{P}T_{v_{i2}}S \rangle = x_{i0}$.

Now by the construction ν is projective quasifinite morphism, hence it is finite ([3], III, ex. 11.2). Thus, since X is smooth and, hence, normal, and ν is birational, the morphism ν is the normalization morphism (this is one of the equivalent definitions of the normalization – see, e.g., [6], p. 160). This gives the proof of the assertions 1) and 2).

For 3) remark, that the equality $M_2 = \sum_{i=1}^{M} (y_{i1} + y_{i2})$ evidently outcomes from the equalities

$$dp_*(T_{y_{i1}}X) \cap dp_*(T_{y_{i2}}X) = \{0\}, \qquad 1 \le i \le M$$

(the intersections are taken in $T_{x_{i0}}\mathbb{P}^{10}$). But these are equivalent to the equalities (ii) above, since, in view of (i) above, one evidently has $T_{x_i}\langle PT_{u_{ij}}S, PT_{v_{ij}}S\rangle = dp_*(T_{y_{ij}}X)$, $1 \le i \le M$, $1 \le j \le 2$. The proposition is proved.

Corollary. *The polynomial $\delta_4(x, y, z, w)$ is given by the formula* (3).

2. Computation of the class $f_*[X]$ in $A_*(\mathbb{P})$

In this section we compute the first half $\int_X f^* f_*[X]$ of the formula (3). Since by the projection formula

$$\int_X f^* f_*[X] = \int_P (f_*[X])^2, \tag{19}$$

we have in fact to find the class $f_*[X] = [\widetilde{Ch}]$ in $A_*(P)$. For this purpose consider the diagram (5) and denote $\xi = c_1(\mathcal{O}_{D/S}(1))$, so that $g_*\mathcal{O}_D(\xi) = N^\vee_{S/\mathbb{P}^{10}}$. To compute $[\widetilde{Ch}]$, we use the following

Lemma 4.

$$[\widetilde{Ch}] = \sigma^*[Ch] - j_* \left(\frac{1}{\xi}[Z]\right) \qquad in \quad A_*(P).$$

Proof. In fact, according to [2], 6.7.(e),

$$[\widetilde{Ch}] = \sigma^* x + j_* y \tag{20}$$

for some $x \in A_*(\mathbb{P}^{10})$, $y \in A_*(D)$. Since $D = P(N_{S/\mathbb{P}^{10}})$, one has by [2], 3.3:

$$y = \sum_{i=0}^{2} \xi^{2+i} g^* a_i, \tag{21}$$

for some $a_i \in A_i(S)$, $0 \le i \le 2$. Thus, since evidently $g_* y = \sum_{i=0}^{2} a_i g_* \xi^{2+i} = 0$, one has:

$$\sigma_* j_* y = i_* g_* y = 0. \tag{22}$$

From (20), (22) and the equality $[Ch] = \sigma_*[\widetilde{Ch}]$ in view of the projection formula we have:

$$[Ch] = \sigma_* \sigma^* x + \sigma_* j_* y = x. \tag{23}$$

Next, $j^* \sigma^*[Ch] = g^* i^*[Ch] = 0$, since $i^*[Ch] = (\deg_{\mathbb{P}^{10}} Ch) i^*[\mathbb{P}^5] = 0$. Therefore, since by (18) $[Z] = [\widetilde{Ch} \cap D] = j^*[\widetilde{Ch}]$, from (20) we get in $A_*(D)$: $[Z] = j^* \sigma^*[Ch] + j^* j_* y = -\xi \cdot y$ (here we use [2], 2.6.(c) and the equality $c_1(N_{D/P}) = -c_1(\mathcal{O}_{D/S}(1)) = -\xi$), wherefrom $y = -\frac{1}{\xi}[Z]$. From here and (20) and (23) the assertion of lemma follows.

Now we proceed to the computation of the class $[Z]$ in $A_*(D)$. For this purpose we use the construction and notations from the proof of lemma 2. In particular, one easily sees that in the diagram (9)

$$\Pi = P(E_4),$$

where E_4 is a rank-4 vector bundle on D, fitting into the exact triple

$$0 \longrightarrow g^* J_1 T_S \longrightarrow E_4 \longrightarrow \mathcal{O}_D(-\xi) \longrightarrow 0. \tag{24}$$

Denote

$$
\begin{aligned}
H &= c_1\left(\mathcal{O}_{\mathbb{P}^{10}}(1) \boxtimes i^*\mathcal{O}_{\mathbb{P}^{10}}(-1)\right), & W_{z,S} &= s_2(\Omega_S), \\
\tilde{H} &= \sigma_Y^* H, & W_{w,S} &= K_S^2, \\
R_S &= c_1\left(i^*\mathcal{O}_{\mathbb{P}^{10}}(1)\right), & W_x &= g^* W_{x,S}, \\
K_S &= c_1(\Omega_S), & W_y &= g^* W_{y,S}, \\
R &= g^* R_S, & W_z &= g^* W_{z,S}, \\
K &= g^* K_S, & W_w &= g^* W_{w,S}, \\
\tilde{R} &= \sigma_Y^* pr_s^* R_S, & \tilde{W}_x &= \sigma_Y^* pr_s^* W_{x,S}, \\
\tilde{K} &= \sigma_Y^* pr_s^* K_S, & \tilde{W}_y &= \sigma_Y^* pr_s^* W_{y,S}, \\
W_{x,S} &= R_S^2, & \tilde{W}_z &= \sigma_Y^* pr_s^* W_{z,S}, \\
W_{y,S} &= R_S \cdot K_S, & \tilde{W}_w &= \sigma_Y^* pr_s^* W_{w,S}. \tag{25}
\end{aligned}
$$

Here

$$\int_S W_{x,S} = x, \quad \int_S W_{y,S} = y, \quad \int_S W_{z,S} = z, \quad \int_S W_{w,S} = w. \tag{26}$$

Now, since by (6) $s_i(J_1 T_S) = s_i(T_S)$, one easily finds from (24):

$$r_* \tilde{H}^{k+3} = s_k(E_4) = \xi^k + \xi^{k-1} K + \xi^{k-2} W_z, \quad k \geq 0 \tag{27}$$

(here $\xi^{k-1} \equiv \xi^{k-2} \equiv 0$ for $k \leq 0$).

Next, consider the embedding $i \times id_s : S \times S \hookrightarrow \mathbb{P}^{10} \times S$. Then evidently in $A_*(\mathbb{P}^{10} \times S)$:

$$[S \times S] = x c_1 \left(\mathcal{O}_{\mathbb{P}^{10}}(1) \boxtimes \mathcal{O}_S\right)^8 = x(H + pr_s^* R_S)^8. \tag{28}$$

Let $\mathbb{P}^{10} = P(V)$, $\dim_\mathbb{C} V = 11$. The vector bundle $J_1 \Omega_S = (J_1 T_S)^\vee$ naturally fits into the exact triple

$$0 \longrightarrow N_{S/\mathbb{P}^{10}}^\vee \longrightarrow V^\vee \otimes \mathcal{O}_S(-1) \longrightarrow J_1 \Omega_S \longrightarrow 0. \tag{29}$$

Next, the triple (6) naturally induces the triple

$$0 \longrightarrow S^2 \Omega_S \longrightarrow S^2(J_1 \Omega_S) \longrightarrow J_1 \Omega_S \longrightarrow 0. \tag{30}$$

In view of the assertion 2 of lemma 1 there is defined an epimorphism $N^\vee_{S/\mathbf{P}^{10}} \xrightarrow{\varepsilon} S^2\Omega_S \longrightarrow 0$ such that the following diagram, including (29) and (30), is commutative:

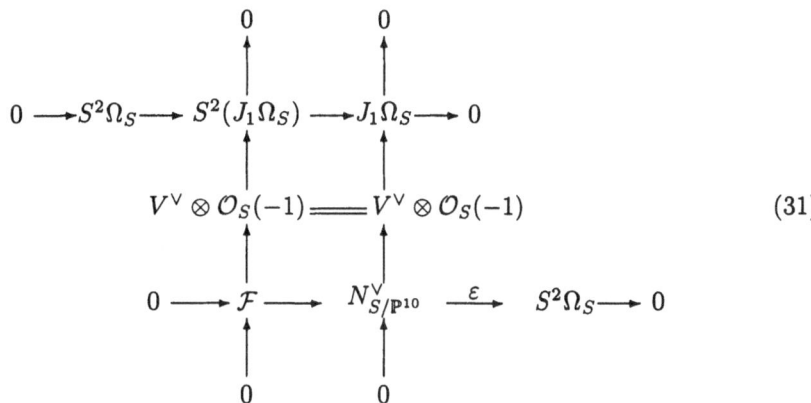

$$(31)$$

where $\mathcal{F} = \mathrm{Ker}(\varepsilon)$. The epimorphism ε defines the embedding

$$P(S^2T_S) \xhookrightarrow{i_1} P(N_{S/\mathbf{P}^{10}}) = D$$

such that

$$i_1\left(P(S^2T_S)\right) = (s)_0, \quad s \in H^0\left(g^*\mathcal{F}^\vee(\xi)\right). \tag{32}$$

Next, the (skew-symmetric) isomorphism $\Omega_S \xrightarrow{\Lambda} \Omega^\vee_S(K_S)$ defines the (symmetric) isomorphism $S^2\Omega_S \xrightarrow{\simeq} S^2\Omega^\vee_S(2K_S)$, i.e., equivalently, the section φ : $\mathcal{O}_S \longrightarrow S^2(S^2\Omega_S)(-2K_S)$, which lifts to the section

$$s_\varphi : \mathcal{O}_{P(S^2T_S)} \longrightarrow \mathcal{O}_{P(S^2T_S)}(2) \otimes \rho_2^*\mathcal{O}_S(-2K_S), \tag{33}$$

where $\rho_2 : P(S^2T_S) \longrightarrow S$ is the structure morphism, and one easily finds that

$$(s_\varphi)_0 \simeq P(T_S) \simeq P(N_{S_\Delta/S\times S}) = \Delta \tag{34}$$

(see diagram (11) for notations). Thus we have a diagram of natural maps

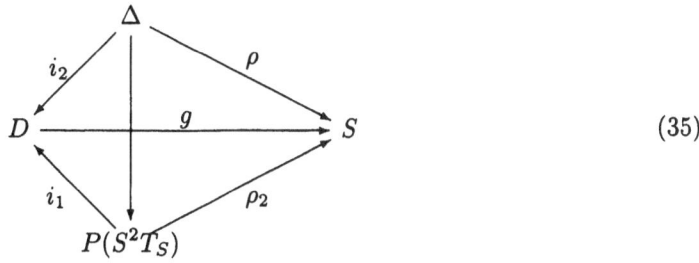

$$(35)$$

Since $\mathcal{O}_{P(S^2 T_S)}(1) = i_1^* \mathcal{O}_D(\xi)$, from (32) and (33) we find in $A_*(D)$:

$$[i_2(\Delta)] = c_5(g^* \mathcal{F}^\vee(\xi)) c_1(\mathcal{O}_D(2\xi - 2K)). \tag{36}$$

Next, the bottom triple in (31) gives:

$$c_t(\mathcal{F}^\vee) = c_t(V \otimes \mathcal{O}_S(1)) s_t(T_S) s_t(S^2 T_S), \tag{37}$$

where

$$s_t(T_S) = 1 + K_S t + W_{z,S} t^2. \tag{38}$$

From (36)–(38) we get in $A_*(D)$:

$$[i_2(\Delta)] = (2\xi - 2K) \left(\xi^5 + \xi^4(11R + 4K) + \xi^3(55W_x + 44W_y + 5W_z + 6W_w) \right). \tag{39}$$

Now consider the constructions and notations introduced in the proof of lemma 2. First, for the embedding $S_\Delta \simeq S \overset{s_0}{\hookrightarrow} Y$ one easily obtains: $s_0(S_\Delta) = (s)_0$, where $s \in H^0(\theta^* T_S \otimes \mathcal{O}_Y(h))$, $h = y^* H$, so that in view of (38) we have in $A_*(Y)$:

$$[s_0(S_\Delta)] = h^2 - h\theta^* K_S + \theta^*(W_{w,S} - W_{z,S}) = c_2(\theta^* T_S(h)). \tag{40}$$

Now consider the embedding

$$\bar{i} : \Delta \hookrightarrow \bar{D} : z \longmapsto (i_2(z), (s_0 \circ \rho)(z)) \tag{41}$$

(we use here the identification (10)), so that in view of (36) and (40)

$$[\bar{i}(\Delta)] = \bar{\theta}^* \left(c_5(g^* \mathcal{F}^\vee(\xi)) c_1(\mathcal{O}_D(2\xi - 2K)) \right) \bar{g}^* c_2(\theta^* T_S(h)).$$

Whence by (39) and (40) we get in $A_*(\bar{D})$:

$$[\bar{i}(\Delta)] = \bar{j}^* A,$$

$$A = 2\tilde{\xi}^6(\tilde{H}^2 - \tilde{H}\tilde{K} + \tilde{W}_w - \tilde{W}_z) + \tilde{\xi}^5[(22\tilde{R} + 6\tilde{K})\tilde{H}^2 - 22\tilde{H}\tilde{W}_y - \tag{42}$$

$$-6\tilde{H}\tilde{W}_w] + \tilde{\xi}^4(110\tilde{W}_x + 66\tilde{W}_y + 10\tilde{W}_z + 4\tilde{W}_w)\tilde{H}^2,$$

where $\tilde{\xi} = r^* \xi$.

Now from the assertion 2 of lemma 1 and the diagrams (9) and (11) we find that the embedding $\bar{i}$ in (41) is included into the commutative diagram

$$
\begin{array}{ccccc}
\Delta & \longrightarrow & \widetilde{S \times S} & \overset{\sigma_\Delta}{\longrightarrow} & S \times S \\
{\scriptstyle \bar{i}} \downarrow & & {\scriptstyle \tilde{i}} \downarrow & & \downarrow {\scriptstyle i \times \mathrm{id}_s} \\
\bar{D} & \overset{\bar{j}}{\longrightarrow} & \Pi & \overset{\sigma_Y}{\longrightarrow} & \mathbb{P}^{10} \times S
\end{array} \tag{43}
$$

hence from (12) we have in $A_*(\Pi)$:

$$[\tilde{i}(\widetilde{S \times S}) \cap \bar{D}] = 2[\tilde{i}(\Delta)].$$

Hence, according to lemma 4 we have from the diagram (43)

$$[\tilde{i}(\widetilde{S \times S})] = \sigma_Y^*[(i \times id_s)(S \times S)] + \bar{j}_* \left(\frac{1}{c_1(N_{\bar{D}/\Pi})} [\tilde{i}(\widetilde{S \times S}) \cap \bar{D}] \right) =$$
(44)
$$= \sigma_Y^*[(i \times id_s)(S \times S)] + 2\bar{j}_* \left(\frac{[\tilde{i}(\Delta)]}{c_1(N_{\bar{D}/\Pi})} \right).$$

Next, from (24) and (25) we find: $\mathcal{O}_\Pi(\bar{D}) = \mathcal{O}_\Pi(\tilde{H} - \tilde{\xi})$, wherefrom

$$c_1(N_{\bar{D}/\Pi}) = \bar{j}^*(\tilde{H} - \tilde{\xi}), \quad c_1\left(\mathcal{O}_\Pi(\bar{D})\right) = \tilde{H} - \tilde{\xi}.$$
(45)

Now, let B be obtained from A in (42) by the substitution $\tilde{\xi} \rightsquigarrow \tilde{H}$:

$$B = 2\tilde{H}^8 + (22\tilde{R} + 4\tilde{K})\tilde{H}^7 + (110\tilde{W}_x + 44\tilde{W}_y + 8\tilde{W}_z)\tilde{H}^6.$$
(46)

Since B is divisible by $\tilde{H}^6$ and evidently $\bar{j}^*\tilde{H}^6 = \bar{g}^*i^*H^6 = 0$, we have

$$\bar{j}^*B = 0.$$
(47)

Now by Bezout theorem $A - B$ is divisible by $\tilde{H} - \tilde{\xi}$. Denoting the corresponding quotient by Q, we get:

$$A - B = (\tilde{H} - \tilde{\xi})Q,$$
(48)

wherefrom by (42), (45) and (47)

$$[\tilde{i}(\Delta)] = \bar{j}^*[(\tilde{H} - \tilde{\xi})Q] = \bar{j}^*Q \cdot c_1(N_{\bar{D}/\Pi}).$$

Whence by (48) and [2], 2.6,

$$\bar{j}_* \left(\frac{[\tilde{i}(\Delta)]}{c_1(N_{\bar{D}/\Pi})} \right) = \bar{j}_*\bar{j}^*Q = (\tilde{H} - \tilde{\xi})Q = A - B.$$
(49)

Next, from (28)

$$\sigma_Y^*[(i \times id_s)(S \times S)] = x(\tilde{H} + \tilde{R})^8 = x(\tilde{H}^8 + 8\tilde{H}^7\tilde{R} + 28\tilde{H}^6\tilde{W}_x).$$
(50)

Besides, from (27) and (42) it follows that

$$r_*A = 0.$$
(51)

Next, remark, that in view of (12) and (18) one has in $A_*(D)$: $r_*[\tilde{i}(\widetilde{S \times S})] = [Z]$. From here and (44), (49)–(51) we get:

$$[Z] = r_* \left(x(\tilde{H} + \tilde{R})^8 - 2B \right) = \alpha\xi^5 + \beta\xi^4 + \gamma\xi^3,$$
(52)

where

$$\alpha = (x - 4)[D],$$

$$\beta = (8x - 44)R + (x - 12)K,$$ (53)

$$\gamma = (28x - 220)W_x + (8x - 132)W_y + (x - 20)W_z - 8W_w.$$

From (52) and lemma 4 we now obtain:

$$[\widetilde{Ch}] = \sigma^*[Ch] - j_*(\alpha\xi^4 + \beta\xi^3 + \gamma\xi^2).$$

Whence according to [2], 6.7, the self-intersection formula for $[\widetilde{Ch}]$ in $A_*(P)$ gives:

$$[\widetilde{Ch}]^2 = \sigma^*[Ch]^2 - j_* \left(\xi(\alpha\xi^4 + \beta\xi^3 + \gamma\xi^2)^2 + 2(\alpha\xi^4 + \beta\xi^3 + \gamma\xi^2)g^*i^*[Ch] \right),$$

i.e.

$$[\widetilde{Ch}]^2 = \sigma^*[Ch]^2 - j_* \left(\alpha^2\xi^9 + 2\alpha\beta\xi^8 + (2\alpha\gamma + \beta^2)\xi^7 \right),$$ (54)

since evidently $i^*[Ch] = 0$ and $j_*\xi^k = 0$ for $k \leq 6$.

Next, since $s_t(N^\vee_{S/\mathbb{P}^{10}}) = s_t(\Omega_{\mathbb{P}^{10}} \big|_S)c_t(N_S) =$
$= (1 + Kst + (W_{w,S} - W_{z,S})t^2)/(1 - Rst)^{11}$, we easily find, using (26):

$$\int_D \xi^9 = \int_S s_2(N^\vee_{S/\mathbb{P}^{10}}) = 66x + 11y - z + w,$$

$$\int_D R\xi^8 = -11x - y, \quad \int_D K\xi^8 = -11y - w,$$ (55)

$$\int_D W_x\xi^7 = x, \quad \int_D W_y\xi^7 = y, \quad \int_D W_z\xi^7 = z, \quad \int_D W_w\xi^7 = w.$$

Besides, by the projection formula:

$$\int_P j_* \left(\alpha^2\xi^9 + 2\alpha\beta\xi^8 + (2\alpha\gamma + \beta^2)\xi^7 \right) =$$

$$= \int_D \left(\alpha^2\xi^9 + 2\alpha\beta\xi^8 + (2\alpha\gamma + \beta^2)\xi^7 \right).$$ (56)

On the other hand, by the double-point formula for the projection of S into $\mathbb{P}^4$,

$$\int_P \sigma^*[Ch]^2 = \int_{\mathbb{P}^{10}} [Ch]^2 = (\deg_{\mathbb{P}^{10}} Ch)^2 = \left(\frac{1}{2}(x^2 - 10x - 5y - z) \right)^2.$$ (57)

Substituting (53) into (56) and using (55), we get from (54), (56), (57) and (19) the following result

Proposition 2.

$$\int_X f^* f_*[X] = \int_P [\widetilde{Ch}]^2 = \frac{1}{4}(x^4 - 60x^3 - 30x^2 y - 6x^2 z +$$

$$+996x^2 + 868xy + 180xz + 64xw + 25y^2 +$$

$$+10yz + z^2 - 3520x - 3520y - 576z - 512w).$$

3. Chern polynomial of the normal sheaf of the immersion f

In this section we compute the second half of formula (3). For this we need the Chern polynomial $c_t(f^* T_P / T_X) = c_t(f^* T_P) s_t(T_X)$ of the normal sheaf $f^* T_P / T_X$ of the immersion $f : X \longrightarrow P$. First, consider the diagram (see (5) and proposition 1)

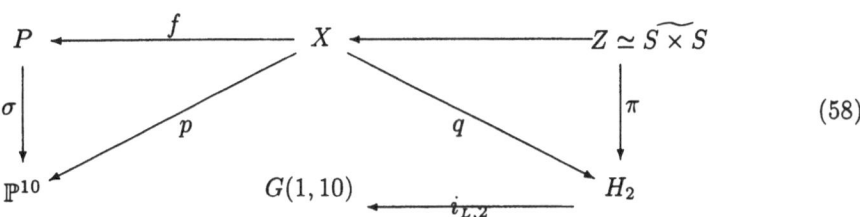

(58)

where $\pi : Z \longrightarrow H_2$ is the double covering ramified in the divisor $\Delta \simeq P(T_S) \subset Z$, $\bar{X} = X \underset{H_2}{\times} Z$. The diagram (58) gives the diagram

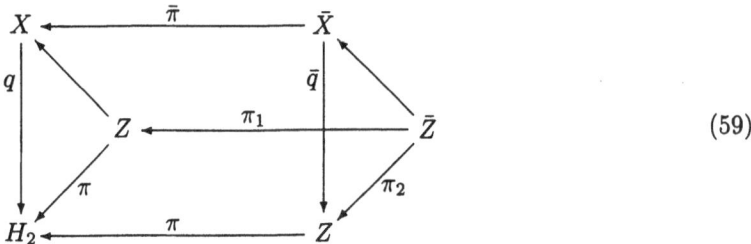

(59)

where

$$\bar{Z} = Z \underset{H_2}{\times} Z = Z_1 \cup Z_2 \tag{60}$$

and Z_1, Z_2 are the two copies of Z, such that there are isomorphisms $\pi_j : Z_i \xrightarrow{\simeq} Z$, $i, j = 1, 2$; moreover, one can identify Z_1, Z_2 in such a way that, say,

$$\pi_1 \left(\pi_2 \big|_{Z_1} \right)^{-1} = id_z, \quad \pi_1 \left(\pi_2 \big|_{Z_2} \right)^{-1} = i_z, \tag{61}$$

where $i_z : Z \xrightarrow{\;\simeq\;} Z$ is the involution on $Z \simeq \widetilde{S \times S}$ induced by the permutation of factors.

Next, consider the diagram of natural projections (see (11))

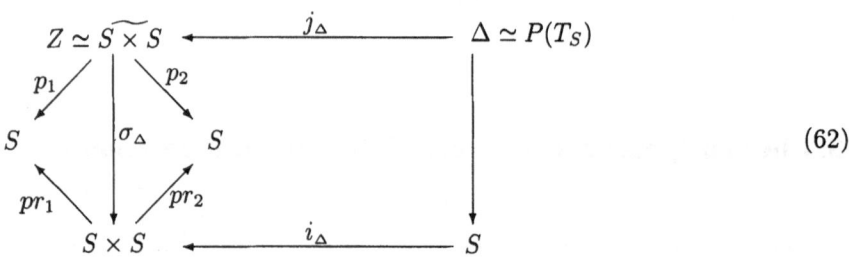

$$\tag{62}$$

and denote

$$
\begin{aligned}
D_i &= p_i^* R_S, & R_x &= c_1 \left(p^* \mathcal{O}_{\mathbb{P}^{10}}(1) \right), \\
W_{i,z} &= p_i^* W_{z,S}, & \bar{R} &= \bar{\pi}^* R_x, \\
W_{i,w} &= p_i^* W_{w,S}, \quad i = 1, 2, & \bar{D}_i &= \bar{q}^* D_i, \\
H_0 &= i_{L,2}^* c_1 \left(\mathcal{O}_{G(1,10)}(1) \right), & \bar{K}_i &= \bar{q}^* p_i^* K_S, \\
H_z &= \pi^* H_0, & \bar{W}_{i,z} &= \bar{q}^* W_{i,z}, \\
\bar{H} &= \bar{q}^* H_z, & \bar{W}_{i,w} &= \bar{q}^* W_{i,w}, \quad i = 1, 2, \\
\bar{\Delta} &= \bar{q}^* \Delta, & Q &= i_{L,2}^* Q_G, \tag{63}
\end{aligned}
$$

where Q_G is the antitautological rank-2 vector bundle on $G(1, 10)$. Since $\pi : Z \longrightarrow H_2$ is the double covering ramified in Δ, one easily checks that $\pi^* \Delta = 2\bar{\Delta}$ and $\mathcal{O}_X(Z) = \mathcal{O}_X \left(2R_x + q^* (\frac{\Delta}{2} - H_0) \right)$, hence in Pic $\bar{X}$ we have:

$$\bar{Z} = 2\bar{R} + \bar{\Delta} - \bar{H}. \tag{64}$$

Next, one has the exact triple on Z:

$$0 \longrightarrow \mathcal{O}_Z(D_2 - \Delta) \longrightarrow q^* Q \big|_Z \longrightarrow \mathcal{O}_Z(D_1) \longrightarrow 0 \tag{65}$$

and the equalities $q^* Q \big|_Z = \pi^* Q$, $c_1(Q) = H_0$, from where in view of (63)

$$H_z = D_1 + D_2 - \Delta, \quad \bar{H} = \bar{D}_1 + \bar{D}_2 - \bar{\Delta}. \tag{66}$$

Besides, (65) defines the embedding $Z_1 \xrightarrow{\;\pi_1\;}_{\simeq} Z \simeq P(\mathcal{O}_Z(-D_1)) \hookrightarrow P\left(q^* Q^\vee \big|_Z \right) = \bar{X}$, coinciding in view of (60) and (61) with the natural embedding $Z_1 \hookrightarrow \bar{X}$; thus, since $\mathcal{O}_{\bar{X}}(\bar{R}) = \mathcal{O}_{P(q^* Q^\vee |_Z)}(1)$, we get: $Z_1 = (s)_0$, $s \in H^0 \left(\mathcal{O}_{\bar{X}}(\bar{R} + \bar{\Delta} - \bar{D}_2) \right)$, wherefrom, in Pic $\bar{X}$

$$Z_1 = \bar{R} + \bar{\Delta} - \bar{D}_2. \tag{67}$$

From (60), (64), (66) and (67) we also get in Pic $\bar{X}$:

$$Z_2 = \bar{R} + \bar{\Delta} - \bar{D}_1. \tag{68}$$

Now consider the diagram (see (5))

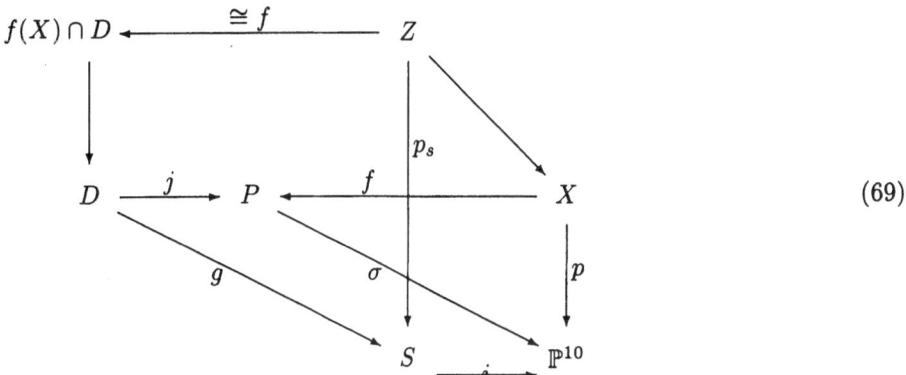

$$\tag{69}$$

Lemma 5.
$$c_t(f^*T_P) = c_t\big((p^*T_{\mathbb{P}^{10}})(-Z)\big) c_t(p_s^*T_S) c_t\big(\mathcal{O}_X(Z)\big). \tag{70}$$

Proof. From the diagram (69) one immediately has:

$$\begin{aligned}
f^*\mathcal{O}_P(D) &= \mathcal{O}_X(Z), \\
f^*g^*T_S &= p_s^*T_S, \\
f^*g^*i^*T_{\mathbb{P}^{10}} &= p_s^*i^*T_{\mathbb{P}^{10}} = p^*T_{\mathbb{P}^{10}}\big|_Z\,.
\end{aligned} \tag{71}$$

Next, for the blowing-up $\sigma : P \longrightarrow \mathbb{P}^{10}$ the following triples are exact (see [2], lemma 15.4):

$$0 \longrightarrow T_P \longrightarrow \sigma^*T_{\mathbb{P}^{10}} \longrightarrow j_*F \longrightarrow 0, \tag{72}$$

$$0 \longrightarrow \mathcal{O}_{D/S}(-1) \longrightarrow g^*N_{S/\mathbb{P}^{10}} \longrightarrow F \longrightarrow 0, \tag{73}$$

where $\mathcal{O}_{D/S}(-1) = \mathcal{O}_P(D)\big|_D$. Besides, there are exact triples

$$0 \longrightarrow \mathcal{O}_P \longrightarrow \mathcal{O}_P(D) \longrightarrow \mathcal{O}_{D/S}(-1) \longrightarrow 0, \tag{74}$$

$$0 \longrightarrow T_S \longrightarrow i^*T_{\mathbb{P}^{10}} \longrightarrow N_{S/\mathbb{P}^{10}} \longrightarrow 0.$$

From (73) and (74) we find $c_t(F) = c_t(g^*i^*T_{\mathbb{P}^{10}})/c_t\big(\mathcal{O}_P(D)\big) c_t(g^*T_S)$, hence by (71)

$$s_t(f^*j_*F) = c_t(p_s^*T_S) c_t\big(\mathcal{O}_X(Z)\big)/c_t(p_s^*i^*T_{\mathbb{P}^{10}}). \tag{75}$$

Now applying to (72) the functor f^* and taking into account the equality $f^*\sigma^*T_{\mathbf{P}^{10}} = p^*T_{\mathbf{P}^{10}}$ and the exact triple

$$0 \longrightarrow p^*T_{\mathbf{P}^{10}}(-Z) \longrightarrow p^*T_{\mathbf{P}^{10}} \longrightarrow p_s^* i^* T_{\mathbf{P}^{10}} \longrightarrow 0,$$

outcoming from (69) and (71), we get:

$$c_t(f^*T_P) = c_t\big((p^*T_{\mathbf{P}^{10}})(-Z)\big)c_t(p_s^* i^* T_{\mathbf{P}^{10}})s_t(f^* j_* F). \tag{76}$$

The equality (70) now follows from (75) and (76), q.e.d.

Now from diagrams (59), (62), (69) and isomorphisms (61) we have:

$$(\pi_1|_{Z_1})^* p_s^* T_S = \bar{q}^* p_1^* T_S |_{Z_1}, \qquad (\pi_1|_{Z_2})^* p_s^* T_S = \bar{q}^* p_2^* T_S |_{Z_2}, \tag{77}$$

$$Z_1 \cap Z_2 = (\bar{q}|_{Z_2})^{-1}(\Delta), \qquad \mathcal{O}_{\bar{X}}(Z_1)|_{Z_2} = \mathcal{O}_{Z_2}(\bar{\Delta}), \tag{78}$$

$$\mathcal{O}_{\bar{X}}(\bar{R})|_{Z_1} = \mathcal{O}_{\bar{X}}(\bar{D}_1)|_{Z_1}, \qquad \mathcal{O}_{\bar{X}}(\bar{R})|_{Z_2} = \mathcal{O}_{\bar{X}}(\bar{D}_2)|_{Z_2}. \tag{79}$$

From (78), (59) and (60) it follows that the exact triple

$$0 \longrightarrow \mathcal{O}_{\bar{X}}(-Z_1)|_{Z_2} \longrightarrow \mathcal{O}_{\bar{Z}} \longrightarrow \mathcal{O}_{Z_1} \longrightarrow 0$$

coincides with the triple

$$0 \longrightarrow (\pi_1|_{Z_2})^* \mathcal{O}_Z(-\Delta) \longrightarrow \pi_1^* \mathcal{O}_Z \longrightarrow (\pi_1|_{Z_1})^* \mathcal{O}_Z \longrightarrow 0. \tag{80}$$

Applying now the functor π_1^* to $p_s^* T_S$, from (77) and (80) we get the exact triple

$$0 \longrightarrow (\bar{q}^* p_2^* T_S |_{Z_2})(-\bar{\Delta}) \longrightarrow \pi_1^* p_s^* T_S \longrightarrow \bar{q}^* p_1^* T_S |_{Z_1} \longrightarrow 0.$$

From here and evident exact triples:

$$0 \longrightarrow \bar{q}^* p_1^* T_S(-Z_1) \longrightarrow \bar{q}^* p_1^* T_S \longrightarrow \bar{q}^* p_1^* T_S |_{Z_1} \longrightarrow 0,$$

$$0 \longrightarrow \bar{q}^* p_2^* T_S(-\bar{\Delta} - Z_2) \longrightarrow \bar{q}^* p_2^* T_S(-\bar{\Delta}) \longrightarrow (\bar{q}^* p_2^* T_S |_{Z_2})(-\bar{\Delta}) \longrightarrow 0,$$

we find:

$$\pi_1^* c_t(p_s^* T_S) = \frac{c_t(\bar{q}^* p_1^* T_S) c_t\big((\bar{q}^* p_2^* T_S)(-\bar{\Delta})\big)}{c_t\big(\bar{q}^* p_1^* T_S(-Z_1)\big) c_t\big(\bar{q}^* p_2^* T_S(-Z_2 - \bar{\Delta})\big)}. \tag{81}$$

Now we proceed to the computation of $\bar{\pi}^* c_t(T_X)$. From the exact triple

$$0 \longrightarrow T_{X/H_2} \longrightarrow T_X \longrightarrow q^* T_{H_2} \longrightarrow 0$$

and the equality $T_{X/H_2} = \mathcal{O}_X(2R_x - q^* H_0)$ we obtain:

$$\bar{\pi}^* c_t(T_X) = \bar{\pi}^* c_t(q^* T_{H_2})\big(1 + (2R_x - q^* H_0)t\big). \tag{82}$$

Now find $\bar{\pi}^* c_t(q^* T_{H_2})$. For this, remark that:

$$j_{\Delta *} \pi^* N_{\Delta/H_2} = j_{\Delta *} \pi^* \big(\mathcal{O}_{H_2}(\Delta)|_{\Delta}\big) = j_{\Delta *}\big(\mathcal{O}_Z(2\Delta)|_{\Delta}\big), \tag{83}$$

and the following triple is exact:

$$0 \longrightarrow T_Z \longrightarrow \pi^* T_{H_2} \longrightarrow j_{\Delta *} \pi^* N_{\Delta / H_2} \longrightarrow 0. \tag{84}$$

Next, similar to (72) and (73) we obtain from the diagram (62) the exact triples

$$0 \longrightarrow T_Z \longrightarrow \sigma_\Delta^* T_{S \times S} \longrightarrow j_{\Delta *} F \longrightarrow 0,$$

$$0 \longrightarrow \mathcal{O}_{\Delta / S}(-1) \longrightarrow \rho^* T_S \longrightarrow F \longrightarrow 0, \tag{85}$$

where $\mathcal{O}_{\Delta / S}(-1) = \mathcal{O}_Z(\Delta) |_\Delta$, and

$$0 \longrightarrow \mathcal{O}_Z \longrightarrow \mathcal{O}_Z(\Delta) \longrightarrow \mathcal{O}_{\Delta / S}(-1) \longrightarrow 0,$$

$$0 \longrightarrow \mathcal{O}_Z(\Delta) \longrightarrow \mathcal{O}_Z(2\Delta) \longrightarrow \mathcal{O}_Z(2\Delta) |_\Delta \longrightarrow 0. \tag{86}$$

From (83)–(86) it follows that

$$\pi^* c_t (T_{H_2}) = \frac{\sigma_\Delta^* c_t (T_{S \times S}) c_t \left(\mathcal{O}_Z(2\Delta) |_\Delta \right)}{c_t \left(j_{\Delta *} F \right)} = \frac{\sigma_\Delta^* c_t (T_{S \times S}) c_t \left(\mathcal{O}_Z(2\Delta) \right)}{c_t \left(j_{\Delta *} \rho^* T_S \right)}. \tag{87}$$

Next, in view of (62) $j_{\Delta *} \rho^* T_S = p_2^* T_S |_\Delta$, so that there is the exact triple:

$$0 \longrightarrow p_2^* T_S(-\Delta) \longrightarrow p_2^* T_S \longrightarrow j_{\Delta *} \rho^* T_S \longrightarrow 0 \tag{88}$$

and, moreover, the equality:

$$\sigma_\Delta^* c_t (T_{S \times S}) = c_t(p_1^* T_S) c_t(p_2^* T_S). \tag{89}$$

From (87)–(89) we have

$$\pi^* c_t (T_{H_2}) = c_t(p_1^* T_S) c_t \left(p_2^* T_S(-\Delta) \right) c_t \left(\mathcal{O}_Z(2\Delta) \right),$$

wherefrom in view of the diagram (59):

$$\bar{\pi}^* c_t (q^* T_{H_2}) = c_t \left(\bar{q}^* \pi^* T_{H_2} \right) = c_t(\bar{q}^* p_1^* T_S) c_t \left(\bar{q}^* p_2^* T_S(-\bar{\Delta}) \right) c_t \left(\mathcal{O}_{\bar{X}}(2\bar{\Delta}) \right).$$

From here and (81) and (70), using the Euler exact triple

$$0 \longrightarrow \mathcal{O}_{\bar{X}}(-\bar{Z}) \longrightarrow V_{11} \otimes \mathcal{O}_{\bar{X}}(\bar{R} - \bar{Z}) \longrightarrow \bar{\pi}^* \left(p^* T_{\mathbf{P}^{10}}(-Z) \right) \longrightarrow 0$$

we find:

$$\bar{\pi}^* c_t (f^* T_P / T_X) = \frac{\bar{\pi}^* c_t(f^* T_P)}{\bar{\pi}^* c_t(T_X)} =$$

$$= \frac{c_t \left(\bar{\pi}^* p^* T_{\mathbf{P}^{10}}(-Z) \right) \bar{\pi}^* c_t(p_s^* T_S) c_t \left(\mathcal{O}_{\bar{X}}(\bar{Z}) \right)}{\left(1 + (2\bar{R} - \bar{H})t \right) c_t(\bar{q}^* p_1^* T_S) c_t \left(\bar{q}^* p_2^* T_S(-\bar{\Delta}) \right) c_t \left(\mathcal{O}_{\bar{X}}(2\bar{\Delta}) \right)} =$$

$$= \frac{\left(1 + (\bar{R} - \bar{Z})t\right)^{11}(1 + \bar{Z}t)}{\left(1 + (2\bar{R} - \bar{H})t\right)(1 - \bar{Z}t)c_t\left(\bar{q}^*p_1^*T_S(-Z_1)\right)c_t\left(\bar{q}^*p_2^*T_S(-Z_2 - \bar{\Delta})\right)(1 + 2\bar{\Delta}t)}.$$

Substituting here the formulas:

$$c_t\left(\bar{q}^*p_1^*T_S(-Z_1)\right) = 1 - (2Z_1 + \bar{K}_1)t + (\bar{W}_{1,w} - \bar{W}_{1,z} + \bar{K}_1 Z_1 + Z_1^2)t^2,$$

$$c_t\left(\bar{q}^*p_2^*T_S(-Z_2 - \bar{\Delta})\right) =$$

$$= 1 - (2Z_2 + \bar{K}_2 + 2\bar{\Delta})t + \left(\bar{W}_{2,w} - \bar{W}_{2,z} + \bar{K}_2(Z_2 + \bar{\Delta}) + (Z_2 + \bar{\Delta})^2\right)t^2,$$

and using (64), (66) and (68), we get

$$\bar{\pi}^* c_t\left(f^*T_P/T_X\right) = \left(1 + (\bar{D}_1 + \bar{D}_2 - 2\bar{\Delta} - \bar{R})t\right)^{11}\left(1 + (2\bar{R} + 2\bar{\Delta} - \bar{D}_1 - \bar{D}_2)t\right) \times$$

$$\times \frac{1}{\left(1 + (2\bar{R} + \bar{\Delta} - \bar{D}_1 - \bar{D}_2)t\right)\left(1 + (\bar{D}_1 + \bar{D}_2 - 2\bar{\Delta} - 2\bar{R})t\right)(1 + 2\bar{\Delta}t)} \times$$

$$\times \frac{1}{1 - (2\bar{R} + 2\bar{\Delta} - 2\bar{D}_2 + \bar{K}_1)t + \left(\bar{W}_{1,w} - \bar{W}_{1,z} + \bar{K}_1(\bar{R} + \bar{\Delta} - \bar{D}_2) + (\bar{R} + \bar{\Delta} - \bar{D}_2)^2\right)t^2} \times$$

$$\times \frac{1}{1 - (2\bar{R} + 4\bar{\Delta} - 2\bar{D}_1 + \bar{K}_2)t + \left(\bar{W}_{2,w} - \bar{W}_{2,z} + \bar{K}_2(\bar{R} + 2\bar{\Delta} - \bar{D}_1) + (\bar{R} + 2\bar{\Delta} - \bar{D}_1)^2\right)t^2}.$$

$$(90)$$

Expanding (90) as a t-series we find the polynomial

$$\bar{\pi}^* c_5\left(f^*T_P/T_X\right) \in \mathbb{Z}[\bar{R}, \bar{D}_1, \bar{D}_2, \bar{K}_1, \bar{K}_2, \bar{W}_{1,z}, \bar{W}_{2,z}, \bar{W}_{1,w}, \bar{W}_{2,w}, \bar{\Delta}]. \qquad (91)$$

Remark, that, since $\bar{\pi}$ is the double covering, one has

$$\int_X c_5\left(f^*T_P/T_X\right) = \frac{1}{2}\int_{\bar{X}} \bar{\pi}^* c_5\left(f^*T_P/T_X\right). \qquad (92)$$

Now taking (26) into account we obtain the following table of intersection indices in $\bar{X}$ (we omit the symbol $\int_{\bar{X}}$ here):

$$\bar{R}\bar{D}_1^2\bar{D}_2^2 = x^2, \qquad\qquad\qquad \bar{R}\bar{K}_1^2\bar{D}_2\bar{\Delta} = \bar{R}\bar{K}_1^2\bar{K}_2\bar{\Delta} = 0,$$

$$\bar{R}\bar{D}_1^2\bar{D}_2\bar{K}_2 = \bar{R}\bar{D}_2^2\bar{D}_1\bar{K}_1 = xy, \qquad\qquad \bar{R}\bar{K}_1^2\bar{\Delta}^2 = -w,$$

$$\bar{R}\bar{D}_1^2\bar{K}_2^2 = xw, \qquad\qquad\qquad \bar{R}\bar{K}_1\bar{D}_2^2\bar{\Delta} = 0,$$

$$\bar{R}\bar{D}_1^2(\bar{W}_{2,w} - \bar{W}_{2,z}) = x(w - z), \qquad \bar{R}(\bar{W}_{1,w} - \bar{W}_{1,z})\bar{D}_2^2 = x(w - z),$$

$$\bar{R}\bar{D}_1^2\bar{D}_2\bar{\Delta} = \bar{R}\bar{D}_1^2\bar{K}_2\bar{\Delta} = 0, \qquad \bar{R}(\bar{W}_{1,w} - \bar{W}_{1,z})\bar{D}_2\bar{K}_2 = y(w - z),$$

$$\bar{R}\bar{D}_1^2\bar{\Delta}^2 = \bar{R}\bar{D}_2^2\bar{\Delta}^2 = -x, \qquad \bar{R}(\bar{W}_{1,w} - \bar{W}_{1,z})\bar{K}_2^2 = w(w - z),$$

$$\bar{R}\bar{D}_1\bar{K}_1\bar{D}_2\bar{K}_2 = y^2, \qquad \bar{R}(\bar{W}_{1,w} - \bar{W}_{1,z})(\bar{W}_{2,w} - \bar{W}_{2,z}) = (w - z)^2.$$

$$\bar{R}\bar{D}_1\bar{K}_1\bar{K}_2^2 = yw,$$
$$\bar{R}\bar{D}_1\bar{K}_1(\bar{W}_{2,w} - \bar{W}_{2,z}) = y(w - z),$$
$$\bar{R}\bar{D}_1\bar{K}_1\bar{D}_2\bar{\Delta} = 0,$$
$$\bar{R}\bar{D}_1\bar{K}_1\bar{K}_2\bar{\Delta} = 0,$$
$$\bar{R}\bar{D}_1\bar{K}_1\bar{\Delta}^2 = -y,$$
$$\bar{R}\bar{K}_1^2\bar{D}_2^2 = xw,$$
$$\bar{R}\bar{K}_1^2\bar{D}_2\bar{K}_2 = yw,$$
$$\bar{R}\bar{K}_1^2\bar{K}_2^2 = w^2,$$
$$\bar{R}\bar{K}_1^2(\bar{W}_{2,w} - \bar{W}_{2,z}) = w(w - z),$$

$$\bar{R}(\bar{W}_{1,w} - \bar{W}_{1,z})\bar{D}_2\bar{\Delta} = 0,$$
$$\bar{R}(\bar{W}_{1,w} - \bar{W}_{1,z})\bar{K}_2\bar{\Delta} = 0,$$
$$\bar{R}(\bar{W}_{1,w} - \bar{W}_{1,z})\bar{\Delta}^2 = z - w,$$
$$\bar{R}\bar{\Delta}^2\bar{D}_2^2 = -x,$$
$$\bar{R}\bar{\Delta}^2\bar{D}_2\bar{K}_2 = -y,$$
$$\bar{R}\bar{\Delta}^3\bar{D}_2 = y,$$
$$\bar{R}\bar{\Delta}^3\bar{K}_2 = w,$$
$$\bar{R}\bar{\Delta}^4 = -z,$$

$$\bar{R}^2\bar{D}_1^2\bar{D}_2 = x^2,$$
$$\bar{R}^2\bar{D}_1^2\bar{K}_2 = xy,$$
$$\bar{R}^2\bar{D}_2^2\bar{\Delta} = x,$$
$$\bar{R}^2\bar{D}_1\bar{K}_1\bar{D}_2 = xy,$$
$$\bar{R}^2\bar{D}_1\bar{K}_1\bar{K}_2 = y^2,$$
$$\bar{R}^2\bar{D}_2\bar{K}_1^2 = xw,$$
$$\bar{R}^2\bar{K}_1^2\bar{K}_2 = yw,$$
$$\bar{R}^2(\bar{W}_{1,w} - \bar{W}_{1,z})\bar{D}_2 = x(w - z),$$
$$\bar{R}^2(\bar{W}_{1,w} - \bar{W}_{1,z})\bar{K}_2 = y(w - z),$$
$$\bar{R}^2\bar{D}_1\bar{D}_2\bar{\Delta} = x,$$
$$\bar{R}^2\bar{D}_1\bar{K}_1\bar{\Delta} = y,$$
$$\bar{R}^2\bar{D}_1\bar{K}_2\bar{\Delta} = y,$$
$$\bar{R}^2\bar{K}_1\bar{D}_2\bar{\Delta} = y,$$
$$\bar{R}^2\bar{K}_1^2\bar{\Delta} = w,$$
$$\bar{R}^2\bar{K}_1\bar{K}_2\bar{\Delta} = w,$$
$$\bar{R}^2(\bar{W}_{1,w} - \bar{W}_{1,z})\bar{\Delta} = w - z,$$
$$\bar{R}^2\bar{D}_1\bar{\Delta}^2 = -2x - y,$$
$$\bar{R}^2\bar{D}_2\bar{\Delta}^2 = -2x - y,$$

$$\bar{R}^2\bar{K}_1\bar{\Delta}^2 = -2y - w,$$
$$\bar{R}^2\bar{\Delta}^3 = 2y + z,$$
$$\bar{R}^3\bar{D}_1^2 = x^2 - x,$$
$$\bar{R}^3\bar{D}_1\bar{D}_2 = x^2 - x,$$
$$\bar{R}^3\bar{D}_1\bar{K}_1 = xy - y,$$
$$\bar{R}^3\bar{D}_1\bar{K}_2 = xy - y,$$
$$\bar{R}^3\bar{K}_1^2 = xw - w,$$
$$\bar{R}^3\bar{K}_1\bar{K}_2 = y^2 - w,$$
$$\bar{R}^3(\bar{W}_{1,w} - \bar{W}_{1,z}) = (x - 1)(w - z),$$
$$\bar{R}^3\bar{D}_1\bar{\Delta} = 3x + y,$$
$$\bar{R}^3\bar{K}_1\bar{\Delta} = 3y + w,$$
$$\bar{R}^3\bar{\Delta}^2 = -3x - 3y - z,$$
$$\bar{R}^4\bar{D}_1 = x^2 - 4x - y,$$
$$\bar{R}^4\bar{K}_1 = xy - 4y - w,$$
$$\bar{R}^4\bar{\Delta} = 6x + 4y + z,$$
$$\bar{R}^5 = x^2 - 10x - 5y - z.$$

$$\text{(91)}$$

Proposition 3.

$$\int_X c_5\left(f^*T_P/T_X\right) = \frac{1}{2}(-100x^2 - 100xy - 20xz - 25y^2 - 10yz - z^2 +$$
$$+ 2200x + 2761y + 684z + 422w).$$

Now the result of the theorem follows from propositions 2 and 3 and formula (3).

References

[1]   M.BELTRAMETTI, A.J.SOMMESE. _Zero cycles and k-th order embeddings of smooth projective surfaces_, Symposia Mathematica, vol.XXXII (1991), 33-48.

[2]   W.FULTON. _Intersection theory_, Springer-Verlag, 1984.

[3]   R.HARTSHORNE. _Algebraic geometry_, Springer-Verlag, 1977.

[4]   A.S.TIKHOMIROV. _Standard bundles on a Hilbert scheme of points on a surface_, Present collection.

[5]   A.N.TYURIN. _On periods of quadratic differentials_, Russian math.surveys, v.XXXIII, 6(1978), 149-195 (Russian).

[6]   I.R.SHAFAREVICH. _Basic algebraic geometry_, Moscow, 1972 (Russian).

Almost Canonical Polynomials of Algebraic Surfaces

Andrej N. Tyurin

0 Introduction

Let S be a smooth simple connected algebraic surface over $\mathbb{C}$, Pic (S) be the group of classes of divisors,

$$L \subset \text{Pic}\,(S) \otimes \mathbb{R} \tag{0.1}$$

be the light cone,

$$L^+ = \text{light cone}/\mathbb{R}^+ \tag{0.2}$$

be the Lobachevski space and

$$C^+ = C_S/\mathbb{R}^+ \tag{0.3}$$

be closed subset of L^+ of rays of polarizations.

For any point $H \in C^+$ let

$$\overline{M^H(2, c_1, c_2)} \tag{0.4}$$

be the Gieseker closure of moduli space of H-stable vector bundles on S (see [G]).

Now the topological type of vector bundles $(2, c_1, c_2)$ (2 is rank) defines a collection of hyperplanes $\{W_e\}$ in K^+ (so called "walls") orthogonal to the collections of vectors $\{e\}$, by the conditions:

$$e \in \text{Pic}\,(S), \tag{0.5}$$

$$e = c_1 \bmod 2,$$

$$c_1^2 - 4c_2 \le e^2 \le 0.$$

This collection of walls divides the Lobachevski space L^+ and C^+ in chambers.

Under a deformation $H_t, t \in [-\varepsilon, \varepsilon]$ $\overline{M^H(2, c_1, c_2)}$ is changing when H_t is going through the wall W_e (say at the moment $t = 0$).

How does $\overline{M^{H_t}(2, c_1, c_2)}$ change?

Very rough picture is:

some vector bundle E becomes semistable and obtaines a line subbundle M, i.e. admits the triple

$$0 \to M \to E \to M' \to 0, \tag{0.6}$$
$$M.H_0 = M'.H_0.$$

But then E isn't distinguished from direct sum

$$M \oplus M' \in \overline{M^{H_0}(2, c_1, c_2)}. \tag{0.7}$$

But

$$\mathbb{P}Ext^1(M', M) \subset \overline{M^{H-\varepsilon}(2, c_1, c_2)}, \overline{M^{H_\varepsilon}(2, c_1, c_2)} \supset \mathbb{P}Ext^1(M, M')$$

and we have the following picture

$$\mathbb{P}\,Ext^1(M',M) \qquad \subset \qquad M^\varepsilon \qquad\qquad M^{-\varepsilon} \qquad \supset \qquad \mathbb{P}\,Ext^1(M,M')$$

$$\searrow \text{blow down} \qquad \searrow \qquad \underset{M^{H_0}}{} \qquad \swarrow \qquad \swarrow \text{blow down}$$

$$\cup$$
$$(M \oplus M')$$

What is a realistic picture? Almost the same:

Let

$$GAM_0(2, M + M', c_2) = \{0 \to M \to E \to J_\xi \otimes M' \to 0\}/\mathbb{C}^* \tag{0.8}$$

be set of all non-trivial extensions up to homotheties, where J_ξ is the ideal sheaf of a 0-dimensional subscheme ξ (of a cluster ξ for short) and

$$\pi : GAM_0(2, M + M', c_2) \to \text{Hilb}^{c_2 - M.M'} S$$

be the natural projection given by sending the extension to the cluster ξ as element of the Hilbert scheme.

A fibre of this projection is

$$\pi^{-1}(\xi) = \mathbb{P}Ext^1(J_\xi \otimes M', M) = \mathbb{P}H^1(J_\xi(M - M' + K))^*$$

by Serre-duality.

And the almost realistic picture (GAM-picture for short) is:

$$GAM_0(2, M + M', c_2) \quad \subset \quad M^\varepsilon \qquad\qquad M^{-\varepsilon} \quad \supset \quad GAM_0(2, M' + M, c_2)$$

$$\searrow \text{blow down} \qquad \searrow \qquad \underset{M^{H_0}}{} \qquad \swarrow \qquad \swarrow \text{blow down} \qquad (0.9)$$

$$\cup$$
$$\text{Hilb} S$$

We need to remark that our situation is a slice of more general situation if $b_2^+(S) = 1$ (see [P-T] or [T-2] specially for Θ). Namely,

we can consider $H^2(S, \mathbb{R})$ instead of Pic (S), the big "light cone" instead of (0.2) and so on, and the moduli spaces of instantons instead of the moduli spaces of stable vector bundles. Then we have the same problem (see [D-K], [K]).

Warning. The natural compactification of moduli spaces of instantons is a different from the Gieseker's compactification (see [Mo] but these compactifications are much more close if we consider moduli spaces of stable pairs (see [T 2])).

Moreover the case of **Higgs-bundles** or **vortex bundles** (stable pairs for short) is much more interesting from this point of view.

In this case we have the space of stable pairs (s,E) where s is for example some section of E (as homomorphism $s : \mathcal{O}_S \to E$) up to $\mathbb{C}^*$-action, with stability conditions

$$M \to E, \quad M.H \leq \tfrac{1}{2}c_1(E).H - \sigma, \quad \text{if } M \neq \text{im } s, \qquad (0.10)$$

$$c_1(\text{im } s).H \leq \tfrac{1}{2}c_1(E).H + \sigma,$$

where σ is non-negative real number (see [B-D], [B] or [Th]). Then we have the family of moduli spaces M^σ beginning with the ordinary moduli space when $\sigma = 0$ and a moduli space is changing when the parameter σ goes through an integer.

All moduli spaces are given as **quotients in the Geometrical Invariant Theory (GIT for short)**

or

results of Symplectic Reductions (SR for short)

or

extremal subset in Equivariant Morse Theory (EMT for short).

The following table gives some survey of the experience in construction of original moduli spaces

GIT	SR	EMT
D.Mumford [M]	V.Arnold [A]	S.Donaldson [D 2]
F.Kirwan [Ki]	N.Hitchin [H]	
P.Newstead [N]	F.Kirwan [Ki]	

and elementary birational transformations

P.Newstead	N.Hitchin [H]	S.Donaldson [D 3]
$(Q_1 \cap Q_2 \in \mathbb{P}^5 \to \mathbb{P}^3)$	V.Guillemin,	
	S.Sternberg [G-S]	

The next table gives some survey of the experience in construction of moduli spaces of stable pairs

M.Thaddeus [Th]	N.Hitchin [H]	V.Pidstrigach,
(D.Huybrechts,	S.Bradlow,	A.Tyurin [P-T]
M.Lehn [H-L])*	G.Daskalopoulos [B-D]	

and elementary birational transformations of moduli spaces of pair

M.Thaddeus [Th]	(S.Bradlow.	V.Pidstrigach [P]
A.Bertram [B]	G.Daskalopoulos	A.Tyurin [T 2]
	Wentworth)*	

In spite of the fact that a general theory of elementary birational modifications – flips is ready for $\mathbb{C}^*$-and torus-actions only (see [Hu]), there is a very active discussion on general reductive group action with great expectations.

But for us (and for some applications) the following question (in the style of classical question about 3-folds: **how does a flip transformation change K^3?**) is much more interesting:

Namely every moduli space M and the standard definition of the slant-homomorphism in the algebro-geometric context give polynomial $\gamma_M \in S^{\dim M} H^2(M, \mathbb{Z})$:

in algebraic geometrical situation we have algebraic polynomial $a\gamma^{ch}(2, c_1, c_2) \in S^d H^2(M, \mathbb{Z})$ (see [T 1] or [O'G]) or canonical Spin-polynomial $a\gamma^{ac,-K}(2, K, c_2) \in S^d H^2(M, \mathbb{Z})$ (see next section);

in almost Kählerian setup we have polynomial $\gamma^{\omega}(2, c_1, c_2) \in S^d H^2(M, \mathbb{Z})$ (see [T 2]);

in EMT we have the original Donaldson polynomial $\gamma^g(2, c_1, c_2) \in S^d H^2(M, \mathbb{Z})$ (see [D 1]);

for moduli spaces of pairs or jumping instantons we provided by Spin-polynomials (see [T 2]).

Our question is **to compute the increment of any polynomial**

$$incr_e(2, c_1, c_2) = \gamma^{+\varepsilon}(2, c_1, c_2) - \gamma^{-\varepsilon}(2, c_1, c_2), \tag{0.11}$$

when the parameter is going through the wall W_e (see [T 2] for some details).

Let me recall that in EMT setup by The Homotopy Conjecture the increment depends only on the homotopy types of the involved objects (see [K-M] , [D 3] and [T 2]).

We can hope that combinatoric description of (0.11) shines light to the geometrical foundation of this conjecture.

The aim of my talk is to show some method (the method of geometrical approximation) to describe elementary transformations (flips for short) of moduli spaces of pairs and to show that the increment (0.11) for Spin-polynomial can be computed purely in terms of effective curves lying on S.

But beforehand I would like to show the new features and difficulties of applications of general theory (for example GIT) to moduli space on non-rational surface.

First and main difficulty is the following:

in GIT there is one case when the theory is perfect – when big variety T where our group G acts (in our situation it is the space of Gieseker's tensors, see [G]) has positive anticanonical system that is when T is Fano-variety. In that case the Kähler cone C^+ is a rational convex polytop (C_S is a rational convex polyhedral cone). It is a corollary of Mori theorem.

For example it is the case of moduli spaces of vector bundles with fixed c_1 on curve and in this case we have the perfect description of birational transformations – flips (or antiflips more precisely) in [B] and [Th].

But in our situation if S isn't rational then the space of Gieseker's tensors T isn't rational too.

Hence our first step is to distinguish "positive" and "negative" parts as in Mori-theory precisely. Daniel Huybrechts tries to do it theoretically and I would like to show how to do this practically, because for applications we need a consideration of surfaces of general type.

The second feature is more unusual (at least for Mori theory):

Every moduli space M admits the family of so called Poincare bundles:

that is the restrictions of the projectivization of the universal bundle on $S \times M$ to $(point) \times M$: $\mathbb{P} \to M$ and we have synchronous elementary transformations on both of levels

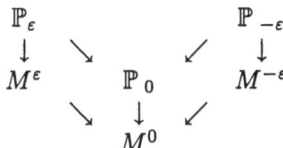

What is a classical analogue of this telescopical flip?

Returning to concret our problem, we need to remark that in situation of algebraic surfaces of general type there is one distinguished (so called almost canonical) chamber. Hence our plan is:

1) to construct all of our objects for this canonical chamber;

2) to reconstruct these objects when a polarization intersects some wall.

Of course we can do it precisely in same simple cases only (the extremal simple case is our **main example**).

1 The canonical and almost canonical polynomials of non-rational surface

Let S as usual be a simple connected smooth, compact non-rational algebraic surface. Then the map

$$m : S \to S_{min} \tag{1.1}$$

of S to the minimal model S_{min} of S is determined uniquely. Let

$$K_{min} = m^*(K_{S_{min}}) \tag{1.2}$$

be a preimage of the canonical class of minimal model.

Then $K_{min} \in H^2(S, \mathbb{R})$ is contained by the closure of the Kähler cone.

Definition 1.1. A polarization H is called an almost canonical polarization (ac-polarization for short) if the ray $\mathbb{R}^+.H$ in the projectivization of the Kähler cone C_S (0.3) is close to the ray $\mathbb{R}^+.K_{min}$ in Lobachevski metric.

There exists the case when we not need to distinguish almost canonical polarizations. Namely, when

$$K_{min}^2 \text{ is odd} \iff b_2^+(S_{min}) \text{ is even.} \tag{1.3}$$

In the other case we must give less compressed explanations.

First of all in this case we must consider the set of non-empty linear systems

$$|C_1|, \dots \text{ such that} \tag{1.4}$$

$$2C.K_{min} = K_{min}^2 \iff C^2 = (K_{min} - C)^2$$

and (it is very important)

$$C^2 = (K_{min} - C)^2 > 0.$$

(It means for example that $\chi(\mathcal{O}_S(C)) = (p_g(S) + 1) - (C^2)^2$ or an arithmetical genus of C

$$p_a(C) = \frac{2C^2 + K_{min}^2}{4} + 1$$

is a positive integer.)

Definition 1.2. The collection (1.4) is called the collection of the canonical walls. On the set W_K of all canonical walls as on a subset of the 2-cohomology lattice

$$W_K \in H^2(S_{min}, \mathbb{Z}) \tag{1.5}$$

the positive semigroup P_{-2} generated by the finite collection of all (-2)-curves

$$C_1, \dots, C_{N_{-2}} \tag{1.6}$$

on S_{min} acts naturally.

Consider the pluricanonical model of $\phi_{|NK_{min}}(S_{min}) = S_{NK}$ given as the image of the pluricanonical map

$$\phi_{|NK_{min}|} : S_{min} \longrightarrow \mathbb{P}^{\frac{1}{2}N(N-1)K_{min}^2 + p_g} \tag{1.7}$$

for $N \gg 0$. Then S_{NK} contains finite set

$$|C_1'|, ..., |C_{N_w}'| \tag{1.8}$$

of curves of degree $\frac{1}{2}N.K_{min}^2$ and the collection of their geometrical preimages

$$|C_1|, ..., |C_{N_w}| \tag{1.8'}$$

is the subcollection of the collection of the canonical walls (1.4).

Let $U_K \in K_S^+$ be a small neighbourhood of the point $\mathbb{R}^+.K_{min}$ inside of the Kähler cone of S. Then the finite set of hyperplanes

$$\{W_i\}, W_i = (K_{min} - 2C_i)^\perp, \quad i = 1, ..., N_w \tag{1.9}$$

divides U_K into the finite set of chambers

$$U_K = \bigcup ch_n. \tag{1.10}$$

Now we not need to distinguish almost canonical polarizations inside one chamber ch and let the symbol ac_{ch} denote any polarization inside ch. Then the symbol $M^{ch}(r, c_1, c_2)$ denotes the moduli space of ac_{ch} - slope stable bundles on S of rank r with Chern classes c_1, c_2.

On the analogy of Jacobian of algebraic curve we proposed (see (4.16) from [T 3]) the following

Definition 1.3. The Gieseker closure (see [G])

$$\overline{M^{ch}(2, K_S, c_2(S))} = J_{ch}(S) \tag{1.11}$$

is called a jacobian of S.

Remark. We will see later that all jacobians are birationally equivalent (up to components) and we will drop the chambers index as long as there is no danger of confusion.

A jacobian $J(S)$ contains the distinguished point

$$T^*S = \Omega S \in J(S) \tag{1.12}$$

– the cotangent bundle of S whicn is stable by Bogomolov theorem.

By Noether formula the constant

$$\mu(S) = \frac{3c_2(S) - K_S^2}{4} \tag{1.13}$$

is an integer. ($\mu(S)$ alias MIAYAOKA NUMBER of S.)

The virtual (expected) dimension of jacobian of S

$$v.\dim J(S) = 5\mu(S). \tag{1.14}$$

This constant is an anologue of genus of algebraic curve (see (4.16) from [T 3]).

The standard definition of the slant-homomorphism in the algebro-geometric context (see [T 1] or [O'G]) gives the polynomial

$$a\gamma_S \in S^{5\mu(S)}H^2(S,\mathbb{Z}) \tag{1.15}$$

(if jacobian $J(S)$ has the expected dimension it is clear but otherwise we need to use some trick as for example in [T 2]).

Now, for any positive integer l we can consider the subspace of $J(S)$:

$$W^l(S) = \{F \in J(S) | h^0(F) \geq l+1\}. \tag{1.16}$$

On the anology of Riemann Theorem in the case of algebraic curves we denote first subspace

$$W^0(S) = \Theta(S). \tag{1.17}$$

The collection of these subspaces defines the filtration:

$$J(S) \supseteq \Theta(S) \supseteq W^1(S) \supseteq ... \supseteq W^l(S) \supseteq ... \tag{1.18}$$

If $F \in W^l - W^{l+1}$ and the family of torsion free sheaves is in "general position" near F, then the fibre of the normal bundle to $W^l(S)$ at F is given by

$$(N_{W^l \subset J(S)})_F = Hom(H^0(F), H^1(F)) \tag{1.19}$$

with $H^0(F) = \mathbb{C}^{l+1}$, $H^1(F) = \mathbb{C}^{l+1-\chi(F)}$ (if $\chi(F)$ is not positive).

Thus the virtual (expected) codimension of $W^l(S)$

$$v.\mathrm{codim}W^l(S) = (l+1)(l+1-\chi(T^*S)) = (l+1)(l+1+\frac{c_2(S)}{3}+\frac{2\mu(S)}{3}), \tag{1.20}$$

and the virtual (expected) dimension of $W^l(S)$

$$v.\dim W^l(S) = \frac{(35-4l)c_2(S)-13K^2}{12} - (l+1)^2. \tag{1.21}$$

In particular,

$$v.\dim \Theta(S) = \frac{35c_2(S)-13K^2}{12} - 1 = 4\mu(S) - (p_g(S)+1) - 1. \tag{1.22}$$

This integer is non-negative iff the inequality

$$(2,696969....)c_2(S) > K_S^2 \tag{1.23}$$

is true.

Remark. Unfortunately we can't repair by this inequality the "watershed conjecture"of Bogomolov.

At any case the standard definition of the slant-homomorphism in the algebro-geometric context (see [T 1] or [O'G]) gives the polynomials

$$a\gamma^{ch}_{\Theta(S)} \in S^{\dim \Theta} H^2(S, \mathbb{Z}), \tag{1.24}$$

$$a\gamma_{\Theta(S)} = \sum_{ch} a\gamma^{ch}_{\Theta(S)},$$

and the polynomials

$$a\gamma^{ch}_l \in S^{v.\dim W^l} H^2(S, \mathbb{Z}), \tag{1.25}$$

$$a\gamma_l = \sum_{ch} a\gamma^{ch}_l.$$

Definition 1.4. The polynomials $a\gamma_S$ (1.15) and $a\gamma_{\Theta(S)}$ (1.25) are called the canonical polynomials of S.

(Actually we can define these polynomials without regularity conditions (see [T 2]).)

Now to have enough such polynomials we consider the analogical constructions with any c_2.

Definition 1.5. The Gieseker closure (see [G])

$$\overline{M^{ac}(2, K_S, c_2(S) + k)} = J_k(S) \tag{1.26}$$

is called a k-th jacobian of S.

On the analogy of (1.3') the k-th jacobian contains the distinguished subspace:

$$\{\Omega S\}_k = \{F \in J_k(S) | F^{**}\} = \Omega S, \tag{1.27}$$

that is the subset of the torsion free sheaves which have the cotangent bundle of S as the reflexive envelopment of ones. The structure of $\{\Omega S\}_k$ can be described below as a projectivization of the standard vector bundle on the Hilbert scheme of S (see below).

The virtual (expected) dimension of k-th jacobian of S

$$v.\dim J_k(S) = 5\mu(S) + 4k. \tag{1.28}$$

Now, for any positive integer l, we can consider the subspace of $J_k(S)$:

$$W^l_k(S) = \{F \in J_k(S) | h^0(F) \geq l + 1\}. \tag{1.29}$$

We denote first subspace

$$W^0_k(S) = \Theta_k(S) \tag{1.30}$$

again.

The collection of these subspaces defines the filtration:

$$J_k(S) \supseteq \Theta_k(S) \supseteq W^1_k(S) \supseteq \cdots \supseteq W^l_k(S) \supseteq \cdots \tag{1.31}$$

The description of the normal bundle is completely the same as in (1.10) and thus the virtual (expected) codimension of $W_k^l(S)$

$$v.\text{codim} W_k^l(S) = (l+1)(l+1-\chi(T^*S)+k) = (l+1)(l+1+\frac{c_2(S)}{3}+\frac{2\mu(S)}{3}+k), \quad (1.32)$$

and the virtual (expected) dimension of $W_k^l(S)$

$$v.\dim W^l(S) = \frac{(35-4l)c_2(S)-13K^2}{12} - (l+1)^2 + (3-l)k. \quad (1.33)$$

In particular,

$$v.\dim \Theta_k(S) = \frac{35c_2(S)-13K^2}{12} + 3k - 1 = 4\mu(S) - (p_g(S)+1) + 3k - 1. \quad (1.34)$$

The standard definition of the slant-homomorphism gives the polynomials

$$a\gamma_{\Theta_k(S)} \in S^{v.\,\dim\,\Theta_k(S)} H^2(S,\mathbb{Z}) \quad (1.35)$$

and the polynomials

$$a\gamma_l^k \in S^{v.\,\dim\,W_k^l} H^2(S,\mathbb{Z}). \quad (1.36)$$

Recall. We take the sum over all chambers again (see (1.24-25)).

Definition 1.6. 1) The subspaces W_k^l are called the almost canonical jumping subspaces (canonical, if $k=0$). (BN-locus for short, theta-locus for $l=0$.)

2) The polynomials $a\gamma_l^k$ and $a\gamma_{\Theta(S)_k}$ are called the almost canonical Spin-polynomials of S.

Remark. BN-locus alias "BRILL-NOETHER locus" of course.

Since our surface S is regular, the intersection

$$\{\Omega S\}_k \bigcap \Theta_k(S) = \emptyset \quad (1.37)$$

always.

Now we need to explain why we consider this very one construction.

2 Why $c_1 = K_S$?

Recall that according to EMT if the underlying smooth manifold of an algebraic surface S is equipped with a Riemannian metric g then for every $SU(2)$ or $SO(3)$–bundle E of a topological type $(2,0,c_2)$ or $(2,w_2,p_1)$ the gauge - orbit space $B(E) = \mathcal{A}^*{}_h(E)/\mathcal{G}$ of irreducible connections contains the subspace $\mathcal{M}^g(2,0,c_2)$ or $\mathcal{M}^g(2,w_2(S),p_1) \subset B(E)$

of antiselfdual connections with respect to the Riemannian metric g, oriented by the choice of the lift to the canonical class K_S of the Stiefel-Whitney class $w_2(S)$ and an orientation of a maximal positive subspace in $H^2(S, \mathbb{R})$. This space provides by slant product-construction the homogeneous polynomials

$$\gamma^g(2, 0, c_2) \quad \text{or} \quad \gamma^g(2, K_S, p_1) \in S^d H^2(S, \mathbb{Z}). \tag{2.1}$$

These polynomials behave naturally under diffeomorphisms of S, namely, if $p_g > 0$ for any $\sigma \in H^2(S, \mathbb{Z})$ and any $\phi \in \mathrm{Diff}\, S$, preserving the orientation of a maximal positive subspace in $H^2(S, \mathbb{R})$, we have

$$\gamma^g(2, 0, c_2)(\sigma) = \gamma^g(2, 0, c_2)(\phi(\sigma)) \tag{2.2}$$

or

$$\gamma^g(2, K_S, p_1)(\sigma) = \gamma^g(2, K_S, p_1)(\phi(\sigma)). \tag{2.2'}$$

These properties provide some information about the structure of the polynomials for example for surfaces with "big monodromy" or for surfaces with blown up point.

Now new system of invariants of such type of the underlying differentiable structure of an algebraic surface was proposed in [P-T], [T 2]. These so called Spin-polynomial invariants can be used to prove that the canonical class of a simply connected non-rational algebraic surface S is defined by its underlying differentiable structure.

Recall that if we consider the anticanonical class $-K_S$ as a $Spin^C$-structure on S (see [P-T] or [T 2]) equipped with a Riemannian metric g then for every $U(2)$–bundle E of a topological type $(2, c_1, c_2)$ the gauge - orbit subspace $\mathcal{M}^g(E) \subset \mathcal{B}(E)$ of antiselfdual connections contains the subspace:

$$\mathcal{M}_1^{g, -K_S}(E) = \{(a) \in \mathcal{M}^g(E) | rk\, ker D_a^{-K_S, \nabla_0} \geq 1\}, \tag{2.3}$$

where $D_a^{-K_S, \nabla_0}$ is a coupled Dirac operator (see [T-P]) with an Hermitian connection ∇_0 on the line bundle with the first Chern class K_S (see [P-T], [T 2]).

This space provides by slant product the homogeneous polynomial

$$\gamma_1^{g, -K_S}(2, c_1, c_2) \in S^{d_1} H^2(S, \mathbb{Z}) \tag{2.4}$$

so called Spin-polynomials (see [T 2]).

Actually, such polynomial $\gamma_1^{g, C}(2, c_1, c_2) \in S^{d_1} H^2(S, \mathbb{Z})$ is defined for any $Spin^C$-structure $C \in H^2(S, \mathbb{Z})$ of S given by lifting up $w_2(S)$ to an integer class and there exists simple rule of a behaviour of polynomials under changing of $Spin^C$-structure:

$$\gamma_1^{g_1, C+2\delta}(2, c_1, c_2) = \gamma_1^{g_1, C}(2, c_1 + 2\delta, c_2 + c_1.\delta + \delta^2). \tag{2.5}$$

On the other hand if $p_g(S) \geq 1$ then for every pair of regular metrics g_1, g_2

$$\gamma_1^{g_1, C}(2, c_1, c_2) = \gamma_1^{g_2, C}(2, c_1, c_2). \tag{2.6}$$

Hence in the last case we can not distinguish first upper index and hence for any $\sigma \in H^2(M, \mathbb{Z})$ and any $\phi \in \text{Diff} S$ we have

$$\gamma_1^{g,C}(2, -C, c_2)(\sigma) = \gamma_1^{g,C}(2, -C, c_2)(\phi(\sigma)) \tag{2.7}$$

up to a sign.

Indeed $\gamma_1^{g,C}(2, -C, c_2)(\phi^{-1}(\sigma)) = \gamma_1^{g,\phi(C)}(2, -\phi(C), c_2)(\sigma) = \gamma_1^{g,C}(2, -\phi(C) + \phi(C) - C, c_2 + \frac{\phi(C)-C}{2}(\frac{\phi(C)-C}{2} + C))$. Now

$$(C + 2\delta)^2 = C^2 \Leftrightarrow \delta(\delta + C) = 0 \tag{2.8}$$

and c_2 is preserved too (see [T 2]).

So if $p_g(S) > 0$ then the polynomial $\gamma_1^{H,C}(2, -C, c_2)$ is an invariant (up to sign of course) of the smooth structure of S.

For an algebraic surface S there exists the canonical $Spin^C$-structure given by the anticanonical class $-K_S$ (we will drop low index as long as there is no danger of confusion) and the main fact is that we can compute these polynomials for $C = -K_S$ precisely in algebraic geometical terms (see [T 2]).

Thus to receive the differential geomerical information we must restrict ourself by the equality $c_1 = K_S$ and investigate the geometry of the moduli spaces of stable vector bundles and torsion free sheaves of the topological type $(2, K_S, c_2)$ that is Θ_k and the almost canonical Spin-polynomials.

For example, in the simplest situation when

$$\text{Pic }(S) = \mathbb{Z}.K_S \tag{2.9}$$

we will prove that for $k \gg 0$ almost canonical polynomial 1) isn't vanishing;

2) is divizible by K_S as a linear form on $H^2(S, \mathbb{Z})$;

3) any form which divides the canonical polynomial is proportional to K_S.

What do we need to add to prove that the canonical class is an invariant of the underlying differentiable structure of S?

We must prove following John Morgan [Mo] that our algebraic almost canonical Θ-polynomial is equal to the differential geometric Spin-polynomial. It was shown in [T 2] that in this case Uhlenbeck compactification of Θ_k is much more close to Gieseker compactification of one then in the case of k-th jacobian. We postpone this step to the special article.

3 Geometrical approximation of Θ

First of all we remind the construction of $GA\Theta_k$ ($= GAM_0(2, K_S, c_2(S) + k)$ in the notations of [T 2] and (0.8-9)).

This variety is a complete intersection in the projectivization of some standard (canonical again) vector bundle on the Hilbert scheme of S.

Let me recall this construction:

Let

$$Z_d \subset S \times \text{Hilb}^d S, \quad \text{with} \quad Z_d \cap (S \times \xi) = \xi, \tag{3.1}$$

be the universal subscheme ($d = c_2(S) + k$ for short) and

$$
\begin{array}{ccc}
 & Z_d & \\
\bar{p}_S \swarrow & & \searrow \bar{p}_H \\
S & & \text{Hilb}^d S
\end{array}
\tag{3.2}
$$

be two projections of it, induced by the projections p_H and p_S of the direct product $S \times \text{Hilb}^d S$.

Consider the vector bundles

$$\mathcal{E}_{2K} = R^0\bar{p}_H(\bar{p}_S^* \mathcal{O}_S(2K)), \tag{3.3}$$

$$\mathcal{E}_{0,2K} = \mathcal{E}_{2K} \oplus H^1(\mathcal{O}_S(2K)) \otimes \mathcal{O}_{\text{Hilb}}, \tag{3.4}$$

$$\mathcal{E}_{2K} = R^0\bar{p}_H(\bar{p}_S^* \mathcal{O}_S(2K)).$$

These sheaves are locally free because the canonical homomorphism is surjective. (Remind that $h^1(\mathcal{O}_S(2K))$ is the number of (-1)-exceptional curves on S).

Let H be the divisor class of the Grothendieck sheaf $\mathcal{O}_{\mathbb{P}\mathcal{E}_{0,2K}^*}(1)$ (or $\mathcal{O}_{\mathbb{P}\mathcal{E}_{2K}^*}(1)$) on the projective bundle $\mathbb{P}\mathcal{E}_{2K}^* \to \text{Hilb}S$) associated with $\mathcal{E}_{0,K}^d$.

Each section s of $\mathcal{E}_{0,K}^d$ corresponds to a section $\bar{s}$ of the Grothendieck sheaf $\mathcal{O}_{\mathbb{P}\mathcal{E}_{0,2K}^*}(1)$.

The restriction homomorphism $H^0(\mathcal{O}_S(2K)) \xrightarrow{res} H^0(\mathcal{E}_{0,2K})$ can be written as the composite

$$H^0(\mathcal{O}_S(2K)) \otimes \mathcal{O}_{\text{Hilb}} \to H^0(\mathcal{E}_{0,K}^d) \otimes \mathcal{O}_{\text{Hilb}} \xrightarrow{wc} \mathcal{E}_{0,K}^d,$$

that is, any section $s \in H^0(\mathcal{O}_S(2K))$ defines a section $res(s)$ of the vector bundle $\mathcal{E}_{0,2K}$ and a section $\overline{res(s)}$ of the Grothendieck sheaf $\mathcal{O}_{\mathbb{P}\mathcal{E}_{0,2K}^*}(1)$.

If $s_1, \ldots, s_{p_2(S)}$ is a basis of the space of sections $H^0(\mathcal{O}_S(2K))$ then the intersection of divisors

$$\bigcap_{i=1}^{p_2(S)} H_{\overline{res(s_i)}} \subset \mathbb{P}\mathcal{E}_{0,2K}^* = GA\Theta_{d-c_2(S)}. \tag{3.5}$$

It is easy to see that if the hypersurfaces (3.5) are in general position then

$$\dim GA\Theta_k = v.\dim\Theta_k \tag{3.6}$$

(see (1.34)).

Now $GA\Theta_k$ is a base of the universal family of torsion free sheaves given on the direct product $S \times GA\Theta$ (we will drop indices as long as there is no danger of confusion) as the universal extension

$$0 \to p_1^*\mathcal{O}_{GA\Theta}(H) \to \mathbb{E} \to (id \times \pi)^*(\mathcal{I}_Z \otimes p_S^*\mathcal{O}_S(K)) \to 0 \tag{3.7}$$

(for description of the cocycle of this extension see the diagram (4.25) and (4.28) from [T 2]).

Thus $GA\Theta$ is the set all nontrivial extensions of the type

$$0 \to \mathcal{O}_S \to E \to J_\xi(K) \to 0. \tag{3.8}$$

As any family of sheaves $\mathbb{E}$ defines according to the slunt-construction some polynomial $ga\gamma_\Theta$: namely we can consider the (2,2)-Künneth component $\mu_\mathbb{E}$ of the class

$$4.c_2(\mathbb{E}) - c_1^2(\mathbb{E}) \tag{3.9}$$

as a cohomological correspondence

$$\mu_\mathbb{E} : H^2(S, \mathbb{Z}) \to H^2(GA\Theta, \mathbb{Z}). \tag{3.10}$$

From (3.8) we have for any $\sigma \in H^2(S, \mathbb{Z})$

$$\mu_\mathbb{E}(\sigma) = 4\pi^*(\tilde{\sigma}) + 2(\sigma.K)H, \tag{3.11}$$

where π is the standard projection of $GA\Theta$ to Hilb, $H = p_{GA\Theta}^*(H)$ for short and we need to recall that the universal subscheme (3.1) as an algebraic correspondence defines the cohomological correspondence

$$\mu_{\text{Hilb}} : H^2(S, \mathbb{Z}) \to H^2(\text{Hilb}, \mathbb{Z}), \tag{3.12}$$

$$\tilde{\sigma} = \mu_{\text{Hilb}}(\sigma).$$

Roughly speaking, if a fundamental 2-cycle σ is given as a smooth oriented surface Σ then the fundamental class $\tilde{\sigma}$ of $\mu_{\text{Hilb}}(\sigma)$ is given as

$$\tilde{\sigma} = \mu_{\text{Hilb}}(\sigma) = [\mu_{\text{Hilb}}(\sigma)] = \{\xi \in \text{Hilb}|Supp\xi \cup \Sigma \neq \emptyset\},$$

that is $\tilde{\sigma}$ contains the clusters $\xi = p_1 + ... + p_d$ such that at least one point p_i is contained in Σ.

Now the value of our polynomial of σ is an intersection number

$$ga\gamma_{\Theta_k}(\sigma) = (4\pi^*(\tilde{\sigma}) + 2(\sigma.K).H)^{v.\dim GA\Theta}.H^{p_2(S)} \tag{3.13}$$

of 2-cocycles on $\mathbb{P}\mathcal{E}_{0,K}^{d*}$.

Definition 3.1. The polynomial (3.13) is called a geometrical approximation of almost canonical Spin-polynomial.

Before computing of a value of this polynomial we should like to prove

Proposition 3.1. *For $k \geq 2p_2(S) - c_2(S)$ one has*

$$\dim GA\Theta_k = (v.d) = v. \dim \Theta_k$$

and $GA\Theta_k$ is irreducible.

Proof. First of all under our conditions the natural projection

$$\pi : GA\Theta \rightarrow \text{Hilb}S \qquad (3.14)$$

is surjective. Moreover the fibre $\pi^{-1}(\xi)$ has right dimension if $h^0(J_\xi(2K)) = 0$. The subvariety of the Hilbert scheme

$$\text{Hilb}|2K| = \{\xi \in \text{Hilb}^d S | h^0(J_\xi(2K)) > 0\} \qquad (3.15)$$

has dimension $\leq d + h^0(\mathcal{O}_S(2K)) - 1$.

On the other hand the dimension of every fibre $\leq d + h^1(\mathcal{O}_S(2K))$. Thus

$$\dim \pi^{-1}(\text{Hilb}|2K|) < v. \dim GA\Theta$$

and we have required equality. To prove irreducibility we need to remark that $GA\Theta$ is the intersection of $h^0(\mathcal{O}_S(2K))$ divisors (3.5). Hence a dimesion of any component is more or equal to $v. \dim \Theta$. We are done.

Main example. In the situation (2.9) we have

Proposition 3.2. *For $k \gg 0$*

$$ga\gamma_{\Theta_k} = a\gamma_{\Theta_k},$$

that is for big k an almost canonical Spin-polynomial coincides with the geometrical approximation of one.

Proof. It is easy to see that every non-trivial extension of the type (3.8) is Gieseker-stable. Hence there is a morphism

$$can : GA\Theta_k \rightarrow \Theta_k,$$

and the family $\mathbb{E}$ (3.7) is induced by this morphism. Hence the comological correspondence μ_E is a composition $\mu_\Theta.can^*$. It provides a coincidence of polynomials and we are done.

Remark. The fibre of the morphism can over $F \in \Theta_k$ is $\mathbb{P}H^0(F)$.

Now using (3.13), the projection formula

$$(\pi^*(\widetilde{\sigma})^j.H^{\dim \mathbb{P}\mathcal{E}-j} = \widetilde{\sigma}^j.\pi_*H^{\dim \mathbb{P}\mathcal{E}-j} \qquad (3.16)$$

and the definition of the Segre classes

$$s_n(\mathcal{E}) = \pi_*H^{rank\mathcal{E}-1+n}$$

we have in the final analysis

Theorem 3.1. *If $GA\Theta_k$ has expected dimension (v.d) and $k > K^2 + (p_g(S) + 1) - c_2(S)$ then*

$$ga\gamma_{\Theta_k}(\sigma) =$$

$$= 2^{c_2(S)+k}(K.\sigma)^{(c_2+k-K^2-p_g-2)} \sum_{j=0}^{2d} 2^j \binom{2c_2+2k}{j} (K.\sigma)^{2d-j} . s_{2c_2+2k-j}(\mathcal{E}_{2K})\tilde{\sigma}^j$$

$$(3.17)$$

(of course $s_{2c_2+2k-j}(\mathcal{E}_{0,2K}) = s_{2c_2+2k-j}(\mathcal{E}_{2K})$).

You can see that if $k \gg 0$ the geometrical approximation of the almost canonical Spin-polynomial is divided by the linear form K.

Main example . In the situation (2.9) the almost canonical Spin-polynomial is divided by K and only by K because the cohomological correspondence (3.10) given by cycle (3.9) of Hodge type (2,2) preserves Hodge type. Hence the polynomial has Hodge type $(v.d, v.d)$ and if any line form L divides this polynomial then L as a cohomology class has Hodge type $(1,1)$. Thus subjecting to (2.9) it is proportional to K.

Moreover you can see that $ga\gamma_{\Theta_k}$ is a superposition of the standard polynomials

$$p_k^j(S)(\sigma) = s_{2c_2+2k-j}(\mathcal{E}_{2K})\tilde{\sigma}^j.$$

$$(3.18)$$

Now we can release the principal part of this polynomial.

For this we consider the following geometrical construction:

Let S^d be d-th direct power of our surface, p_i be projection to the i-th component, $S^{(d)}$ be the symmetric power of S, $f : S^d \to S^{(d)}$ be the quotient morphism and $g : \text{Hilb}^d S \to S^{(d)}$ be the natural morphism sending a zero-dimensional subscheme to the cycle. Consider the natural diagram

$$
\begin{array}{ccc}
& S^{[d]} & \\
\bar{g} \nearrow & & \searrow \bar{f} \\
S^d & & \text{Hilb}^d \\
\searrow f & & g \nearrow \\
& S^{(d)} &
\end{array}
\qquad (3.19)
$$

Now we have

$$d!p_k^j(S)(\sigma) = s_{2c_2+2k-j}(\bar{f}^* \mathcal{E}_{2K})(\bar{f}^* \tilde{\sigma})^j.$$

$$(3.20)$$

Let Δ be the preimage of the (large) diagonal of $S^{(d)}$. Then

$$\bar{f}^*\tilde{\sigma} = \sum_{i=0}^{d} p_i^*(\sigma) - \Delta.$$

$$(3.21)$$

On the other hand by the definition we have the following exact sequence

$$0 \to \bar{f}^* \mathcal{E}_{2K} \to \bigoplus_{i=0}^{d} p_i^*(\mathcal{O}_S(K)) \to \mathcal{L} \to 0$$

$$(3.22)$$

and the support of $\mathcal{L}$ is Δ.

From this the total Segre class of $\bar{f}^* \mathcal{E}_{2K}$ is

$$s(\overline{f}^* \mathcal{E}_{2K}) = \prod_{i=0}^{d} (1 + p_i^*(K))^{-1} . c(\mathcal{L}), \tag{3.23}$$

where $c(\mathcal{L})$ is the total Chern class of $\mathcal{L}$.

Hence

$$d! p_k^j(S)(\sigma) = [\prod_{i=0}^{d} (1 + p_i^*(K))^{-1} . c(\mathcal{L})]_{2d-j} (\sum_{i=0}^{d} p_i^*(\sigma) - \Delta)^j. \tag{3.24}$$

Consider the term

$$T_k^j(S) = [\prod_{i=0}^{d} (1 + p_i^*(K))^{-1}]_{2d-j} . (\sum_{i=0}^{d} p_i^*(\sigma))^j. \tag{3.25}$$

To compute one we need to remark that

$$p_i^*(K)^3 = p_i^*(\sigma)^3 = 0, \qquad p_i^*(K)^2 = K^2 . (S^{d-1}),$$

$$p_i^*(K) . p_i^*(\sigma) = (K.\sigma)(S^{d-1}), \quad p_i^*(\sigma)^2 = \sigma^2 . (S^{d-1}). \tag{3.26}$$

Let as usual q_S be the intersection quadratic form of S on $H^2(S, \mathbb{Z})$. Then any non-vanishing term of (3.25) has a form

$$a_n q_S(\sigma)^n . (K.\sigma)^{2d-j}, \tag{3.27}$$

where a_n is concrete integer depended on K^2 only.

Hence the principal part (3.25) has the form

$$T_k^j(S) = \sum_{i=0}^{c_2(S)+k} a_i^{k,j} q_S^i . K^{2c_2(S)+2k-2i}, \tag{3.28}$$

where integer constants $a_i^{k,j}$ depend on the topology of S only.

Thus $GA\Theta$, $ga\gamma_\Theta$ and the picture (0.9) are some approximation of the real objects and procedures. To receive the almost canonical Spin-polynomial or the right picture of biratio-nal transformations (0.9) we have to apply the chain of elementary transformations to our objects and procedures. In the next section we will describe the elementary transformations for the case of Spin-canonical polynomials. We hope that it gives the hint what we need to do in other situations.

4 Through the walls

Non-rational algebraic surface S defines the finite set of non-empty complete linear systems

$$\{|C_i|\}, \quad i = 0, 1, ..., N, \qquad 2C_i . K_{min} \le K_{min}. \tag{4.1}$$

These linear systems define the collection of walls for Θ-case. More precisely any class C defines the wall $e = K - 2C$ (see (0.5)) when we go out from an almost canonical chamber.

The set (4.1) admits the following distinguishing:

$$|0|; \{E_i\}, i = 1, ..., N_{-1}; \{E'_j\}, j = 1, ..., N_{-2}; \tag{4.2}$$

where E_i is an exceptional (-1)-curve; E'_j is an exceptional (-2)-curve from minimal model. It is easy to see, that the exceptional curves are determinated by the equality $C_i.K_{min} = 0$.

Moreover, the positive semigroup $P_{-2,-1}$ generated by the finite collection of all exceptional curves (2.2) acts naturally on the set (2.1). Up to this action we have the subset (1.8') of the canonical walls

$$\{|C_n|\}_w, n = 0, 1, ..., N_w; \tag{4.2'}$$

and the finite set

$$\{|C_m|\}_-, m = 1, ..., N_p \tag{4.2''}$$

of the preimages of curves of degree less then $\frac{1}{2}N.K^2_{min}$ on the pluricanonical model S_{NK} (1.7).

Suppose that **for every curve** C **from the set (4.1)** $C^2 < 0$ (in particular, the set of the canonical walls (1.8') is empty). It is easy to see that in this case for $k \gg 0$ $GA\Theta$ is irreducible (see Proposition 3.1) and birationally equivalent to Θ.

Consider the following subvariety of $GA\Theta_k$:

$$\Delta_k = \{F \in GA\Theta | F \text{ isn't ac} - \text{semistable}\}. \tag{4.3}$$

Consider any point $E \in \Delta_k$ given as non-trivial extension

$$0 \to \mathcal{O}_S(-K) \to E(-K) \to J_\xi \to 0. \tag{4.4}$$

Then the hypothetical destabilizing line bundle must be of type $\mathcal{O}_S(-C)$, where C is an effective curve, the cluster ξ must be supported on this effective curve and the nonstability conditions are equivalents to inequality (4.1) that is this curve C is from our system (4.1).

Hence we have a decomposition of the locus (4.3)

$$\Delta_k = \bigcup_{C \in (4.1)} \Delta_C, \tag{4.5}$$

where

$$\Delta_C = \{E | E \text{ is destabilised by } \mathcal{O}_S(-C)\}. \tag{4.6}$$

It is easy to see that if C and C' have not common components and $k \gg 0$, then

$$\Delta_C \bigcap \Delta_{C'} = \emptyset.$$

Moreover, the image of the standard projection π is

$$\pi(\Delta_C) \subset S^{c_2(S)+k} C \subset \text{Hilb}^{c_2+k} S, \tag{4.7}$$

where $S^{c_2+k}C$ is "the symmetric product" of the curve C that is

$$S^{c_2+k}C = \mathrm{Hilb}C \subset \mathrm{Hilb}^{c_2+k}S,$$

which is defined in general case by the following construction: the section $s \in H^0(\mathcal{O}_S(C))$ which defines C as its zero-set defines the section $\bar{s}$ of the standard vector bundle $\mathcal{E}_C$ on $\mathrm{Hilb}S$ (see (3.4)) and the zero scheme of this section

$$(\bar{s})_0 = \mathrm{Hilb}C \tag{4.8}$$

is the subvariety of $\mathrm{Hilb}S$ containing the clusters lying on the curve C.

Now if $|C|$ is a linear system, then

$$\mathrm{Hilb}|C| = \bigcup_{C' \in |C|} \mathrm{Hilb}C' \tag{4.9}$$

is a union of all clusters lying on curves of the linear system $|C|$ and

$$\Delta_{|C|} = \bigcup_{C' \in |C|} \Delta_{C'}. \tag{4.10}$$

Moreover, as a cohomology class

$$[\mathrm{Hilb}C] = c_{top}(\mathcal{E}_C) \tag{4.11}$$

and

$$[\mathrm{Hilb}|C|] = c_{rank\mathcal{E}_C - h^0(\mathcal{O}_S(C)) - 1}(\mathcal{E}_C). \tag{4.12}$$

Now to describe a fibre of Δ over $\xi \in \mathrm{Hilb}C$ we need to recall that any homomorphism $\phi_\xi : \mathcal{O}_S(-C) \to J_\xi$ can be lifted to a homomorphism $\mathcal{O}_S(-C) \to E(-K)$ in the short exact sequence (4.4) if and only if the element $e \in Ext^1(J_\xi, \mathcal{O}_S(-K))$ which defines the extension (4.4) is contained in the kernel of the homomorphism $Ext^1(J_\xi, \mathcal{O}_S(-K)) \xrightarrow{\phi^1} Ext^1(\mathcal{O}_S(-C), \mathcal{O}_S(-K))$ induced by the homomorphism ϕ_ξ. But this kernel

$$ker\phi_\xi^1 = Ext^1(\mathcal{O}_C(-\xi), \mathcal{O}_S(-K)) \tag{4.13}$$

and by the Serre-duality $Ext^1(\mathcal{O}_C(-\xi), \mathcal{O}_S(-K))^* = Ext^1(\mathcal{O}_S(-K), \mathcal{O}_C(K|_C - \xi)) = H^1(\mathcal{O}_C(2(K)|_C - \xi))) = H^0(\mathcal{O}_C((C)^2 + (\xi - K|_C)))^*$. Thus

$$\pi^{-1}(\xi) = |(C)^2 - K.C + \xi|. \tag{4.14}$$

The destabilizing homomorphism of line bundle $\mathcal{O}_S(-C)$ to the extension $E(-K)$ gives new representation $E(-K)$ (4.4) as an extension and we have the diagram

$$
\begin{array}{ccccc}
0 & & 0 & & 0 \\
\uparrow & & \uparrow & & \uparrow \\
0 \to \mathcal{O}_S(-K) & \xrightarrow{\quad} & J_\eta(C-K) & \to & \mathcal{O}_C(-\xi) \to 0 \\
\uparrow & {\phi_\xi(C-K) \nearrow} & \uparrow & & \uparrow \\
0 \to \mathcal{O}_S(-K) & \xrightarrow{\quad} & E(-K) & \to & J_\xi \to 0 \\
\uparrow & & \uparrow & {\phi_\xi \nearrow} & \uparrow \\
0 & \xrightarrow{\quad} & \mathcal{O}_S(-C) & \to & \mathcal{O}_S(-C) \to 0 \\
& & \uparrow & & \uparrow \\
& & 0 & & 0
\end{array}
\tag{4.15}
$$

It is easy to see that under the identification (4.14) the horisontal extension is given by the element

$$\eta \in |(C)^2 - K.C + \xi|. \tag{4.16}$$

It means that we can describe Δ_C purely in terms of the geometry of curve C:

$$\pi \quad c) \subset \text{Hilb}C, \eta \in |(C)^2 - K\!\!\!\!\!\not{C} + \xi|. \tag{4.17}$$

Of course for the case of curve all of our constructions are classical and are the subject of almost all monografies about the geometry of curves (see for example the end of [ACGH]). Namely, for a curve C there exists the universal subscheme

$$Z \subset C \times \text{Hilb}C$$

and its two projections

$$
\begin{array}{ccc}
& Z & \\
\bar{p}_C \swarrow & & \searrow \bar{p}_H \\
C & & \text{Hilb}C
\end{array}
\tag{4.18}
$$

induced by the projections p_H and p_C of the direct product $C \times \text{Hilb}C$. Any divisor class $D \in \text{Pic } C$ induces a standard rank d vector bundle

$$\mathcal{E}_D^d = R^0 \bar{p}_H(\bar{p}_C^* \mathcal{O}_C(D)) \tag{4.19}$$

on $\text{Hilb}^d C$, with fibre $H^0(\mathcal{O}_\eta(D))$ at $\eta \in \text{Hilb}^d C$.

The analogue of one for surface that is the standard vector bundle on $\text{Hilb}S$

$$\mathcal{E}_D = \mathcal{E}_D^d = R^0 \bar{p}_H(\bar{p}_S^* \mathcal{O}_S(D)) \tag{4.20}$$

can be restricted to $\text{Hilb}C$ if $C \subset S$ and it is easy to see that for $D \in \text{Pic } S$ we have the equality

$$\mathcal{E}_{D.C}^d = \mathcal{E}_D^d|_{\text{Hilb}C}. \tag{4.21}$$

Now Δ_C over $\text{Hilb}C$ admits the same description as GAM over $\text{Hilb}S$. Namely, consider the diagram

$$
\begin{array}{ccccccccc}
& & & & & & 0 & & \\
& & & & & & \downarrow & & \\
& & 0 & & 0 & & \mathcal{O}_C(-\xi) & & \\
& & \downarrow & & \downarrow & & \downarrow & & \\
0 & \rightarrow & \mathcal{O}_S(-C) & \rightarrow & \mathcal{O}_S & \rightarrow & \mathcal{O}_C & \rightarrow & 0 \\
& & \downarrow & & \downarrow & & \downarrow & & \\
0 & \rightarrow & J_\xi & \rightarrow & \mathcal{O}_S & \rightarrow & \mathcal{O}_\xi & \rightarrow & 0 \\
& & \downarrow & & \downarrow & & \downarrow & & \\
& & \mathcal{O}_C(-\xi) & & 0 & & 0 & & \\
& & \downarrow & & & & & & \\
& & 0 & & & & & &
\end{array}
\tag{4.22}
$$

Apply the functor $\mathcal{E}xt_{\mathcal{O}_S}(\ , \mathcal{O}_S(-K))$ to this diagram we have got the diagram

$$
\begin{array}{ccccccc}
0 & \leftarrow & H^1(\mathcal{O}_S(2K))^* & \leftarrow & H^1(\mathcal{O}_S(2K))^* & \leftarrow & 0 \\
& & \downarrow & & \downarrow & & \downarrow \\
0 & \leftarrow & H^1(\mathcal{O}_S(2K-C))^* & \leftarrow & Ext^1(J_\xi, \mathcal{O}_S(-K)) & \leftarrow & H^1(\mathcal{O}_C((2K)|_C - \xi))^* \\
& & \downarrow & & \downarrow & & \downarrow \\
0 & \leftarrow & H^0(\mathcal{O}_C((2K)|_C)^* & \leftarrow & H^0(\mathcal{O}_\xi(2K))^* & \leftarrow & H^1(\mathcal{O}_C((2K)|_C - \xi))^* \\
& & \downarrow & & \downarrow & & \downarrow \\
0 & \leftarrow & H^0(\mathcal{O}_S(2K))^* & \leftarrow & H^0(\mathcal{O}_S(2K))^* & \leftarrow & 0 \\
& & \downarrow & & \downarrow & & \\
& & H^0(\mathcal{O}_S(2K-C))^* & \leftarrow & H^0(J_\xi(2K))^* = 0 & &
\end{array}
$$
$$\text{(4.23)}$$

Remark. This diagram is provided by the clockwise rotation of the diagram (4.22). The zeroes in the left row are provided by the growing of $\deg \xi$. From this we can see that Δ_C is the base subscheme of the complete linear system on

$$
\mathbb{P}(\mathcal{E}_{2K}^{c_2+k}|_{\mathrm{Hilb}C}) = \mathbb{P}(\mathcal{E}_{(2K).C}^{c_2+k} \oplus H^1(\mathcal{O}_S(2K)) \otimes \mathcal{O}_{\mathrm{Hilb}C}),
$$

given by the rational map

$$
\mathbb{P}(\mathcal{E}_{(2K).C}^{c_2+k} \oplus H^1(\mathcal{O}_S(2K))^* \otimes \mathcal{O}_{\mathrm{Hilb}}) \to \mathbb{P}(H^0(\mathcal{O}_S(2K))^*), \qquad \text{(4.24)}
$$

which is the restriction of the map of $\mathbb{P}\mathcal{E}_{0,2K}^{c_2+k}$.

Hence,

$$
\Delta_C = \pi^{-1}(\mathrm{Hilb}C) \cap GA\Theta, \qquad \text{(4.25)}
$$
$$
N_{\Delta_C \subset GA\Theta} = \pi^* N_{\mathrm{Hilb}C \subset \mathrm{Hilb}S} = \pi^* \mathcal{E}_C|_{\mathrm{Hilb}C},
$$
$$
\Delta_{|C|} = \pi^{-1}(\mathrm{Hilb}|C|) \cap GA\Theta,
$$
$$
N_{\Delta_{|C|} \subset GA\Theta} = \pi^* N_{\mathrm{Hilb}|C| \subset \mathrm{Hilb}S}.
$$

Now the variety $\Delta_{|C|}$ admits two projections to Hilb:

$$
\begin{array}{ccc}
 & \Delta_{|C|} & \\
\pi \swarrow & & \searrow \pi' \\
\mathrm{Hilb}^{c_2+k}|C| & & \mathrm{Hilb}^{(C-K).C+c_2+k}C
\end{array}
\qquad \text{(4.26)}
$$

where the left side is given by (4.7) and the right side is sending a vector bundle $E(-K)$ of the diagram (4.15) to the cluster η.

It is the main observation which we need for the description of the partial modification of $GA\Theta$.

The partial modification of the component $GA\Theta$ is the blow up of the subvariety $\Delta_{|C|}$ in $\mathrm{Hilb}S$, an "elementary transformation" of the lift of the universal family $\mathbb{E}$ (3.7) and recomputation of a cohomological correspondence (3.10) and of new polynomial (more precisely, of the correction term of one).

The geometrical description of blow up of Δ is very easy. We need to use the description of $\text{Hilb}|C|$ as some Chern class of the standard vector bundle on Hilb. Namely, let

$$\sigma_C : \widetilde{\text{Hilb}}S \rightarrow \text{Hilb}S \tag{4.27}$$

be the blow up of $\text{Hilb}|C|$ in $\text{Hilb}S$.

We need to consider the projectivization of dual to the standard vector bundle $\mathcal{E}_C$ and the base set $GAM_0(2, C - K, d)$ of the linear system on $\mathbb{P}\mathcal{E}_C^*$ given by the canonical projection

$$\mathbb{P}\mathcal{E}_C^* \xrightarrow{can} \mathbb{P}H^0(\mathcal{O}_S(C))^*,$$

that is an analogue of $GA\Theta$ (see (3.5)). Then the exceptional divisor is the intersection

$$\sigma_C^{-1}(\text{Hilb}|C|) = \widetilde{\text{Hilb}}|C| = \pi^{-1}(\text{Hilb}|C|) \cap GAM_0(2, C - K, d) \tag{4.28}$$

in the projectivization $\mathbb{P}\mathcal{E}_C^*$.

The blow up σ_C (4.27) induces the blow up

$$\sigma_{0,C} : \widetilde{\mathbb{P}\mathcal{E}}_{0,2K}^* \rightarrow \mathbb{P}\mathcal{E}_{0,2K}^*,$$

$$\widetilde{\mathbb{P}\mathcal{E}}_{0,2K}^* = \mathbb{P}\pi^*(\mathcal{E}_{0,2K}^*), \tag{4.29}$$

and the blow up with the same notation

$$\sigma_{0,C} : \widetilde{GA\Theta} \rightarrow GA\Theta \tag{4.30}$$

of the subvariety $\Delta_{|C|} \subset GA\Theta$. The exceptional divisor is

$$\widetilde{\Delta}_{|C|} = \sigma_{0,C}^{-1}(\Delta_{|C|}) =$$

$$= (\pi^{-1}(\text{Hilb}|C|)) \cap GAM_0(2, C - K, d) \underset{\text{Hilb}|C|}{\times} (\pi^{-1}(\text{Hilb}|C|) \cap GA\Theta). \tag{4.31}$$

We will denote the preimage of the Grothendieck generator H from $\text{Pic } \mathbb{P}\mathcal{E}_{0,2K}^*$ (see (3.4)) by the symbol H_m. Let $\widetilde{H_{\overline{res(s_i)}}}$ be a geometrical preimage of the hypersurface $H_{\overline{res(s_i)}}$. Then the result of the blow up of $GA\Theta$ along $\Delta_{|C|}$ is

$$\widetilde{GA\Theta} = \overset{h^0(\mathcal{O}_S(2K))}{\underset{i=1}{\bigcap}} \widetilde{H_{\overline{res(s_i)}}} \subset \widetilde{\mathbb{P}\mathcal{E}}_{0,2K}^*. \tag{4.32}$$

Now doing the "elementary transformation" of the lift of the universal family (3.7) we realize $\widetilde{GA\Theta}$ as a base of family of bundles and torsion free sheaves: simplifying the notations let

$$\text{Hilb}^{c_2(S)+k+C(C-K)}S = \text{Hilb}'S \tag{4.33}$$

be the target of the right arrow π' in (4.26). Then we have the morphism

$$(\sigma_{0,C}.\pi') : \widetilde{\Delta}_{|C|} \rightarrow \text{Hilb}'S \tag{3.34}$$

and the morphism

$$id \times (\sigma_{0,C}.\pi') : S \times \tilde{\Delta}_{|C|} \to S \times \text{Hilb}'S. \tag{4.35}$$

Let Z' be the preimage of the universal subscheme Z (3.1) with respect to this morphism. Then on the divisor $S \times \tilde{\Delta}_{|C|}$ of the variety $S \times \widehat{GA\Theta}$ we have the torsion free sheaf $(id \times (\sigma_{0,C}.\pi'))^*(\mathcal{I}_{Z'} \otimes p_S^* \mathcal{O}_S(C - K))$ and the epimorphism of the restriction of the twisted universal extension (3.7) to it:

$$(id \times \sigma_{0,C})^* \mathbb{E}(-K)|_{S \times \tilde{\Delta}_{|C|}} \to (id \times (\sigma_{0,C}.\pi'))^*(\mathcal{I}_{Z'} \otimes p_S^* \mathcal{O}_S(C - K)) \to 0$$

(see the diagram (4.15)).

We can consider this epimorphism as an epimorphism of the twisted universal sheaf $\mathbb{E}(-K)$ to a torsion sheaf on $S \times \widehat{GA\Theta}$ which provides the exact sequence and:

$$0 \to \mathbb{E}' \to (id \times (\sigma_{0,C}.\pi'))^* \mathbb{E}(-K) \to (id \times (\sigma_{0,C}.\pi'))^*(\mathcal{I}_{Z'} \otimes p_S^* \mathcal{O}_S(C - K)) \to 0 \tag{4.36}$$

and its kernel $\mathbb{E}'$ is the "elementary transformation" which we need.

The exact sequence (4.36) provides

$$\mu_{\mathbb{E}'}(\sigma) = 4\tilde{\sigma} + 2(K.\sigma).H_m - 2((K + 2C).\sigma).\tilde{\Delta}_{|C|}, \tag{4.37}$$

and we have a new polynomial

$$ga_C\gamma_\Theta(\sigma) = (4\pi^*(\tilde{\sigma}) + 2(\sigma.K).H_m - 2((K+2C).\sigma).\tilde{\Delta}_{|C|})^{v.d}.(H_m - .\tilde{\Delta}_{|C|})^{h^0(\mathcal{O}_S(2K))}, \tag{4.38}$$

and the binomial decomposition provides

$$ga_C\gamma_\Theta(\sigma) = ga\gamma_\Theta(\sigma)+ \tag{4.39}$$

$$\tilde{\Delta}_{|C|}[\sum_{i=0}^{h^0(\mathcal{O}_S(2K))} \sum_{j=1}^{v.d} \binom{h^0(\mathcal{O}_S(2K))}{i} \binom{v.d}{j}((K + 2C).\sigma)^j.\tilde{\Delta}_{|C|}^{i+j-1}.H_m^{h^0(\mathcal{O}_S(2K))-i} \times$$

$$\times (4\pi^*(\tilde{\sigma}) + 2(\sigma.K).H_m)^{v.d-j}].$$

The correcting term of polynomial is the same as the increment of the almost canonical Spin-polynomial (see (0.11)) and we can compute one by restricting all our divisors to

$$\tilde{\Delta}_{|C|} = (\pi^{-1}(\text{Hilb}C) \cap GAM_0(2, C - K, d) \underset{\text{Hilb}|C|}{\times} (\pi^{-1}(\text{Hilb}C) \cap GA\Theta), \tag{4.40}$$

that is in terms of the theory of curves.

Let H' be the Grothendieck generator of Pic $\mathbb{P}\mathcal{E}_{C^2}$, then $\tilde{\Delta}_{|C|}^2 = -H'$ and

$$\tilde{\sigma}.\text{Hilb}C = (C.\sigma).H'. \tag{4.41}$$

Now the computation of the correcting sum in (4.39) we do on the relative product (4.31). It easy to see that the correcting sum has the form

$$\sum a_i(C^2, C.K, D.C)(C.\sigma)^i(D.\sigma)^{d_1-i}, \qquad (4.42)$$

where the constants $a_i(x, y, z)$ arise from the products of Segre classes of $\mathcal{E}_{C^2}$ and $\mathcal{E}^*_{2K.C}$ on Hilb$C$ and binomial coefficients. We can compute these precisely as in Example 14.4.17 of [F] or the beautiful computations in [ACGH] or using the analogue of the diagram (3.19) for the case of curves.

At last to receive the almost canonical Spin-polynomial (or the right picture of birational transformation (0.9)) we must use the described correcting procedure step by step ordering the collection of curves (4.1) by divisibility property. It isn't very surprising that for surfaces this procedure is as simple as for curves, because there exists very simple but extremely useful trick for embedding our situation into (now classical) situation of stable pair on an algebraic curve (see [B] and [Th]): there exists (see [T 4]) the smooth curve C_{big} on S such that

1) the restriction map

$$res : M^{ac}(2, K, c_2(S) + k) \rightarrow M_{C_{big}} \qquad (4.43)$$

of moduli space of ac-stable vector bundles to stable vector bundles with $c_1 = K.C_{big}$ on the curve C_{big} is an embedding;

2) for every $|C_i|$ from the set (4.1) and $GA\Theta - \Delta_{|C_i|}$ the restriction map (4.43) is an embedding;

3) for every pair of curves C_1 and C_2 from collection (4.1)

$$C_1|_{C_{big}} = C_2|_{C_{big}} \Longleftrightarrow C_1 = C_2. \qquad (4.44)$$

Now we can restrict our procedure to C_{big}.

Why doesn't this trick reduce our problem to the perfect understood (actually, Fano-) case of Thaddeus and Bertram?

The obstacle is the following: the restriction map (4.43) gives a new compactification of $M^{ac}(2, K, c_2(S) + k)$ which is much more dense then Gieseker compactification and even is denser then Uhlenbek-compactification, but a priori this compactification depends on C_{big} (see [T 4]).

References

[A]   V.Arnold. *Mathematical Methods of Classical Mechanics*, Springer-Verlag, New York, 1978.

[ACGH] E. Arbarello, M. Cornalba, P.A. Griffits, J. Harris. *Geometry of Algebraic Curves*, Springer-Verlag, New York, 1985.

[B]   A. Bertram. *Moduli of rank 2 vector bundles, theta divisors and the geometry of curves in projective space*, J. Diff. Geom., v. 35, (1992), 429-469.

[B-D]   S. Bradlow and G.Daskalopoulos. *Moduli of stable pair for holomorphic bundles over Riemann surfaces*, Int. J. Math., v. 2, (1991), 477-513.

[D 1]   S.Donaldson. *Polynomial invariants for smooth 4-manifolds*, Topology, v. 29, (1990), 257-315.

[D 2]   S.Donaldson. *Instantons and Geometric Invariant Theory* , Comm. Math. Phys., v. 93, (1984), 453-460.

[D 3]   S.Donaldson. *Irrationality and h-cobordism theory*, J. Diff. Geom., v. 26, (1987), 141-168.

[D-K]   S.Donaldson and P.Kronheimer. *The Geometry of Four-Manifolds*, Clarendon Press, Oxford, 1990.

[F]   W.Fulton. *Intersection Theory*, Springer-Verlag, New York, 1987.

[G]   D.Gieseker. *On the moduli of vector bundles on an algebraic surface*, Ann. of Math., v. 106, (1977), 45-60.

[O'G]   K.G.O'Grady. *Algebro-geometric analogues of Donaldson polynomials*, Inv. Math., v. 107, (1992), 351-395.

[G-S]   V. Guillemin, S. Sternberg. *Birational equivalence in symplectic category*, Invent. Math., v. 97, (1989), 485-522.

[H]   N.Hitchin. *The self duality equations on a Riemann surface*, Proc. London Math. Soc., v. 55, (1987), 59-126.

[Hu]   Y.Hu. *The geometry and topology of quotient varieties of torus actions*, Duke Math. Journal, v. 68, (1992), 151-183.

[H-L]   D. Huybrechts, M. Lehn. *Stable pairs on curves and surfaces*, Preprint MPI, (1992-1993).

[Ki]   F. Kirvan. *Cohomology of quotient in symplectic and algebraic geometry*, Springer-Verlag, New York, 1987.

[K-M]   D.Kotschick and J.Morgan. $SO(3)$*-invariants for 4-manifolds with* $b_2^+ = 1$. *II*, in print.

[Mo]   J.Morgan. *Comparison of the Donaldson polynomial invariants with their algebro-geometric analogues*, in print.

[M]   D. Mumford. *Geometric invariant theory*, Springer-Verlag, New York, 1965.

[N]   P. Newstead. *Introduction to Moduli Problems and Orbit Spaces*, Tata Inst. Lectures 51, Springer-Verlag, Heidelberg, 1978.

[P]   V.Pidstrigach. *Spin polynomial invariants and walls structure*, Preprint (to appear).

[P-T]   V.Pidstrigach and A.Tyurin. *Invariants of the smooth structures of an algebraic surfaces arising from Dirac operator*, Iz. AN SSSR, v. 52:2, (1992), 279-371 (Russian); English transl. in Warwick preprint, (1992), v. 22.

[Th]   M.Thaddeus. *Stable pairs, linear system and the Verlinde formula*, Preprint, (1992).

[T 1]   A.Tyurin. *Algebraic-geometrical aspects of smooth structures I. Donaldson polynomials*, Russian Math. Surveys, v. 44:3, (1989), 117-178.

[T 2]   A.Tyurin. *The Spin-polynomial invariants of the smooth structures of algebraic surfaces*, Iz. AN SSSR, v. 57:2, (1993), 279-371 (Russian).

[T 3]   A.Tyurin. *Cycles, curves and vector bundles on algebraic surface*, Duke Math. J., v. 52, N 4, (1987), 813-851.

[T 4]   A.Tyurin. *The moduli spaces of vector bundles on threefolds, surfaces and curves I.*, Preprint of Math. Inst. of Univ. Erlangen-Nürnberg, (1990).

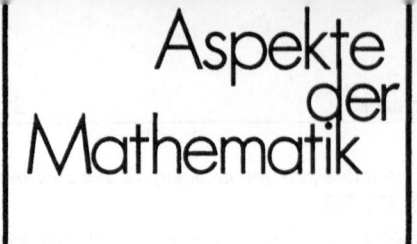

Aspekte der Mathematik

Edited by Klas Diederich

Band D 1: H. Kraft: Geometrische Methoden in der Invariantentheorie

Band D 2: J. Bingener: Lokale Modulräume in der analytischen Geometrie 1

Band D 3: J. Bingener: Lokale Modulräume in der analytischen Geometrie 2

Band D 4: G. Barthel/F. Hirzebruch/T. Höfer: Geradenkonfigurationen und Algebraische Flächen*

Band D 5: H. Stieber: Existenz semiuniverseller Deformationen in der komplexen Analysis

Band D 6: I. Kersten: Brauergruppen von Körpern

*A Publication of the Max-Planck-Institut für Mathematik, Bonn